Series Editor
John M. Walker
School of Life Sciences
University of Hertfordshire
Hatfield, Hertfordshire, AL10 9AB, UK

For further volumes:
http://www.springer.com/series/7651

Transgenic Cotton

Methods and Protocols

Edited by

Baohong Zhang

Department of Biology, East Carolina University, Greenville, NC, USA

Editor
Baohong Zhang
Department of Biology
East Carolina University
Greenville, NC
USA

ISSN 1064-3745   ISSN 1940-6029 (electronic)
ISBN 978-1-62703-211-7   ISBN 978-1-62703-212-4 (eBook)
DOI 10.1007/978-1-62703-212-4
Springer New York Heidelberg Dordrecht London

Library of Congress Control Number: 2012951131

Preface

Cotton is the most important textile and cash crop and is widely cultivated in more than 70 countries, including the United States, China, and India. Because its long life cycle and complicated genetic background, it is hard to improve cotton using traditional breeding techniques; although, it has made much progress in the last several decades. Currently, transgenic techniques have become a powerful tool to improve cotton. Transgenic cotton is among the first commercially genetically modified crops. Since it was adopted by the cotton farmers in the middle of 1990s, transgenic cotton has been widely adopted around the world. Transgenic cotton not only provides huge benefits to cotton farmers, including increasing yield and reducing cost and labor, but also brings lots of environmental and societal impacts, such as reducing environmental pollution by reducing usage of pesticides. Transgenic cotton is also employed in basic research, such as investigating cellulose biosynthesis as well as gene expression and regulation.

This book provides a comprehensive collection of the cutting-edge methods for creating and monitoring transgenic cotton and its application on agricultural and basic research. Worldwide experts contributed to this book and presented their firsthand methods in the field of transgenic cotton. This book is divided into five major parts. The introduction part describes the current status and perspectives of transgenic cotton. The transformation part presents the principle and methods for making transgenic cotton. The detection part provides a comprehensive collection on the methods for detecting foreign gene copy and expression in transgenic cotton plants. The application part describes the improvement of cotton using transgenic technology. The risk assessment part presents the method for monitoring the potential impact of transgenic cotton on environment, including gene flow. This book provides a good resource for scientists as well as graduate students who work on transgenic plants, plant genetics, molecular biology, and agricultural sciences.

I greatly appreciate all the authors who have contributed excellent chapters to this book. Their expertise makes the book a valuable resource for scientists and aspiring graduate students interested in transgenic plants, particularly in transgenic cotton. I also want to express our sincere appreciation to Professor John M. Walker, the Methods in Molecular Biology Series Editor, and Mr. David Casey from Humana Press for their help, support, and commitment during its preparation.

Greenville, NC, USA   *Baohong Zhang*

Contents

Contributors

MUHAMMAD ARSHAD • *Gene Transformation Lab, Agricultural Biotechnology Division (ABD), National Institute for Biotechnology and Genetic Engineering (NIBGE), Faisalabad, Pakistan*

SHAHEEN ASAD • *Gene Transformation Lab, Agricultural Biotechnology Division (ABD), National Institute for Biotechnology and Genetic Engineering (NIBGE), Faisalabad, Pakistan*

RASHMI CHHABRA • *National Research Centre on DNA Fingerprinting, National Bureau of Plant Genetic Resources, New Delhi, India*

WANGZHEN GUO • *National Key Laboratory of Crop Genetics and Germplasm Enhancement, Nanjing Agricultural University, Nanjing, P.R. China*

YAN HONG • *Temasek Life Sciences Laboratory, National University of Singapore, Singapore*

SUCHITRA KAMLE • *Molecular Genomics Laboratory, National Institute of Immunology, New Delhi, India*

HEE JIN KIM • *Cotton Fiber Bioscience Research Unit, USDA-ARS, Southern Regional Research Center, New Orleans, LA, USA*

ARVIND KUMAR • *Faculty of Science, Molecular Biology and Immunology Lab, School of Biotechnology, Banaras Hindu University, Varanasi, Uttar Pradesh, India*

CHENGQI LI • *Henan Institute of Sciences and Technology, Xinxiang, Henan, China*

FANG LIU • *Cotton Research Institute, The Chinese Academy of Agricultural Sciences, Anyang, Henan, China*

JIANFENG LIU • *College of Life Sciences, Hebei University, Baoding, China*

ZHI LIU • *National Key Laboratory of Crop Genetics and Germplasm Enhancement, Cotton Research Institute, Nanjing Agricultural University, Nanjing, P.R. China*

ZHIYING MA • *North China Key Laboratory for Crop Germplasm Resources of Education Ministry, Agricultural University of Hebei, Baoding, China*

WEIGUO MIAO • *College of Environment and Plant Protection, Hainan University, Haikou, People's Republic of China*

ABHISHEK OJHA • *Structural and Computational Biology Lab, International Centre for Genetic Engineering and Biotechnology, New Delhi, India*

XIAOPING PAN • *Department of Biology, East Carolina University, Greenville, NC, USA*

KANNIAH RAJASEKARAN • *Southern Regional Research Center, USDA-ARS, New Orleans, LA, USA*

GURINDER JIT RANDHAWA • *National Research Centre on DNA Fingerprinting, National Bureau of Plant Genetic Resources, New Delhi, India*

MONIKA SINGH • *National Research Centre on DNA Fingerprinting, National Bureau of Plant Genetic Resources, New Delhi, India*

YINGCHUAN TIAN • *State Key Laboratory of Plant Genomics, Institute of Microbiology, Chinese Academy of Sciences, Beijing, China*

HONGMEI WANG • *Cotton Research Institute, The Chinese Academy of Agricultural Sciences, Anyang, Henan, China*

JINGSHENG WANG • *College of Plant Protection, Nanjing Agricultural University, Nanjing, People's Republic of China*

MIN WANG • *Beijing Key Laboratory of Plant Resources Research and Development, Department of Biotechnology, School of Science, Beijing Technology and Business University, Haidian District, Beijing, People's Republic of China*

QINGLIAN WANG • *Henan Institute of Sciences and Technology, Xinxiang, Henan, China*

XINGFEN WANG • *North China Key Laboratory for Crop Germplasm Resources of Education Ministry, Agricultural University of Hebei, Baoding, China*

JIAHE WU • *State Key Laboratory of Plant Genomics, Institute of Microbiology, Chinese Academy of Sciences, Beijing, China*

CHENGXIN YI • *JOil (S) Pte Ltd, 1 Research Link, National University of Singapore, Singapore*

YUSUF ZAFAR • *Gene Transformation Lab, Agricultural Biotechnology Division (ABD), National Institute for Biotechnology and Genetic Engineering (NIBGE), Faisalabad, Pakistan*

BAOHONG ZHANG • *Department of Biology, East Carolina University, Greenville, NC, USA*

BAOLONG ZHANG • *National Key Laboratory of Crop Genetics and Germplasm Enhancement, Cotton Research Institute, Nanjing Agricultural University, Nanjing, People's Republic of China*

JUN ZHANG • *Cotton Research Center, Shandong Academy of Agricultural Sciences, Jinan, Shandong, P.R. China*

TIANZHEN ZHANG • *National Key Laboratory of Crop Genetics and Germplasm Enhancement, Cotton Research Institute, Nanjing Agricultural University, Nanjing, People's Republic of China*

XIN ZHANG • *Henan Institute of Sciences and Technology, Xinxiang, Henan, China*

ZHEN ZHU • *Genetics and Developmental Biology Institute, Chinese Academy of Sciences, Beijing, China*

Part I

Introduction

Chapter 1

Transgenic Cotton: From Biotransformation Methods to Agricultural Application

Baohong Zhang

Abstract

Transgenic cotton is among the first transgenic plants commercially adopted around the world. Since it was first introduced into the field in the middle of 1990s, transgenic cotton has been quickly adopted by cotton farmers in many developed and developing countries. Transgenic cotton has offered many important environmental, social, and economic benefits, including reduced usage of pesticides, indirect increase of yield, minimizing environmental pollution, and reducing labor and cost. *Agrobacterium*-mediated genetic transformation method is the major method for obtaining transgenic cotton. However, pollen tube pathway-mediated method is also used, particularly by scientists in China, to breed commercial transgenic cotton. Although transgenic cotton plants with disease-resistance, abiotic stress tolerance, and improved fiber quality have been developed in the past decades, insect-resistant and herbicide-tolerant cotton are the two dominant transgenic cottons in the transgenic cotton market.

1. Introduction

Cotton is the most important fiber crop; it is also the major resource of plant protein and edible oil. Since it was first cultivated between 5,000 and 10,000 years ago, cotton production has enormously influenced global economic development (1). Currently, cotton is cultivated commercially in more than 70 countries around the world, including the USA, China, and many developing countries (2); it is becoming an important cash crop and it is estimated that more than 180 million people are associated with the worldwide cotton fiber industry, which annually produces 20–30 billion dollars worth of raw cotton (3). Thus, improving cotton has attracted much attention from scientists and breeders in the past decades. One of the biggest contributions is to improve cotton agronomic performance, particularly on cotton resistance to insects and tolerance to herbicide, using transgenic technologies.

Baohong Zhang (ed.), *Transgenic Cotton: Methods and Protocols*, Methods in Molecular Biology, vol. 958, DOI 10.1007/978-1-62703-212-4_1, © Springer Science+Business Media New York 2013

The first transgenic cotton was obtained in 1987 by two independent groups (4, 5), in which they transformed a reporter gene into cotton genome through an *Agrobacterium*-mediated genetic transformation. Although it is among the first crops, including model species, obtaining transgenic plants at the early stage of transgenic plant techniques, few progress had been made in the first several years majorly due to the fact that plant regeneration via somatic embryogenesis remained extremely difficult for cotton. As we know, the most successful method for plant transformation is *Agrobacterium*-mediated genetic transformation, which requires two important steps: (1) transfer and integrate the foreign genes into a plant genome and (2) obtain an entire plant from the single transformed cell. Although any cell contains a complete set of genetic information and allows a cell to have the potential to become an entire plant, the tissue culture technique is not mature enough to induce any cell to differentiate into somatic embryo or shoot. Thus, plant regeneration is still the bottleneck for the transformation of many plant species, including cotton. Since then, many scientists from different laboratories have investigated the effect of different factors on cotton somatic embryogenesis and plant regeneration (6–33). Due to the improvement of cotton tissue culture and plant regeneration as well as new transformation technology development for cotton, transgenic cotton has made significant progress in the past two decades. Transgenic techniques have been widely used in cotton, from basic research to agricultural application. Transgenic cotton is among the first commercial transgenic crops and currently it is widely used around the world.

2. Biotransformation Method

Since Firoozabady (4) and Umbeck (5) obtained the first transgenic cotton plants using *Agrobacterium*-mediated genetic transformation, many transformation methods have been developed and/or adopted to transfer foreign genes into cotton genome. Among these methods, *Agrobacterium*-mediated genetic transformation, biolistic particle delivery system, and pollen tube pathway-mediated method are three major methods for cotton transformation; currently almost all transgenic cotton plants are obtained through these three methods.

2.1. Agrobacterium-Mediated Genetic Transformation

Agrobacterium-mediated genetic transformation is the most widely used and also most successful transformation technique, particularly for dicot plant species. *Agrobacterium tumefaciens* is a natural tool for genetic transformation, in which *Agrobacterium tumefaciens* contains a plasmid, called tumor-induced plasmid (Ti plasmid).

Fig. 1. Flowchart for obtaining transgenic cotton via *Agrobacterium*-mediated transformation.

Agrobacterium uses a complicated mechanism to transfer and insert the transfer DNA (T-DNA) into plant genome (34). One characteristic of T-DNA is that T-DNA can accept any foreign gene through recombinant DNA technique (34). Through the T-DNA, the foreign gene is transferred and integrated into a plant genome.

Agrobacterium-mediated genetic transformation of cotton is a process containing multiple complicated steps, in which it is involved in plant tissue culture. *Agrobacterium*-mediated genetic transformation starts with coculture of *Agrobacterium* with wounded cotton explants (such as cotyledon and hypocotyls) and then obtains transgenic cotton plants after screening the transgenic cells and somatic embryogenesis (Fig. 1). *Agrobacterium*-mediated transformation is also the major transformation method for cotton. The fist transgenic cotton plants were obtained through *Agrobacterium*-mediated method (35, 36). Since then, *Agrobacterium*-mediated transformation of cotton has been established in many research groups (37–55).

Many factors influence the successful rate of *Agrobacterium*-mediated transformation in cotton. Currently, several *Agrobacterium* strains were successfully employed for obtaining transgenic cotton. Among them, the most commonly used strains are LBA 4404 and EHA105. Although each strain works, studies show that stain LBA 4404 is significantly better than stain EHA 105 (51) or C58C3 (56). Using a model cotton transformation genotype Coker 312, the transformation efficiency of the strain LBA4404 was more than twofold higher than that of EHA105 (51). Although coculture was carried out at any temperature between 21 and 28°C and transgenic cotton plants were also obtained from those, low temperature for coculture was better than high temperature (51).

Different explants also affected the transformation rate; currently the widely used explants are cotyledon and hypocotyls. Adding acetosyringone (AS) into the coculture medium or pre-culture of *Agrobacterium* significantly enhances *Agrobacterium*-mediated transformation of cotton (51, 56, 57). AS is a phenolic natural product, particularly, in relation to wounding and other physiologic changes; it is well studied that AS functions as signaling molecule, induces the expression of *Agrobacterium vir* genes, and then initiates the transformation process (58–61).

Genotype is the major constraint for *Agrobacterium*-mediated transformation of cotton due to the fact that only a limited number of genotypes can be cultured to obtain somatic embryogenesis and plant regeneration. To avoid this issue, several laboratories have developed new strategies for obtaining transgenic cotton plants via *Agrobacterium*-mediated transformation, which include transforming some tissues or cells which have high potentials for obtaining an entire plant; these tissues or cells include embryogenic callus (37, 49, 56, 57, 62) and apex (63).

2.2. Biolistic Particle Delivery System

Biolistic particle delivery system is an alternative transformation technique employed for obtaining transgenic cotton. There are several different names for biolistic particle delivery system, such as gene gun and particle bombardment, which uses a device for injecting cell with genetic information. McCabe and Martinell (1993) employed the gene gun delivery system to directly deliver foreign genes into the meristematic tissue of excised cotton embryonic axes using high-velocity gold beads coated with DNA. Their results show that foreign genes were stably integrated and transmitted to progeny in a Mendelian fashion (64). Since then, several other researches also transformed reporter genes and target genes into cotton genome by transformed different explants (64–71).

The major advantage of biolistic particle delivery system is that it can be used to transform any tissue types of any cotton cultivars. However, biolistic particle delivery system may also results in high rate of mutant and multiple copies.

2.3. Pollen Tube Pathway-Mediated Transformation

During plant fertilization, pollens are dropped on the stigma, and then they germinate, grow into the style, and finally arrive at embryo sacs and fuse with the eggs. During this process, a long distance of pollen tube pathway is formed. It is possible that foreign gene is delivered through this pathway to embryo sacs and transfer into the fertilized eggs. Based on this principle, one research group led by Dr. Zhou transferred foreign genomic DNAs into upland cotton and observed many mutants from the transformants (72). Since then, this transformation technique has been successfully employed to obtain transgenic cotton (72–74), watermelon (75, 76), soybean (77–79), wheat (80–83), papaya (84), and corn (75, 85). Some of them have been widely used in the field.

Pollen tube pathway-mediated genetic transformation method is a genotype-independent transformation method.

3. Agricultural Application

Since transgenic insect-resistant Bt cotton was adopted by the cotton farmers in the middle of 1990s, transgenic cotton has been widely adopted around the world (86, 87). Currently, transgenic insect-resistant *Bt* cotton and transgenic herbicide-tolerant cotton are two major types of transgenic cotton in the field.

3.1. Development of Commercial Transgenic Cotton Cultivars

It is a long process to obtain a commercial transgenic cotton cultivar after obtaining a transgenic cotton plant. Usually, the transgenic plants need to be monitored for their agronomic traits as well the expression levels and stability of the foreign genes. Backcrossing is always taken to transfer the foreign gene into a commercial cultivar with good agronomic characteristics. Thus, it may take at least a couple of years before a transgenic cotton plant can be used in the field.

Recently, scientists also try to develop hybrid transgenic cotton for commercialization, in which transgenic cotton is crossed with a non-transgenic commercial cotton cultivar with desirable agronomic characteristics. This technology will allow quick usage of transgenic technology in the field. However, unlike other crops, cotton hybrid is normally produced by manually removing the male part and crossfertilization. Thus, it is lab extensive. Currently, transgenic hybrid cotton is only adopted in China and India.

3.2. Commercialization and Global Adoption of Transgenic Cotton

Transgenic cotton was first adopted by cotton farmers in 1994 in China. In 1996, Bt cotton was started to grow in the USA with about 730,000 hectares with additional in Mexico and Australia for a global total of about 0.8 million hectares. Two years later, the area of Bt cotton is doubled to 1.5 million hectares and then increased to 5.4 million hectares in 2003 (86). Currently, almost all major cotton-planted countries have adopted the transgenic cotton.

3.3. Insect-Resistant Cotton

Pests are a major problem in cotton production in any area. It is estimated that pests significantly damage yield by 15–50% and pests also affect fiber quality. Up to 1,326 species of insects have been reported on cotton worldwide. The majority pests causing cotton damage belong to the caterpillar species (Lepidoptera), such as bollworm and budworm. Thus, from the very beginning, development of transgenic cotton with insect resistance is the top priority of improving cotton using transgenic technology.

It is well known that Bt is an effective biological pesticide, which has been used in the field for many years. More important, many pests targeted by Bt are agronomically important, such as bollworm and armyworm (88). Currently, many *Bt* genes have been identified, cloned, and characterized, and some of them have been transferred into plants, including cotton, soybean, maize, and tomato. Field and lab bioassay shows that three major cotton pests, bollworm (*Helicoverpa zea*), tobacco budworm (*H. Virescens*), and the pink bollworm (*Petinophora gossypiella*), are susceptible to *Bt* and much less pests could survive on transgenic Bt cotton plants (89–98). Filed tests demonstrate that Bt cotton can reduce the damage of bollworm by as much as 93–100%, and damage to crops by leaf-feeding Lepidoptera was also reduced. Planting Bt cotton can save more than up to 70% pesticide usage.

3.4. Herbicide-Tolerant Cotton

Transgenic cotton with herbicide tolerance is another successful story for transgenic plants. Weeds are another big issue for growing any crops, including cotton. Weeds caused significant losses in cotton and require carefully field management. There are more than 30 genera of plants grow in cotton field, including several important weed species, such as grass, sedge, and broadleaf plants (99). The most convenient way to kill the weeds is to spray herbicides. However, when herbicides kill the weeds, it also damage cotton plants. Thus, it is necessary to insert the herbicide-resistant genes into cotton and allow cotton plant to have the ability to resist herbicides.

One widely used herbicide is glyphosate, the active ingredient in herbicide Roundup. Glyphosate is a nonselective broad-spectrum herbicide, which tends to kill any weeds in the field including cotton plants. When plant uptakes the herbicide, glyphosate targets 5-enolpyruvylshikimate-3-phosphate synthase (EPSPS) and then blocks aromatic amino acid synthesis (100). Transgenic cotton encoding a glyphosate-tolerant EPSPS from *Agrobacterium* sp. CP4 (CP4 EPSPS) provides tolerance to the herbicide glyphosate (101, 102).

Currently, several commercialized transgenic cottons with herbicide tolerance have been developed and widely used in several developed countries, including the USA and Australia. Transgenic herbicide-tolerant cotton provides an excellent weed control methods for the cotton farmers, which provide significant economic and societal benefits.

3.5. Transgenic Cotton with Other Traits

Except the transgenic cotton with pest resistance and herbicide tolerance, transgenic technology is also used to modify cotton tolerance to abiotic stress (103–107) as well as cellulose biosynthesis and fiber quality (108–111). However, these transgenic cottons are still in the infant stage and there is still a long way to go before it will be commercialized.

4. Three Major Concerns About Transgenic Cotton

Transgenic cotton is a new thing created by transgenic technique. Like all other new things, it has the potential to cause some side effects.

4.1. Is Transgenic Bt Cotton Safe to Our Human and Other Non-targeted Animals?

When Bt cotton is quickly adopted around the world, two questions are arisen consequentially: Is Bt safe to our human beings and nontarget insects, particularly for the beneficial insects? What is the fate of Bt protein in the field and also in the animal GI track?

Bt is a biological pesticide that is produced by soil bacterium *Bacillus thuringiensis*, which has been adopted to control pests for more than 40 years (88). Bt is extremely selective against lepidopteran pests (such as bollworms) and highly safe to humans, domestic animals, wildlife, agriculturally beneficial insects, and environment (86, 112). There are two reasons for this: (1) Bt proteins are not toxic at all; only when they are depredated into the toxic part they become toxic; however, the biodegradation needs basic condition that only happens in insect GI track; for mammals, including human beings, the GI track is in acidic condition; thus, Bt cannot be converted to toxic compounds in human bodies. (2) The action of Bt proteins requires binding to a specific midgut epithelium receptor, which is specific to different insects but other animals do not have this receptor; thus, Bt is extremely selective against certain pests and safe to others.

4.2. Pest Resistance to Bt Cotton

After long-term exposure to transgenic Bt cotton, it is possible that pests evolve resistance to Bt protein and the Bt cotton. This has been observed in laboratory as well as field tests (113–115). To prevent this, several strategies have been developed.

1. Refuge
   The refuge strategy is the commonly used strategy for preventing pests from developing resistance to Bt cotton. Large-scale field tests show that refuge strategy delayed pest resistance to Bt (116–118). Using this strategy, non-transgenic cotton requires to be planted in or near the Bt cotton, in which the resistant pests will be diluted by increasing the number of susceptible pests.

2. Using multiple *Bt* genes for Bt cotton

2A theoretical model shows that *Bt* cotton containing two dissimilar Bt genes (pyramided plants) delays pest resistance to Bt significantly. Usually, Bt cotton with one single Bt gene can be used for 7–10 years before pests develop resistance to it; but for two Bt genes, it can be used for up to 20 years. This has been confirmed by Bt cotton containing Cry1Ac and Cry1C genes in the greenhouse after 24 generations of selection (119).

3. Other strategies

Several other strategies are also used for delaying pest resistance to Bt cotton. These include crop rotation, spraying pesticides, and integrated pest management (IPM).

4.3. Gene Flow

Gene flow means that transgene moves from transgenic plant to non-transgenic plant or from one type of transgenic plant to another one. Gene flow is another major concern with release of transgenic plants into the environment because that transgene flow can cause superweeds and contamination of non-transgenic seeds and food as well as reduction in species fitness and genetic diversity (120, 121). Although cotton is a self-pollinating crop and gene flow is not a big concern compared with other crops, transgene flow was observed in transgenic cotton in China, Australia, and the USA (122–125). The gene flow was also observed between transgenic cotton and its wild species, which may cause a critical issue in cotton biodiversity (126). Insects are the big player causing gene flow in cotton (127).

References

1. Stephens SG, Mosley ME (1974) Early domesticated cottons from archaeological sites in central coastal. Peru Am Antiquity 39:109–122

2. Zhang BH, Feng R (2000) Cotton resistance to insect and pest-resistant cotton. Chinese Agricultural Science and Technology Press, Beijing

3. IAC (1996) Cotton: review of world situation. Monogram by International Advisory Committee, Washington, DC

4. Firoozabady E, Deboer DL, Merlo DJ, Halk EL, Amerson LN, Rashka KE, Murray EE (1987) Transformation of cotton (*Gossypium hirsutum* L.) by *Agrobacterium tumefaciens* and regeneration of transgenic plants. Plant Mol Biol 10:105–116

5. Umbeck P, Johnson G, Barton K, Swain W (1987) Genetically transformed cotton (*Gossypium hirsutum* L.) plants. Bio-Technology 5:263–266

6. Divya K, Anuradha TS, Jami SK, Kirti PB (2008) Efficient regeneration from hypocotyl explants in three cotton cultivars. Biologia Plantarum 52:201–208

7. Han GY, Wang XF, Zhang GY, Ma ZY (2009) Somatic embryogenesis and plant regeneration of recalcitrant cottons (*Gossypium hirsutum*). Afr J Biotechnol 8:432–437

8. Hussain SS, Rao AQ, Husnain T, Riazuddin S (2009) Cotton somatic embryo morphology affects its conversion to plant. Biologia Plantarum 53:307–311

9. Khan T, Singh AK, Pant RC (2006) Regeneration via somatic embryogenesis and organogenesis in different cultivars of cotton (*Gossypium* spp.). Vitro Cell Develop Biol Plant 42:498–501

10. Kouakou TH, Waffo-Teguo P, Kouadio YJ, Valls J, Richard T, Decendit A, Merillon J-M (2007) Phenolic compounds and somatic embryogenesis in cotton (*Gossypium hirsutum* L.). Plant Cell Tissue Organ Cult 90:25–29

11. Wang J, Sun Y, Yan S, Daud MK, Zhu S (2008) High frequency plant regeneration

from protoplasts in cotton via somatic embryogenesis. Biol Plantarum 52:616–620

12. Zhang B, Wang Q, Liu F, Wang K, Frazier TP (2009) Highly efficient plant regeneration through somatic embryogenesis in 20 elite commercial cotton (*Gossypium hirsutum* L.) cultivars. Plant Omics 2:259–268

13. Aydin Y, Talas-Ogras T, Ipekci-Altas Z, Gozukirmizi N (2006) Effects of brassinosteroid on cotton regeneration via somatic embryogenesis. Biologia 61:289–293

14. Ikram-ul H, Zafar Y (2004) High frequency of callus induction, its proliferation and somatic embryogenesis in cotton (*Gossypium hirsutum* L.). J Plant Biotechnol 6:55–61

15. Mishra R, Wang HY, Yadav NR, Wilkins TA (2003) Development of a highly regenerable elite Acala cotton (*Gossypium hirsutum* cv. Maxxa) – a step towards genotype-independent regeneration. Plant Cell Tissue Organ Cult 73:21–35

16. Rao AQ, Hussain SS, Shahzad MS, Bokhari SYA, Raza MH, Rakha A, Majeed A, Shahid AA, Saleem Z, Husnain T, Riazuddin S (2006) Somatic embryogenesis in wild relatives of cotton (Gossypium spp.). J Zhejiang Univ Sci 7:291–298

17. Sakhanokho HF, Ozias-Akins P, May OL, Chee PW (2004) Induction of somatic embryogenesis and plant regeneration in select Georgia and pee dee cotton lines. Crop Sci 44:2199–2205

18. Sun YQ, Zhang XL, Huang C, Guo XP, Nie YC (2006) Somatic embryogenesis and plant regeneration from different wild diploid cotton (Gossypium) species. Plant Cell Rep 25:289–296

19. Sun YQ, Zhang XL, Huang C, Nie YC, Guo XP (2005) Factors influencing in vitro regeneration from protoplasts of wild cotton (G-klotzschianum A) and RAPD analysis of regenerated plantlets. Plant Growth Regul 46:79–86

20. Wu JH, Zhang XL, Nie YC, Jin SX, Liang SG (2004) Factors affecting somatic embryogenesis and plant regeneration from a range of recalcitrant genotypes of Chinese cottons (*Gossypium hirsutum* L.). Vitro Cell Develop Biol Plant 40:371–375

21. Kumria R, Sunnichan VG, Das DK, Gupta SK, Reddy VS, Bhatnagar RK, Leelavathi S (2003) High-frequency somatic embryo production and maturation into normal plants in cotton (*Gossypium hirsutum*) through metabolic stress. Plant Cell Rep 21:635–639

22. Sakhanokho HF, Zipf A, Raiasekaran K, Saha S, Sharma GC (2001) Induction of highly embryogenic calli and plant regeneration in upland (*Gossypium hirsutum* L.) and pima (Gossypium barbadense L.) cottons. Crop Sci 41:1235–1240

23. Zhang BH, Feng R, Liu F, Wang QL (2001) High frequency somatic embryogenesis and plant regeneration of an elite Chinese cotton variety. Bot Bull Acad Sin 42:9–16

24. Zhang BH, Feng R, Liu F, Yao CB (1999) Direct induction of cotton somatic embryogenesis. Chinese Sci Bull 44:766–767

25. Zhang BH, Feng R, Liu F, Zhou DY, Wang QL (2001) Direct somatic embryogenesis and plant regeneration from cotton (*Gossypium hirsutum* L.) explants. Israel J Plant Sci 49:193–196

26. Gonzalez-Benito ME, Carvalho JMFC, Perez C (1997) Cotton (*Gossypium hirsutum* L.) somatic embryogenesis: a comparative study between two cultivars. Phytomorphology 47:375–382

27. Hemphill JK, Maier CGA, Chapman KD (1998) Rapid in-vitro plant regeneration of cotton (*Gossypium hirsutum* L.). Plant Cell Rep 17:273–278

28. Rajasegar G, Rangasamy SRS, Venkatachalam P, Rao GR (1996) Callus induction, somatic embryoid formation and plant regeneration in cotton (*Gossypium hirsutum* L.). J Phytol Res 9:145–147

29. Trolinder NL, Goodin JR (1987) Somatic embryogenesis and plant-regeneration in cotton (Gossypium-hirsutum-L). Plant Cell Rep 6:231–234

30. Trolinder NL, Goodin JR (1988) Somatic embryogenesis in cotton (Gossypium). 1. Effects of source of explant and hormone regime. Plant Cell Tissue Organ Cult 12:31–42

31. Trolinder NL, Goodin JR (1988) Somatic embryogenesis in cotton (Gossypium). 2. Requirements for embryo development and plant-regeneration. Plant Cell Tissue Organ Cult 12:43–53

32. Voo KS, Rugh CL, Kamalay JC (1991) Indirect somatic embryogenesis and plant recovery from cotton Gossypium-hirsutum L. Vitro Cell Develop Biol Plant 27P·117–124

33. Zhang BH, Feng R, Li XH, Li FL (1996) Anther culture and plant regeneration of cotton (Gossypium klotzschianum Anderss). Chinese Sci Bull 41:145–148

34. Gelvin SB (2003) *Agobacterium*-mediated plant transformation: the biology behind the "gene-Jockeying" tool. Microbiol Mol Biol Rev 67:16–37

35. Firoozabady E, Deboer DL, Merlo DJ, Halk EL, Amerson LN, Rashka KE, Murray EE

(1987) Transformation of cotton (*Gossypium hirsutum* L.) by *Agrobacterium tumefaciens* and regeneration of transgenic plants. Plant Mol Biol 10:105–116

36. Umbeck P, Johnson G, Barton K, Swain W (1987) Genetically transformed cotton (*Gossypium hirsutum* L.) plants. Bio-Technology 5:263–266

37. Asad S, Mukhtar Z, Nazir F, Hashmi JA, Mansoor S, Zafar Y, Arshad M (2008) Silicon carbide whisker-mediated embryogenic callus transformation of cotton (*Gossypium hirsutum* L.) and regeneration of salt tolerant plants. Mol Biotechnol 40:161–169

38. Chen TZ, Wu SJ, Zhao J, Guo WZ, Zhang TZ (2010) Pistil drip following pollination: a simple in planta Agrobacterium-mediated transformation in cotton. Biotechnol Lett 32:547–555

39. Hashmi JA, Zafar Y, Arshad M, Mansoor S, Asad S (2011) Engineering cotton (*Gossypium hirsutum* L.) for resistance to cotton leaf curl disease using viral truncated AC1 DNA sequences. Virus Genes 42:286–296

40. Katageri IS, Vamadevaiah HM, Udikeri SS, Khadi BM, Kumar PA (2007) Genetic transformation of an elite Indian genotype of cotton (*Gossypium hirsutum* L.) for insect resistance. Curr Sci 93:1843–1847

41. Kim HJ, Murai N, Fang DD, Triplett BA (2009) Functional analysis of *Gossypium hirsutum* cellulose synthase catalytic subunit 4 promoter in transgenic Arabidopsis and cotton tissues. Plant Sci 180:323–332

42. Li FF, Wu SJ, Chen TZ, Zhang J, Wang HH, Guo WZ, Zhang TZ (2009) Agrobacterium-mediated co-transformation of multiple genes in upland cotton. Plant Cell Tissue Organ Cult 97:225–235

43. Liu JF, Zhao CY, Ma J, Zhang GY, Li MG, Yan GJ, Wang XF, Ma ZY (2009) Agrobacterium-mediated transformation of cotton (*Gossypium hirsutum* L.) with a fungal phytase gene improves phosphorus acquisition. Euphytica 181:31–40

44. Nandeshwar SB, Moghe S, Chakrabarty PK, Deshattiwar MK, Kranthi K, Anandkumar P, Mayee CD, Khadi BM (2009) Agrobacterium-mediated transformation of cry1Ac gene into shoot-tip meristem of diploid cotton Gossypium arboreum cv. RG8 and regeneration of transgenic plants. Plant Mol Biol Rep 27:549–557

45. Wu JH, Luo XL, Zhang XR, Shi YJ, Tian YC (2011) Development of insect-resistant transgenic cotton with chimeric TVip3A*accumulating in chloroplasts. Transgenic Res 20:963–973

46. Wu SJ, Wang HH, Li FF, Chen TZ, Zhang J, Jiang YJ, Ding YZ, Guo WZ, Zhang TZ (2008) Enhanced Agrobacterium-mediated transformation of embryogenic calli of upland cotton via efficient selection and timely subculture of somatic embryos. Plant Mol Biol Rep 26:174–185

47. Zhang J, Cai L, Cheng JQ, Mao HZ, Fan XP, Meng ZH, Chan KM, Zhang HJ, Qi JF, Ji LH, Hong Y (2008) Transgene integration and organization in Cotton (*Gossypium hirsutum* L.) genome. Transgenic Res 17:293–306

48. Ikram Ul H (2004) Agrobacterium-mediated transformation of cotton (*Gossypium hirsutum* L.) via vacuum infiltration. Plant Mol Biol Rep 22:279–288

49. Leelavathi S, Sunnichan VG, Kumria R, Vijaykanth GP, Bhatnagar RK, Reddy VS (2004) A simple and rapid Agrobacterium-mediated transformation protocol for cotton (*Gossypium hirsutum* L.): embryogenic calli as a source to generate large numbers of transgenic plants. Plant Cell Rep 22:465–470

50. Satyavathi VV, Prasad V, Lakshmi BG, Sita GL (2002) High efficiency transformation protocol for three Indian cotton varieties via Agrobacterium tumefaciens. Plant Sci 162:215–223

51. Sunilkumar G, Rathore KS (2001) Transgenic cotton: factors influencing *Agrobacterium*-mediated transformation and regeneration. Mol Breed 8:37–52

52. Tohidfar M, Mohammadi M, Ghareyazie B (2005) Agrobacterium-mediated transformation of cotton (*Gossypium hirsutum*) using a heterologous bean chitinase gene. Plant Cell Tissue Organ Cult 83:83–96

53. Yuceer SU, Koc NK (2006) Agrobacterium-mediated transformation and regeneration of cotton plants. Russian J Plant Physiol 53:413–417

54. Zhao FY, Li YF, Xu PL (2006) Agrobacterium-mediated transformation of cotton (*Gossypium hirsutum* L. cv. Zhongmian 35) using glyphosate as a selectable marker. Biotechnol Lett 28:1199–1207

55. Zhu SW, Gao P, Sun JS, Wang HH, Luo XM, Jiao MY, Wang ZY, Xia GX (2006) Genetic transformation of green-colored cotton. Vitro Cell Dev Biol Plant 42:439–444

56. Jin SX, Zhang XL, Liang SG, Nie YC, Guo XP, Huang C (2005) Factors affecting transformation efficiency of embryogenic callus of Upland cotton (*Gossypium hirsutum*) with *Agrobacterium tumefaciens*. Plant Cell Tissue Organ Cult 81:229–237

57. Wu S-J, Wang H-H, Li F-F, Chen T-Z, Zhang J, Jiang Y-J, Ding Y, Guo W-Z, Zhang T-Z (2008) Enhanced Agrobacterium-mediated transformation of embryogenic calli of upland cotton via efficient selection and timely subculture of somatic embryos. Plant Mol Biol Rep 26:174–185

58. Joubert P, Beaupere D, Lelievre P, Wadouachi A, Sangwan RS, Sangwan-Norreel BS (2002) Effects of phenolic compounds on Agrobacterium vir genes and gene transfer induction – a plausible molecular mechanism of phenol binding protein activation. Plant Sci 162:733–743

59. Lai E-M, Shih H-W, Wen S-R, Cheng M-W, Hwang H-H, Chiu S-H (2006) Proteomic analysis of Agrobacterium tumefaciens response to the vir gene inducer acetosyringone. Proteomics 6:4130–4136

60. Nair GR, Lai X, Wise AA, Rhee BW, Jacobs M, Binns AN (2011) The integrity of the periplasmic domain of the VirA sensor kinase is critical for optimal coordination of the virulence signal response in Agrobacterium tumefaciens. J Bacteriol 193:1436–1448

61. Stachel SE, Messens E, Vanmontagu M, Zambryski P (1985) Identification of the signal molecules produced by wounded plant cells that activate T-DNA transfer in *Agrobacterium tumefaciens.* Nature 318:624–629

62. Wu J, Zhang X, Nie Y, Luo X (2005) High-efficiency transformation of *Gossypium hirsutum* embryogenic calli mediated by Agrobacterium tumefaciens and regeneration of insect-resistant plants. Plant Breed 124:142–146

63. Zapata C, Park SH, El-Zik KM, Smith RH (1999) Transformation of a Texas cotton cultivar by using Agrobacterium and the shoot apex. Theor Appl Genet 98:252–256

64. McCabe DE, Martinell BJ (1993) Transformation of elite cotton cultivars via particle bombardment of meristems. Bio-Technology 11:596–598

65. Chlan CA, Lin JM, Cary JW, Cleveland TE (1995) A procedure for biolistic transformation and regeneration of transgenic cotton from meristematic tissue. Plant Mol Biol Rep 13:31–37

66. Liu JF, Wang XF, Li QL, Li X, Zhang GY, Li MG, Ma ZY (2011) Biolistic transformation of cotton (*Gossypium hirsutum* L.) with the phyA gene from Aspergillus ficuum. Plant Cell Tissue Organ Cult 106:207–214

67. Rech EL, Vianna GR, Aragao FJL (2008) High-efficiency transformation by biolistics of soybean, common bean and cotton transgenic plants. Nat Protoc 3:410–418

68. Banerjee AK, Agrawal DC, Nalawade SM, Krishnamurthy KV (2002) Transient expression of beta-glucuronidase in embryo axes of cotton by Agrobacterium and particle bombardment methods. Biologia Plantarum 45:359–365

69. Dangat SS, Rajput SG, Wable KJ, Jaybhaye AA, Patil VU (2007) A biolistic approach for transformation and expression of cry 1Ac gene in shoot tips of cotton (*Gossypium hirsutum*). Res J Biotechnol 2:43–46

70. Rajasekaran K, Hudspeth RL, Cary JW, Anderson DM, Cleveland TE (2000) High-frequency stable transformation of cotton (*Gossypium hirsutum* L.) by particle bombardment of embryogenic cell suspension cultures. Plant Cell Rep 19:539–545

71. Finer JJ, McMullen MD (1990) Transformation of cotton (*Gossypium hirsutum* L.) via particle bombardment. Plant Cell Rep 8:586–589

72. Zhou G, Weng J, Zheng Y, Huang J, Qian S, Liu G (1983) Introduction of exogenous DNA into cotton embryos. Methods Enzymol 101:433–481

73. Huang GC, Dong YM, Sun JS (1999) Introduction of exogenous DNA into cotton via the pollen-tube pathway with GFP as a reporter. Chinese Sci Bull 44:698–701

74. Ni WC, Guo SD, Jia SR (2000) Cotton transformation with the pollen tube pathway. Rev China Agricult Sci Technol 2:27–32

75. Yang A, Su Q, An L, Liu J, Wu W, Qiu Z (2009) Detection of vector- and selectable marker-free transgenic maize with a linear GFP cassette transformation via the pollen-tube pathway. J Biotechnol 139:1–5

76. Hao J, Niu Y, Yang B, Gao F, Zhang L, Wang J, Hasi A (2011) Transformation of a marker-free and vector-free antisense ACC oxidase gene cassette into melon via the pollen-tube pathway. Biotechnol Lett 33:55–61

77. Hu CY, Wang LZ (1999) In planta soybean transformation technologies developed in China: procedure, confirmation and field performance. Vitro Cell Dev Biol Plant 35:417–420

78. Shou HX, Palmer RG, Wang K (2002) Irreproducibility of the soybean pollen-tube pathway transformation procedure. Plant Mol Biol Rep 20:325–334

79. Yang S, Li G, Li M, Wang J (2011) Transgenic soybean with low phytate content constructed by *Agrobacterium transformation* and pollen-tube pathway. Euphytica 177:375–382

80. Martin N, Forgeois P, Picard E (1992) Investigations on transforming *Triticum aestivum* via pollen tube pathway. Agronomie 12:537–544

81. Qiu Z, Su Q, An L-J (2008) Application of FITC tracing in the optimization of wheat transformation via pollen-tube pathway. Xibei Zhiwu Xuebao 28:611–616

82. Yin J, Yu G-R, Ren J-P, Li L, Song L (2004) Transforming anti-TrxS gene into wheat by means of pollen tube pathway and ovary injection. Xibei Zhiwu Xuebao 24:776–780

83. Zeng JZ, Wang DJ, Wu YQ, Zhang J, Zhou WJ, Zhu XP, Xu NZ (1994) Transgenic wheat obtained with pollen tube pathway method. Sci China Ser B Chem 37:319–325

84. Wei J-Y, Liu D-B, Chen Y-Y, Cai Q-F, Zhou P (2008) Transformation of PRSV-CP dsRNA gene into papaya by pollen-tube pathway technique. Xibei Zhiwu Xuebao 28:2159–2163

85. Zhang YS, Yin XY, Yang AF, Li GS, Zhang JR (2005) Stability of inheritance of transgenes in maize (*Zea mays* L.) lines produced using different transformation methods. Euphytica 144:11–22

86. Zhang BH, Pan XP, Wang QL (2005) Development and commercial use of Bt cotton. Physiol Mol Biol Plants 11:51–64

87. Zhang BH, Liu F, Yao CB, Wang KB (2000) Recent progress in cotton biotechnology and genetic engineering in China. Curr Sci 79: 37–44

88. Zhang BH, Feng R (2000) Cotton-resistance to pests and transgenic pest-resistant cotton. China Agricultural Science and Technology, Beijing

89. Baur ME, Boethel DJ (2003) Effect of Bt-cotton expressing Cry1A(c) on the survival and fecundity of two hymenopteran parasitoids (Braconidae, Encyrtidae) in the laboratory. Biol Contr 26:325–332

90. Mellet MA, Schoeman AS, Broodryk SW, Hofs JL (2004) Bollworm (Helicoverpa armigera (Hubner), Lepidoptera: Noctuidae) occurrences in Bt- and non-Bt-cotton fields, Marble Hall, Mpumalanga, South Africa. African Entomol 12:107–115

91. Li YX, Greenberg SM, Liu TX (2006) Effects of Bt cotton expressing Cry1Ac and Cry2Ab and non-Bt cotton on behavior, survival and development of Trichoplusia ni (Lepidoptera: Noctuidae). Crop Prot 25:940–948

92. Carriere Y, Ellers-Kirk C, Biggs RW, Sims MA, Dennehy TJ, Tabashnik BE (2007) Effects of resistance to Bt cotton on diapause in the pink bollworm, Pectinophora gossypiella. J Insect Sci 7:1–12

93. Ramasundaram P, Vennila S, Ingle RK (2007) Bt cotton performance and constraints in central India. Outlook Agricult 36:175–180

94. Zhao J, Lu M, Fan X, Xie F (1998) Survival and growth of different instar larvae of Helicoverpa armigera (Hubner) on transgenic Bt cotton. Acta Entomol Sin 41:354–358

95. Adamczyk JJ, Gore J (2003) Varying levels of Cry1Ac in transgenic Bacillus thuringiensis Berliner (Bt) cotton leaf bioassays. J Agricult Urban Entomol 20:49–53

96. Parker CD, Mascarenhas VJ, Luttrell RG, Knighten K (2000) Survival rates of tobacco budworm (Lepidoptera: Noctuidae) larvae exposed to transgenic cottons expressing insecticidal protein of Bacillus thuringiensis Berliner. J Entomol Sci 35:105–117

97. Gore J, Leonard BR, Church GE, Cook DR (2002) Behavior of bollworm (Lepidoptera: Noctuidae) larvae on genetically engineered cotton. J Econ Entomol 95:763–769

98. Gore J, Leonard BR, Church GE, Russell JS, Hall TS (2000) Cotton boll abscission and yield losses associated with first-instar bollworm (Lepidoptera: Noctuidae) injury to nontransgenic and transgenic Bt cotton. J Econ Entomol 93:690–696

99. Buchanan GA (1992) Trends in weed control methods. In: Weeds of cotton: characterization and control. Cotton Foundation

100. Steinrucken HC, Amrhein N (1980) The herbicide glyphosate is a potent inhibitor of 5-enolpyruvyl shikimic acid-3-phosphate synthase. Biochem Biophys Res Commun 94: 1207–1212

101. Nida DL, Kolacz KH, Buehler RE, Deaton WR, Schuler WR, Armstrong TA, Taylor ML, Ebert CC, Rogan GJ, Padgette SR, Fuchs RL (1996) Glyphosate-tolerant cotton: genetic characterization and protein expression. J Agric Food Chem 44:1960–1966

102. Riar DS, Norsworthy JK, Griffith GM (2011) Herbicide programs for enhanced glyphosate-resistant and glufosinate-resistant cotton (*Gossypium hirsutum*). Weed Technol 25: 526–534

103. Pasapula V, Shen GX, Kuppu S, Paez-Valencia J, Mendoza M, Hou P, Chen JA, Qiu XY, Zhu LF, Zhang XL, Auld D, Blumwald E, Zhang H, Gaxiola R, Payton P (2011) Expression of an Arabidopsis vacuolar H(+)-pyrophosphatase gene (AVP1) in cotton improves drought- and salt tolerance and increases fibre yield in the field conditions. Plant Biotechnol J 9:88–99

104. Zhang KW, Guo N, Lian LJ, Wang J, Lv SL, Zhang JR (2011) Improved salt tolerance and seed cotton yield in cotton (*Gossypium hirsu-*

tum L.) by transformation with betA gene for glycinebetaine synthesis. Euphytica 181:1–16

105. Zhu CF, Wang YX, Li YB, Bhatti KH, Tian YC, Wu JH (2011) Overexpression of a cotton cyclophilin gene (GhCyp1) in transgenic tobacco plants confers dual tolerance to salt stress and Pseudomonas syringae pv. tabaci infection. Plant Physiol Biochem 49:1264–1271

106. Lv S, Zhang KW, Gao Q, Lian LJ, Song YJ, Zhang JR (2008) Overexpression of an H(+)-PPase gene from Thellungiella halophila in cotton enhances salt tolerance and improves growth and photosynthetic performance. Plant Cell Physiol 49:1150–1164

107. Light GG, Mahan JR, Roxas VP, Allen RD (2005) Transgenic cotton (*Gossypium hirsutum* L.) seedlings expressing a tobacco glutathione S-transferase fail to provide improved stress tolerance. Planta 222:346–354

108. Wang HY, Wang J, Gao P, Jiao GL, Zhao PM, Li Y, Wang GL, Xia GX (2009) Down-regulation of GhADF1 gene expression affects cotton fibre properties. Plant Biotechnol J 7:13–23

109. FeiFei L, ShenJie W, Fenni L, TianZi C, Ming J, HaiHai W, YanJie J, Jie Z, WangZhen G, TianZhen Z (2009) Modified fiber qualities of the transgenic cotton expressing a silkworm fibroin gene. Chinese Sci Bull 54:1210–1216

110. Zhang M, Zheng XL, Song SQ, Zeng QW, Hou L, Li DM, Zhao J, Wei Y, Li XB, Luo M, Xiao YH, Luo XY, Zhang JF, Xiang CB, Pei Y (2011) Spatiotemporal manipulation of auxin biosynthesis in cotton ovule epidermal cells enhances fiber yield and quality. Nat Biotechnol 29:453–458

111. Lee J, Burns TH, Light G, Sun Y, Fokar M, Kasukabe Y, Fujisawa K, Maekawa Y, Allen RD (2010) Xyloglucan endotransglycosylase/hydrolase genes in cotton and their role in fiber elongation. Planta 232:1191–1205

112. McClintock JT, Schaffer CR, Sjoblad RD (1995) A comparative review of the mammalian toxicity of bacillus thuringiensis-based pesticides. Pest Sci 45:95–105

113. Gao YL, Wu KM, Gould F (2009) Frequency of Bt resistance alleles in H-armigera during 2006-2008 in Northern China. Environ Entomol 38:1336–1342

114. Tabashnik BE, Van Rensburg JBJ, Carriere Y (2009) Field-evolved insect resistance to Bt crops: definition, theory, and data. J Econ Entomol 102:2011–2025

115. Tabashnik BE, Carriere Y, Dennehy TJ, Morin S, Sisterson MS, Roush RT, Shelton AM, Zhao JZ (2003) Insect resistance to transgenic Bt crops: lessons from the laboratory and field. J Econ Entomol 96:1031–1038

116. Carriere Y, Ellers-Kirk C, Hartfield K, Larocque G, Degain B, Dutilleul P, Dennehy TJ, Marsh SE, Crowder DW, Li XC, Ellsworth PC, Naranjo SE, Palumbo JC, Fournier A, Antilla L, Tabashnik BE (2012) Large-scale, spatially-explicit test of the refuge strategy for delaying insecticide resistance. Proc Natl Acad Sci USA 109:775–780

117. Frisvold GB, Reeves JM (2008) The costs and benefits of refuge requirements: the case of Bt cotton. Ecol Econ 65:87–97

118. Huang FN, Andow DA, Buschman LL (2011) Success of the high-dose/refuge resistance management strategy after 15 years of Bt crop use in North America. Entomol Exp Appl 140:1–16

119. Zhao JZ, Cao J, Li YX, Collins HL, Roush RT, Earle ED, Shelton AM (2003) Transgenic plants expressing two Bacillus thuringiensis toxins delay insect resistance evolution. Nat Biotechnol 21:1493–1497

120. Chevre AM, Eber F, Baranger A, Renard M (1997) Gene flow from transgenic crops. Nature 389:924

121. Snow AA (2002) Transgenic crops – why gene flow matters. Nat Biotechnol 20:542–542

122. Heuberger S, Ellers-Kirk C, Tabashnik BE, Carriere Y (2011) Pollen- and seed-mediated transgene flow in commercial cotton seed production fields. PLoS One 5

123. Zhang BH, Pan XP, Guo TL, Wang QL, Anderson TA (2005) Measuring gene flow in the cultivation of transgenic cotton (*Gossypium hirsutum* L.). Mol Biotechnol 31:11–20

124. Llewellyn D, Fitt G (1996) Pollen dispersal from two field trials of transgenic cotton in the Namoi Valley, Australia. Mol Breed 2:157–166

125. Umbeck PF, Barton KA, Nordheim EV, McCarty JC, Parrott WL, Jenkins JN (1991) Degree of pollen dispersal by insect from a field test of genetically engineered cotton. J Econ Entomol 84:1943–1950

126. Wegier A, Pineyro-Nelson A, Alarcon J, Galvez-Mariscal A, Alvarez-Buylla ER, Pinero D (2011) Recent long-distance transgene flow into wild populations conforms to historical patterns of gene flow in cotton (*Gossypium hirsutum*) at its centre of origin. Mol Ecol 20:4182–4194

127. Free JB (1970) Insect pollination of crops. In: Insect pollination of crops. p 544

Chapter 2

Genetically Modified Cotton in India and Detection Strategies

Gurinder Jit Randhawa and Rashmi Chhabra

Abstract

India is one of the largest cotton-growing countries. Cotton is a fiber crop with varied applications from making tiny threads to fashionable clothing in the textile sector. In the near future, cotton crop will gain popularity as a multipurpose crop in India. The commercialization of *Bt* cotton in 2002 and consequently the fast adoption of *Bt* cotton hybrids by cotton farmers have enhanced the cotton production in India. Presently, genetically modified (GM) cotton has occupied 21.0 million hectares (mha) that comprise 14% of the global area under GM cultivation. In the coming years, improved cotton hybrids, with stacked and multiple gene events for improved fiber quality, insect resistance, drought tolerance, and herbicide tolerance, would further significantly improve the cotton production in India. With the dramatic increase in commercialization of GM crops, there is an urgent need to develop cost-effective and robust GM detection methods for effective risk assessment and management, post release monitoring, and to solve the legal disputes. DNA-based GM diagnostics are most robust assays due to their high sensitivity, specificity, and stability of DNA molecule.

1. Introduction

Cotton (*Gossypium hirsutum* L.), a fiber crop, is being cultivated in an area of 11.0 mha in India (1), the largest cotton-growing country in the world. Though cotton is a fiber crop, it is regarded as a multipurpose crop in India because of its usage in the form of both cotton lint and cottonseeds. Cotton is used as (1) an edible oil for human consumption, (2) de-oiled cake as an animal feed, and (3) kapas for fiber (2). Keeping in view the economic importance of cotton and major biotic threat to cotton production, due to insect pests, GM cotton for insect resistance was developed. *Bt* cotton expresses insect resistance transgene from *Bacillus thuringiensis*, conferring resistance to bollworm, a lepidopteron insect

Baohong Zhang (ed.), *Transgenic Cotton: Methods and Protocols*, Methods in Molecular Biology, vol. 958,
DOI 10.1007/978-1-62703-212-4_2, © Springer Science+Business Media New York 2013

pest of cotton. In India *Bt* cotton was commercialized for the first time in 2002. Presently, Bt cotton is being cultivated in an area of more than 10.6 mha in India, which is 86% of the total cotton-growing area (1).

1.1. Commercialization of GM Crops in India

So far, cotton is the only GM crop which has been commercialized In India, occupying 15.4% of the global area under GM cultivation. The first *Bt* cotton event, i.e., MON531 (Bollgard® I), was commercialized in India way back in 2002. In 2011, 883 hybrids and 1 variety of six events, i.e., MON531 with *cry1Ac* gene, MON15985 (Bollgard® II) with *cry1Ac* and *cry2Ab* genes, Event1 with *cry1Ac* gene, GFM-*cry1A* with fused *cry1Ab* and *cry1Ac* genes, Dharwad Event (Bt Bikaneri Nerma) in a variety with truncated *cry1Ac* gene, and 9124 Bt cotton with synthetic *cry1C* gene in hybrids, have been commercially released (1) (Fig. 1). Out of these six commercialized events, three events, MON531, MON15985 (Maharashtra Hybrid Seeds Co. Ltd.), and GFM-*cry1A* (Nath Seeds Ltd.), have been imported in the years 1995, 2000, and 2002, respectively, whereas the other three events are indigenously developed, i.e., Event 1 developed in the year 2002 at IIT, Kharagpur, using indigenous *cry1Ac* gene (3) and commercialized by J.K. Agrigenetics Ltd. while BN-Bt (4) was developed and commercialized by Central Institute of Cotton Research (CICR), Nagpur, in 2008 and event 9124 was developed and commercialized by Metahelix Life Sciences, Bangalore, in 2009 (1) (Table 1).

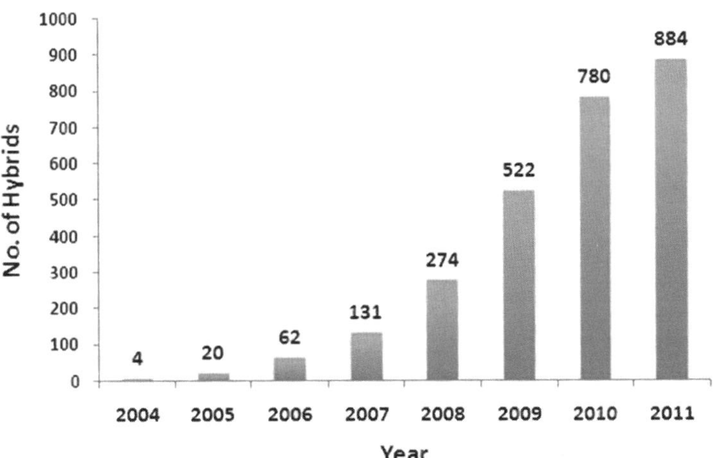

Fig. 1. Year-wise commercialization of Bt cotton hybrids (ISAAA, 2011).

Table 1

Commercially released hybrids/variety of six Bt cotton events in India

S. no	Event	No. of hybrids/variety	Developer
1.	MON-531	215	Mahyco/Monsanto
2.	MON-15985	528	Mahyco/Monsanto
3.	Event-1	41	JK Agri-Genetics
4.	GFM event	96	Nath Seeds
5.	BNLA-601	2[a]	CICR(ICAR) & UAS Dharwad
6.	MLS-9124	2	Metahelix Life Sciences

Source: ISAAA, 2011

[a]Bt Cotton variety

1.2. GM Cotton Events Commercialized, in Field Trials, and Imported Events in India

1.2.1. Commercialized GM Cotton Events

Till date, 883 hybrids and one variety of six events of Bt cotton have been commercialized in India. Out of these hybrids and variety, 59.7% hybrids are of stacked Bt cotton event (Bollgard® II) (Fig. 3).

1. Single gene GM cotton events: Five insect resistance commercialized Bt cotton events are single gene events, viz., (1) MON531 expressing *cry1Ac* gene, (2) Event 1 expressing synthetic *cry1Ac* gene, (3) GFM-cry1A expressing fused *cry1Ac-1Ab* gene, (4) BN-Bt expressing truncated *cry1Ac* gene, and (5) MLS-9124 expressing *cry1C* gene. First three events occupy 24.3%, 4.63%, and 10.85% of total GM cotton area, respectively, whereas cotton hybrids of BN-Bt event and MLS-9124 event collectively cover 0.4% (Fig. 2).

2. Stacked GM cotton events: MON15985 (Bollgard® II), expressing insect resistance *cry1Ac* and *cry2Ab* genes, is the only stacked event amongst the six commercialized events, covering 59.7% of total Bt cotton hybrids (Fig. 3). MON15985 has a trait for enhanced protection against a range of insects, viz., Spodoptera (a leaf-eating tobacco caterpillar), American bollworm, Pink bollworm, and Spotted bollworm. Due to better performance of stacked event of *Bt* cotton, higher profits are being earned due to cost savings associated (1) with lesser sprays for Spodoptera control and (2) increasing yield by 8–10% over single gene *Bt* cotton hybrids.

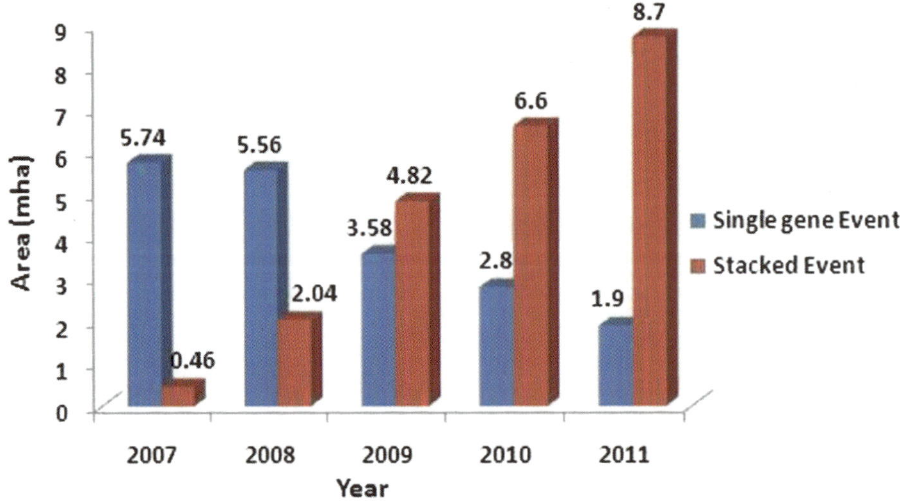

Fig. 2. Area under single gene and stacked gene Bt cotton hybrids (2007–2010). Source: ISAAA, 2011.

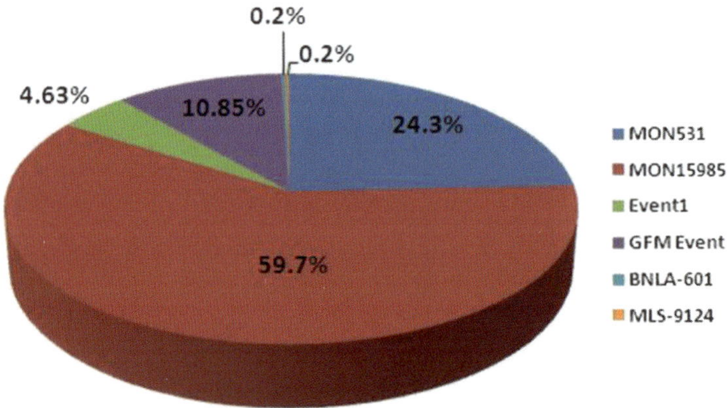

Fig. 3. Commercialized Bt cotton hybrids of six events. Source: ISAAA, 2011.

1.2.2. GM Cotton Events in Field Trials

Amongst the GM cotton events under field trials in India from 2009 to 2011, 54.6% are single gene events and 44.4% of the events are stacked (Table 2).

1. Single gene GM cotton events: Single gene events under field trials express *cry1Ac*, *cry1Ec*, *cry1F*, *cry2Ae*, and *cry1Ab* for insect resistance and *epsps*, *2mepsps*, and *pat* for herbicide tolerance.

2. Stacked GM cotton events: Stacked GM events under field trials include:

   (a) MON15985 x MON88913 (Bollgard®II-Roundup Ready Flex (BGIIRRF®)) of Mahyco, expressing *cry1Ac*, *cry2Ab*, and *epsps* genes for both insect and herbicide resistance.

   (b) Widestrike of Dow Agrosciences, expressing *cry1F* and *cry1Ac* genes for insect resistance.

Table 2
GM cotton in field trials in India (2009–2011)

Event	Gene	Trait	Institute/company
2009			
Event-1	*cry1Ac*	Insect resistance	JK Agrigenetics Ltd, Hyderabad
Event-24	*cry1EC*		
Widestrike	*cry1 Ac* and *cry1F*		Dow Agrosciences India Pvt. Ltd., Mumbai
Event 3006-210-23 x Event 281-24-236			
2010			
WideStrike Event 3006-210-23 x Event 281-24-236	*cry1Ac* and *cry1F*	Insect resistance	Dow AgroSciences India Pvt. Ltd., Mumbai
Event-1	*cry1Ac*		JK Agri Genetics Ltd., Hyderabad
Event-24	*cry1EC*		
ILK-Bt 77-1 to ILK-Bt 77-7	*cry1Ac*		Central Institute for Cotton Research, Nagpur
Anjali-AcBt-1, Anjali-AcBt-2, Anjali-AcBt-3	*cry1Ac*		
Anjali-FBt-1, Anjali-FBt-2			
G-822-Bt (Desi Bt G 822-1 to Desi Bt G-822-4),	*cry1F*		
PA255-Bt (CICR Bt Desi-1 to CICR Desi Bt-6)	*cry1Ac*		
Events MIR-cotton (1–131)	*cry1Ac* and *cry1EC*	Insect resistance	Krishidhan Seeds Ltd., Jalna
SP499 G, SP503 G, SP7017 G, SP7140 G, SP7139 G, SP7152 G, SP7230 G	*2mEPSPS*	Herbicide tolerant	Bayer Bioscience Pvt. Ltd, Hyderabad
2011			
GHB 119	*cry2Ae/ PAT*	Insect resistance and herbicide tolerant	Bayer Bioscience Pvt Ltd., Hyderabad
T304-40	*cry1Ab/PAT*		
MLS9124 or GFM Cry1A event		Insect resistance	Metahelix Life Sciences Pvt. Ltd., Bangalore
MON 88913	*cp4epsps*	Herbicide tolerant	Maharashtra Hybrid Seeds Company Ltd., Jalna

Source: http://www.igmoris.nic.in

(c)   Event-1 x Event-24 of J.K. Agrigenetics, expressing *cry1Ac* and *cry1Ec* genes for insect resistance.

1.2.3. Imported GM Cotton Events

Twenty six (26) GM cotton planting material have been imported for research purposes by various public and private research institutions from the USA, China, and Israel, through National Bureau of Plant Genetic Resources (NBPGR), the nodal organization under Indian Council of Agricultural Research (ICAR) for

Table 3
Details of imported GM cotton planting material for research purposes

S. no	Trait (no. of imports)	Transgenes	Importing institute/company
1.	Insect resistance (14)	*crylAc, cry2Ab, crylAb, cry2Ae, crylF, GFM crylA*	• Syngenta India Pvt. Ltd., Pune • Nath Seeds Ltd., Aurangabad • De-nocil Crop Protection Pvt. Ltd., Mumbai • Proagro, PGS Ltd., Gurgaon • Vikki's Agrotech Pvt. Ltd., Hyderabad • Bayer Bioscience Pvt. Ltd., New Delhi
2.	Herbicide tolerance (5)	*epsps, 2mepsps, bar*	• Maharashtra Hybrid Seeds Co. Ltd., New Delhi • Maharashtra Hybrid Seeds Co. Ltd., Mumbai • Bayer Bioscience Pvt. Ltd., New Delhi
3.	Insect resistance and herbicide tolerance (4) (Stacked Events)	*crylAc, cry2Ab, crylF, epsps*	• Emergent Genetics India Pvt. Ltd., Hyderabad • Monsanto Genetics India Pvt. Ltd., Mumbai
4.	Abiotic stress tolerance (3) Drought tolerance Drought and salinity tolerance	*35 S rol A, B, and C; Mannosyl transferase At A-20, At SOS1, At SOS2, At-ANP1, At-CBF3*	• Nath Seeds Ltd., Aurangabad • Ankur Seeds Pvt. Ltd., Nagpur

import and quarantine processing of transgenic planting material (Table 3).

1. Single gene GM cotton events: More than 50% of these imports are for insect resistance, and other GM events are for herbicide tolerance, for both insect and herbicide resistance, and for abiotic stress tolerance (Fig. 4a, b).

   These imports constitute a range of GM traits with an array of transgenes:

   (a) Insect resistance: *crylAc, vip3A, crylAb–cylAc, crylF, cry2Ab, crylAb,* and *cry2Ae.*

   (b) Herbicide tolerance: *epsps, 2mepsps,* and *bar.*

   (c) Insect Resistance and Herbicide Tolerance (Stacked Events): *crylAc, cry2Ab, crylF,* and *epsps.*

   (d) Abiotic stress tolerance: *35 S-rolA, B,* and *C, Mannosyl transferase, At A-20, At SOS1, At SOS2, At-ANP1,* and *At-CBF-3.*

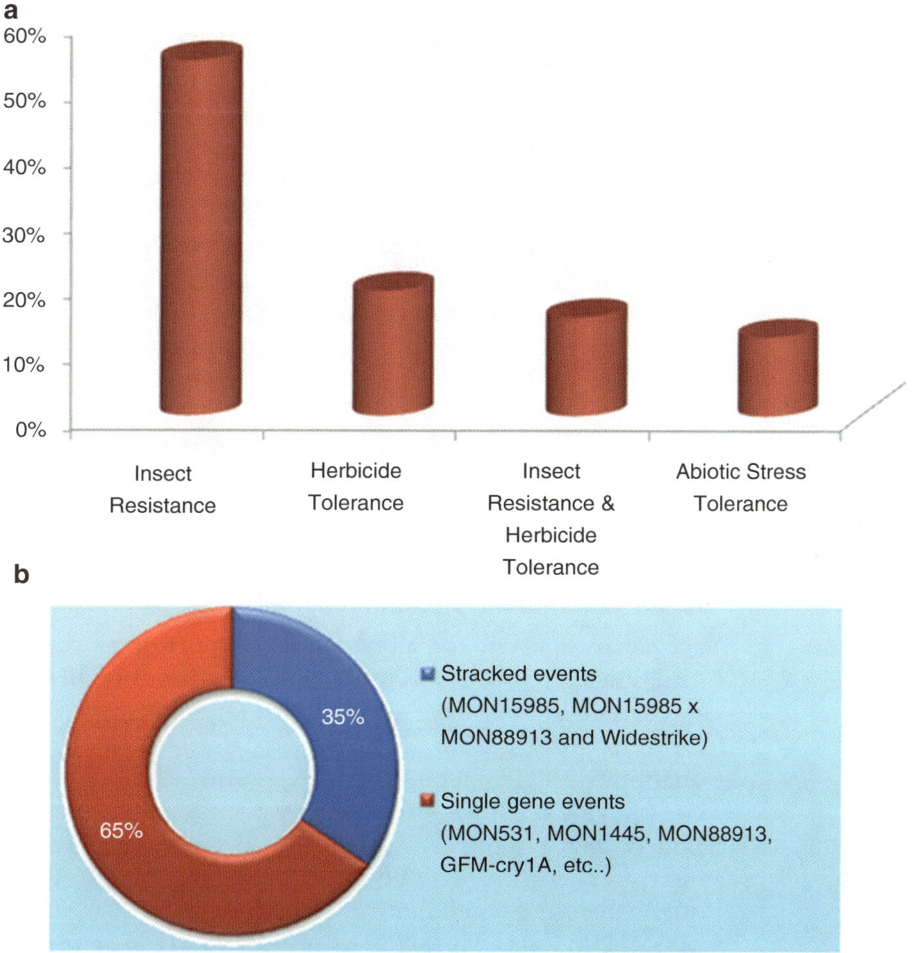

Fig. 4. (**a**) Trait-wise analysis of imported GM cotton planting material in India; (**b**) stacked and single gene events in imported GM cotton in India.

2. Stacked GM cotton events: Out of the 26 imports of GM cotton planting material, 34% of imports are of stacked GM events including Bollgard®II, BGIIRRF® and Widestrike.

The trend of GM cotton commercialized, imported, and under field trials, in India, clearly indicates that the stacked events are going to be much more in demand in the near future.

2. Detection Methods Employed for GM Cotton

The development, commercialization, and deployment of GM crops, both in terms of acreage of cultivated land as well as event/trait diversification, are increasing dramatically. The availability of reliable and robust assays, reference materials, and analytical methods that allow identification and accurate determination of GM content/trait in crops is an important key element to meet the regulatory obligations and legislative requirements as well as to effectively address the biosafety issues pertaining to GM crops.

Among conventional PCR technologies, multiplex PCR is time efficient and cost-effective, which can detect multiple target sequences of inserted gene construct in a single reaction. The multiplex PCR assays for screening of different *Bt* crops either commercialized or under field trials in India have been developed (5). Several multiplex PCR methods have been developed and validated for precise and accurate monitoring, tracing, and regulation of GM cotton (6–9). A decaplex PCR and real-time PCR for identification and differentiation of MON531 and MON15985, two major commercialized events of Bt cotton in India, have also been reported (6) (Table 4).

To increase the accuracy, sensitivity, and reproducibility for detection of GM crops and for automatic and high throughput, multiplex PCRs have been coupled with other methods. In 2009, Nadal et al. developed a multiplex PCR assay coupled to capillary gel electrophoresis for amplicon identification by size and color (multiplex PCR-CGE-SC) for simultaneous detection of cotton species and five events of GM cotton, viz., Bollgard®I, Bollgard®II, Roundup Ready, 3006-210-23, and 281-24-236 (10). Real-time quantitative PCR method for detection of Widestrike GM cotton event 281-24-236/3006-210-23 was developed based on detection of DNA sequences in the junction between the transgene insert and cotton genome (11). Lee et al. (2007) reported the qualitative and quantitative detection of GM cotton events MON15985 and MON88913 using two kinds of specific primer pairs, probes, and one standard plasmid, and confirmed the applicability for practical use by in-house validation experiment (12). Real-time PCR assays have also been developed by our laboratory for quantification of *cry1Ac* and *cry2Ab* genes in two *Bt* cotton events, viz., MON531 and MON15985 of (6).

For rapid screening of GM cotton expressing *chitinase* (*chi*) gene and Bt cotton containing the *cry1A(b)* gene, a visual and rapid loop-mediated isothermal amplification (LAMP) assay was developed by Rostamkhani et al. in 2011 (13). This method can amplify nucleic acids with high specificity, sensitivity, and speed under isothermal conditions (14).

Tohidfar et al. have reported PCR and southern blot analysis to confirm the integration of *cry1Ab* and *nptII* transgenes into the GM cotton genome. Western immunoblot analysis of proteins extracted from leaves of GM cotton revealed the presence of an immunoreactive band with a molecular weight (MW) of approximately 67 kDa in transgenic cotton lines using the anti-Cry1Ab polyclonal antiserum (15).

Table 4

Summary of DNA-based detection systems being employed for GM cotton

GM event/crop	PCR system	Target gene	References
MON531 and MON15985	Decaplex PCR	*cry1Ac* and *cry2Ab* transgenes; *nptII*, *aadA*, and *uidA* marker genes; *CaMV* 35S promoter and *nos* terminator; two construct-specific sequences, i.e., *cry1Ac* transgene construct and *cry2Ab* transgene construct; and endogenous *Sad1* gene	(6)
	Real-time PCR	Quantification of *cry1Ac* and *cry2Ab* genes	
GM cotton (VipCot14, VipCot29)	Multiplex; construct-specific PCR	*vip-s* gene and *vip3A*-like genes, *CaMV* 35 S promoter, *nos* terminator, and *npt II* marker gene	(9)
MON15985	Multiplex PCR	*Cry2Ab*, promoter, terminator, and *nptII* genes	(7)
Widestrike cotton (Event 281-24-236x 3006-210-23)	Quantitative real-time PCR	A cotton-specific endogenous reference gene *SAH7* and event s 281-24-236 and 3006-210-23	(11)
Mon531, GK19, SGK321	Conventional as well as quantitative	Cowpea trypsin inhibitor (*CpTI*) gene of SGK321 cotton and the specific junction DNA sequences containing partial *Cry1A(c)* gene and *NOS* terminator of Mon531, GK19, and SGK321 cotton varieties	(8)
Bollgard I, Bollgard II, Roundup Ready, 3006-210-23, and 281-24-236	Multiplex PCR-CGE-SC	Bollgard I, Bollgard II, Roundup Ready, 3006-210-23, and 281-24-236	(10)
MON15985, MON88913	Event-specific qualitative PCR and quantitative real-time PCR	MON15985, MON88913	(12)
GM Cotton	LAMP	*Chitinase*	(13)
Bt Cotton	PCR, Southern Blot and Western Blot	*Cry1Ah, nptII*	(15)

3. Detection Methods for GM Cotton in India

CICR, Nagpur, has developed three Bt cotton testing kits, namely, Cry1Ac Bt-Quant, an ELISA kit; Cry 1Ac Bt-detect, a dot-blot assay kit; and Cry1Ac Bt express, a dip-stick format, for the detection of Bt toxin. These kits have been effectively deployed to verify

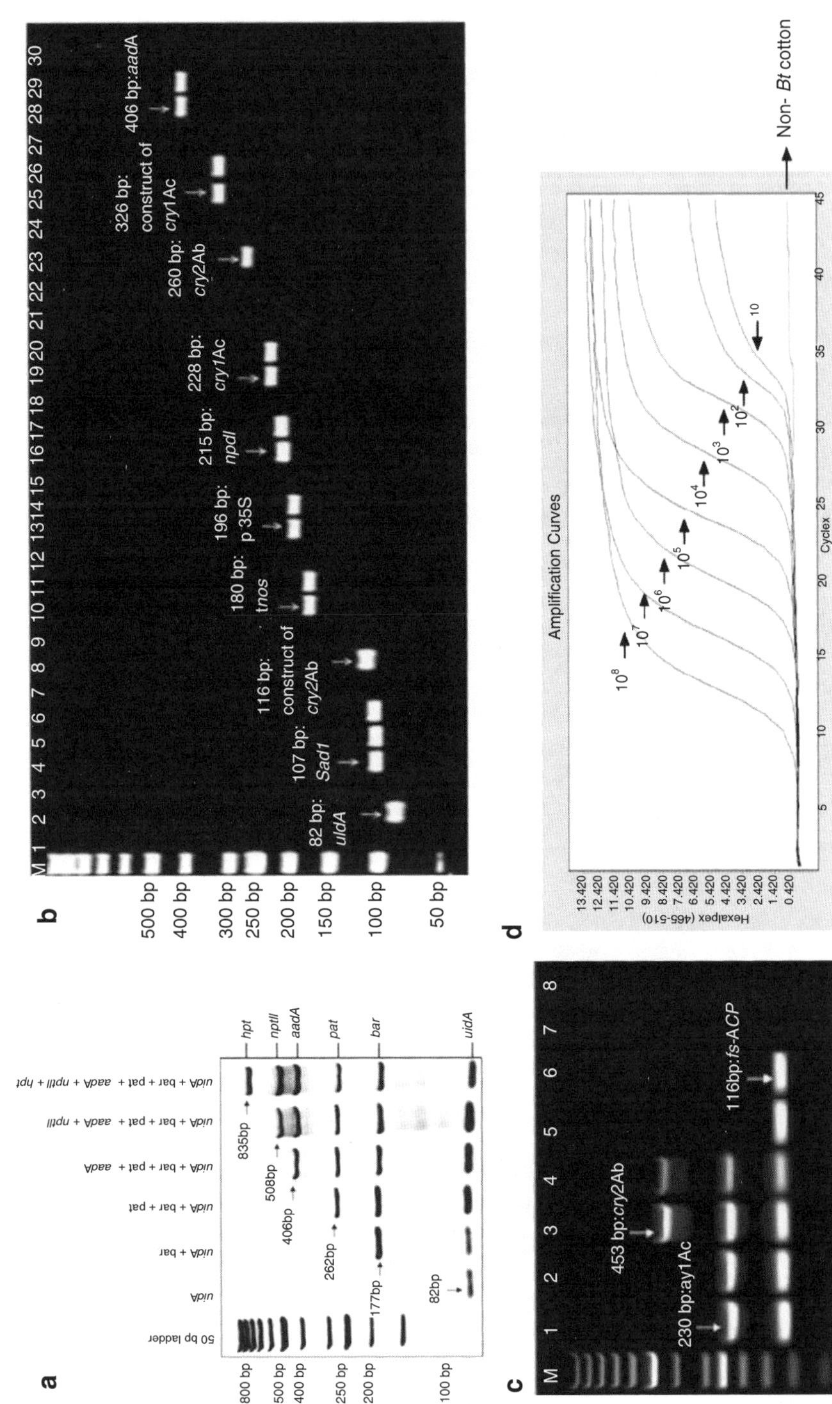

Fig. 5. (**a**) Hexaplex PCR for simultaneous amplification of six commonly used marker genes, i.e., *uidA*, *bar*, *pat*, *aadA*, *nptII*, and *hpt*; (**b**) simplex PCR for amplification of inserted genes, construct-specific sequences, and endogenous gene in two Bt cotton events; i.e., MON531 and MON15985, using primer pairs for *cry1Ac* and *cry2Ab* transgenes, *nptII*, *aadA*, and *uidA*
Fig. 5. (continued) marker genes, *CaMV35S* promoter, *nos* terminator, endogenous *Sad1* gene, and specific gene constructs in MON531/MON15985 and MON15985: (lane M) 50 bp ladder; (lanes 1, 4, 7, 10, 13, 16, 19, 22, 25, 28) samples of MON531 cotton; (lanes 2, 5, 8, 11, 14, 17, 20, 23, 24, 29) samples of MON531/MON15985 cotton; (lanes 3, 6, 9, 12, 15, 18, 21, 24, 27, 30) samples of non-GM cotton; (**c**) triplex PCR to differentiate MON531 and MON15985 Bt cotton events, Lane M: 50 bp ladder, Lanes 1–2: MON531, Lanes 3–4: MON15985, Lanes 5–6: Non-GM Cotton, Lane 7: Water control; (**d**) amplification curves generated for eight serial dilutions of standard plasmid with 10 to 10^8 copies of *cry2Ab* gene.

the purity of Bt seed and ensure the supply of quality Bt hybrid seed to the farming community.

NBPGR, New Delhi, has also developed robust DNA-based GM diagnostics for initial screening and for identification and quantification of GM content in more than ten GM crops. Some of these GM diagnostics, with special reference to GM cotton, are the following:

1. For initial screening of GM crops for checking the GM status of a sample irrespective of crop and trait, PCR assays have been developed targeting commonly used markers, promoter and terminator genes in simplex and multiplex formats.

2. Decaplex and triplex PCR assays have been developed to differentiate between two major commercialized Bt cotton events (covering more than 80% of the total cultivated area for GM cotton) in India, viz., MON531 and MON15985 along with simplex PCRs for each transgene element present in these two Bt cotton events (5, 6) (Fig. 5b).

3. Real-time PCR-based quantitative analysis of *cry1Ac* gene in Bt cotton events, MON531 and MON15985; *cry2Ab* gene in MON15985 has also been developed (6) (Fig. 5c).

4. Rapid and cost-effective diagnostic kits for GM cotton events, viz., Bollgard®I (MON531) and Bollgard®II (MON15985), have also been developed.

   (a) Hexaplex PCR assay has been developed for simultaneous amplification of commonly used six marker genes, i.e., *aadA*, *bar*, *hpt*, *nptII*, *pat*, and *uidA* (16) (Fig. 5a).

   (b) Heptaplex PCR assay simultaneously amplifying a combination of marker genes, *nptII*, *aadA*, *pat*, and *uidA* and regulatory elements, viz., *CaMV* 35 S, *nos* promoters, and *nos* terminator, has also been developed.

References

1. Clives J (2011) Global status of commercialized biotech/GM crops: 2011. ISAAA brief no. 43. Ithaca, NY, ISAAA

2. Choudhary B, Gaur K (2010) *Bt* cotton in India: a multipurpose crop, 2–4

3. Nayak P, Basu D, Das S, Basu A, Ghosh D, Ramakrishnan NA, Ghosh M, Sen SK (1997) Transgenic elite indica rice plants expressing cry1Ac ä-endotoxin of *Bacillus thuringiensis* are resistant against yellow stem borer. Proc Natl Acad Sci USA 94:2111–2116

4. Katageri IS, Vamadevaiah HM, Udikeri SS, Khadi BM, Umar PA (2007) Genetic transformation of an elite genotype of cotton (*Gossypium hirsutum* L.) for insect resistance. Curr Sci 93(12):1843–1847

5. Randhawa GJ, Singh M, Chhabra R, Sharma R (2010) Qualitative and quantitative molecular testing methodologies and traceability systems for Bt crops commercialised or under field trials in India. Food Anal Methods 3(4):295–303

6. Randhawa GJ, Chhabra R, Singh M (2010) Decaplex and real-time PCR based detection of MON531 and MON15985 *Bt* cotton events. J Agric Food Chem 58(18):9875–9881

7. Singh KC, Ojha A, Kachru DN (2007) Detection and characterization of *cry1Ac* transgene construct in *Bt* cotton: multiplex polymerase chain reaction approach. J AOAC Int 90:1517–1525

8. Yang L, Pan A, Zhang K, Guo J, Yin C, Chen J, Huang C, Zhang D (2005) Identification

and quantification of three genetically modified insect resistant cotton lines using conventional and TaqMan real-time polymerase chain reaction methods. J Agric Food Chem 53(16):6222–6229

9. Singh CK, Ojha A, Bhatanagar RK, Kachru DN (2007) Detection and characterization of recombinant DNA expressing vip3A-type insecticidal gene in GMOs-standard single, multiplex and constructspecific PCR assays. Anal Bioanal Chem 390(1):377–387

10. Nadal A, Esteve T, Pla M (2009) Multiplex polymerase chain reaction-capillary gel electrophoresis: a promising tool for GMO screening–assay for simultaneous detection of five genetically modified cotton events and species. J AOAC Int 92(3):765–772

11. Baeumler S, Wulff D, Tagliani L, Song P (2006) A real-time quantitative PCR detection method specific to widestrike transgenic cotton (event 281-24-236/3006-210-23). J Agric Food Chem 54:6527–6534

12. Lee SH, Kim JK, Yi BY (2007) Detection methods for biotech cotton MON15985 and MON88913 by PCR. J Agric Food Chem 55:3351–3357

13. Rostamkhani N, Haghnazari A, Tohidfar M, Moradi A (2011) Rapid identification of transgenic cotton (*Gossypium hirsutum* L.) plants by loop-mediated isothermal amplification. Czech J Genet Plant Breed 47(4):140–148

14. Fu S, Qu G, Guo S, Ma L, Zhang N, Zhang S, Gao S, Shen Z (2010) Applications of loop-mediated isothermal DNA amplification. Appl Biochem Biotechnol 163:845–850

15. Tohidfar M, Ghareyazie B, Mosavi M, Yazdani S, Golabchian R (2008) *Agrobacterium*-mediated transformation of cotton (*Gossypium hirsutum*) using a synthetic cry1Ab gene for enhanced resistance against *Heliothis armigera*. Iranian J Biotechnol 6(3):164–173

16. Randhawa GJ, Singh M, Chhabra R (2009) Multiplex PCR for simultaneous amplification of selectable marker genes and reporter genes for the screening of genetically modified crops. J Agric Food Chem 57:5167–5172

Part II

Transformation

Chapter 3

Agrobacterium-Mediated Transformation of Cotton

Baohong Zhang

Abstract

There are many methods and techniques that can be used to transfer foreign genes into cells. In plant biotechnology, *Agrobacterium*-mediated transformation is a widely used traditional method for inserting foreign genes into plant genome and obtaining transgenic plants, particularly for dicot plant species. *Agrobacterium*-mediated transformation of cotton involves several important and also critical steps, which includes coculture of cotton explants with *Agrobacterium*, induction and selection of stable transgenic cell lines, recovery of plants from transgenic cells majorly through somatic embryogenesis, and detection and expression analysis of transgenic plants. In this chapter, we describe a detailed step-by-step protocol for obtaining transgenic cotton plants via *Agrobacterium*-mediated transformation.

1. Introduction

Transgenic technology is a powerful technique for improving plant yield, quality, and tolerance to abiotic and biotic stress as well for fundamental basic research, such as investigating the gene function. There are many methods and techniques that can be used to transfer foreign genes into plant cells, which include *Agrobacterium*-mediated transformation, particle bombardment (gene gun) with DNA-coated microprojectiles, electroporation, pollen tube pathway, and PEG treatment of protoplasts (1). Of these methods and techniques, *Agrobacterium*-mediated transformation is a widely used traditional method for inserting foreign genes into plant genome and obtaining transgenic plants.

Plant Crown Gall disease is a common disease of many woody plants and some herbaceous plants, which is caused by the bacterium *Agrobacterium tumefaciens*. *Agrobacterium tumefaciens* is a gram-negative soil bacterium that infects the wounded plant through its Ti plasmid. The Ti plasmid transfers and insert its own

Baohong Zhang (ed.), *Transgenic Cotton: Methods and Protocols*, Methods in Molecular Biology, vol. 958,
DOI 10.1007/978-1-62703-212-4_3, © Springer Science+Business Media New York 2013

gene fragment (part of the Ti plasmid, called T-DNA) into plant genome and then causes plant tumor. The transformation and integration of T-DNA from *Agrobacterium* into plant genome is a complicated process with complicated mechanism, in which partial mechanism is well studied (2). Generally speaking, there are six steps involved in this process, which include the following: (1) bacterium and plant cell recognition, (2) wounded plant cell release signal and activating bacterium *vir* genes, (3) T-DNA is excised from Ti-plasmid, transported into (4) the plant cell and then into (5) the nuclei, and finally integrated into plant genome (2). Many plants, particularly dicots, are susceptible to *Agrobacterium*. Thus, *Agrobacterium* is a natural transformation tool for plant transformation.

Agrobacterium-mediated transformation is also the major transformation method for obtaining transgenic cotton. The first transgenic cotton plants were obtained through *Agrobacterium*-mediated method (3, 4), in which two independent research groups transformed reporter gene and selectable markers neomycin phosphotransferase (*nptII*) into cotton and obtained stable transformant cotton plants. Since then, many research groups have established their research strategies for obtaining transgenic cotton using *Agrobacterium*-mediated transformation (5–23). Although there is a little bit of difference among these protocols, all *Agrobacterium*-mediated transformation of cotton involves several important and also critical steps, which include coculture of cotton explants with *Agrobacterium*, induction and selection of stable transgenic cell lines, recovery of plants from transgenic cells majorly through somatic embryogenesis, and detection and expression analysis of transgenic plants. In this chapter, we describe a detailed step-by-step protocol for obtaining transgenic cotton plants via *Agrobacterium*-mediated transformation.

2. Materials

2.1. Coculture of Cotton Explants with Agrobacterium

1. Seeds of upland cotton (*Gossypium hirsutum* L.) cultivar Coker 201 (see Note 1).

2. *Agrobacterium tumefaciens* strain LBA4404 harboring pCNL56. The plasmid pCNL56 binary vector carries *uidA* as reporter gene and the nptII gene as the selectable marker gene (see Note 2).

3. Murashige and Skoog (MS) inorganic salts (24).

4. B$_5$ vitamins (25).

5. Agar.

6. Sucrose.

7. Kanamycin. Store at –20°C.
8. Rifampicin. Store at –20°C.
9. 2,4-D. Store at 4°C.
10. ZT. Store at –20°C.
11. Sterilized deionized water (ddH$_2$O).
12. Bleach.
13. Ethanol.
14. 1 N KOH.
15. 1 N HCL.
16. Glucose.
17. MES.
18. Sodium phosphate buffer pH 5.6.
19. Acetosyringone.
20. ProX 0.22 μm membrane filter.
21. Syringe.
22. Hood.
23. 250×30 mm test tubes.
24. 1.5 mL centrifuge tubes.
25. Filter paper.
26. LB medium.
27. Plant growth incubator with light and temperature control.

2.2. Induction and Selection of Stable Transgenic Cell Lines

1. Carbenicillin. Store at –20°C.
2. IAA.
3. Others are same as in Subheading 2.1.

2.3. Recovery of Plants from Transgenic Cells

1. Charcoal.
2. Others are same as in Subheading 2.2.

2.4. Detection and Expression Analysis of Transgenic Plants

There are many methods, such as Northern blotting and Western blotting, for detection and expression analysis of transgene in transgenic plants. All these methods can be used in this step. Here, we only present a quantitative real-time PCR (qRT-PCR) method.

2.4.1. Isolation of Total RNA

In this protocol, mirVana™ miRNA Isolation Kit is employed to isolate total RNAs from cotton leaves.

1. Leaves from tested cotton plants (see Note 3).
2. *mir*Vana™ miRNA Isolation kit (Ambion, Austin, TX).
   (a) Lysis/Binding Buffer: Store at 4°C.
   (b) miRNA Homogenate Additive: Store at 4°C.

(c) miRNA Wash Solution 1. Before use, add 21 mL 100% ethanol into the Wash Solution 1. This solution contains guanidinium thiocyanate that is a potential biohazard. Thus, wear gloves and handle with care.

(d) Wash Solution 2/3. Before use, add 40 mL 100% ethanol. This solution can be left at room temperature for up to 1 month. For longer storage periods, store at 4°C but warm to room temperature before use.

(e) Collection tubes.

(f) Filter cartridges.

(g) Acid-phenol:Chloroform. Store at 4°C. Phenol is a poison and an irritant and therefore gloves or other protection should be worn when handling this reagent. Dispose of phenol waste appropriately.

(h) Elution Solution or Nuclease-Free Water. Preheated to 95°C when used and stored at 4°C or room temperature.

3. 100% RNase-free ethanol.

4. Liquid nitrogen.

5. RNase-free water.

2.4.2. RT-PCR

1. TaqMan® MicroRNA Reverse Transcription Kit (Applied Biosystems, Foster City, CA). Store at–20°C. All contents should be thawed on ice and centrifuged briefly before using.

(a) 10× RT Buffer.

(b) dNTP mix with dTTP (100 mM).

(c) RNase Inhibitor (20 U/μL).

(d) Multiscribe™ RT enzyme (50 U/μL).

2. Nuclease-Free water.

3. Poly T RT Primers.

2.4.3. qRT-PCR

1. PCR master mix with SYBR green.

2. PCR Primers for nptII gene (Forward primer is TCATCGA CTGTGGCCGGCTG; reversed primer is AAGCGGTCAG CCCATTCGCC).

3. PCR primers for reference genes UBQ7 and GAPDH (Forward primer for UBA7 is GAATGTGGCGCCGGGACCTTC; reversed primer for UBQ7 is ACTCAATCCCCACCAGCC TTCTGG; forward primer for GADPH is TGATGCCAA GGCTGGAATTGCTT; reversed primer for GADPH is GTGTCGGATCAAGTCGATAACACGG) (see Note 5).

4. Nuclease-Free water.

3. Methods

3.1. Coculture of Cotton Explants with Agrobacterium

3.1.1. Establishment of Sterilized Cotton Seedlings

1. Prepare the medium for seed germination and seedling development.
   - (a) Prepare the medium (BMSB) for seed germination and young seedling development. BMSB only contains half strength of MS inorganic salts and B_5 vitamins.
   - (b) Add 2% Sucrose into the BMSB medium.
   - (c) Adjust the pH value to 5.8 using 1 N KOH or 1 N HCl.
   - (d) Add 0.7 g/L agar to the BMSB medium.
   - (e) Aliquot medium to 250×30 mm test tubes. Each tube contains 20 mL medium.
   - (f) Autoclave under 121°C for 15 min.
   - (g) Allow the sterilized medium to be solidified in room temperature.
2. Sterilize and culture cotton seeds.

   All the following operation should be in a sterilized condition:
   - (a) Select the mature cotton seeds.
   - (b) Manually remove seed coats.
   - (c) Soak the seeds in 70% ethanol for 2 min.
   - (d) Wash the seeds twice using sterilized ddH_2O.
   - (e) Seeds are surface-sterilized by soaking the seeds in 10% bleach for 15 min.
   - (f) Wash the seeds for at least three times using sterilized ddH_2O (see Note 4).
   - (g) Dry seeds on a sterilized filter paper.
   - (h) Place seeds on the BMSB medium. Two seeds are placed in each tube (see Note 6).
   - (i) Allow the seed germinate under a 14-h day/10-h night cycle with a light intensity of 3,000 lux at 28 ± 2°C.

3.1.2. Prepare Agrobacterium for Transformation

1. Prepare the medium for *Agrobacterium* culture.
   - (a) Prepare LB medium.
   - (b) Aliquot LB medium to 125 mL flasks. Each tube contains 520 mL LB medium.
   - (c) Autoclave under 121°C for 15 min.
   - (d) Add 100 mg/L kanamycin and 10 mg/L rifampicin (see Note 7).
2. Culture *Agrobacterium*.
   - (a) Pick up two *Agrobacterium* colonies and inoculate them into the LB medium containing antibiotics (see Note 8).

(b) Culture *Agrobacterium* in an incubator at 28°C with 200 rpm shaking.

(c) Culture *Agrobacterium* for 24 h (see Note 9).

3.1.3. Coculture of Cotton Explants with Agrobacterium

1. Prepare the preinduction medium (PIM).

   (a) Prepare PIM. The PIM contains 1% glucose, 7.5 mM MES, and 2 mM sodium phosphate buffer pH 5.6.

   (b) Add acetosyringone to a final concentration of 100 µM just before to coculture the cotton explants and *Agrobacterium* (see Note 10).

2. Prepare coculture medium (MSBC).

   (a) Prepare coculture medium (MSBC) for coculture of *Agrobacterium* and cotton explants. MSBC contains MS inorganic salts and B$_5$ vitamins.

   (b) Add acetosyringone to a final concentration of 100 µM to the MSBC medium (see Note 10).

   (c) Add 0.5 mg/L 2,4-D and 0.1 mg/L ZT into the medium.

   (d) Add 3% sucrose into the MSBC medium.

   (e) Adjust the pH value to 5.8 using 1 N KOH or 1 N HCl.

   (f) Add 0.7 g/L agar to the MSBC medium.

   (g) Autoclave under 121°C for 15 min.

   (h) Aliquot medium to 100×20 mm Petri Dish. Each tube contains 20 mL medium (see Note 11).

   (i) Allow the sterilized medium to be solidified in room temperature.

   (j) Place a sterilized filter paper on the medium.

3. Coculture of *Agrobacterium* and cotton explants.

   (a) Select the well-developed 7–10-day-old young seedlings (see Note 12).

   (b) Cut cotyledons into small pieces with 5–7 mm in one dimension.

   (c) Cut hypocotyls into small segments with about 5–7 mm in length.

   (d) Place the cotyledon disks or hypocotyls segments into the PIM medium with *Agrobacterium*.

   (e) Coculture them for 10 min in PIM medium.

   (f) Place horizontally the hypocotyls segment on the MSBC medium with filter paper.

   (g) Place the cotyledon disks on the MSBC medium with filter paper. The adaxial sides should be faced down on the filter paper.

   (h) Culture them for 48 h in the dark at 22°C (see Note 13).

<table>
<tr><td valign="top">

3.2. Induction and Selection of Stable Transgenic Cell Lines

</td><td valign="top">

1. Prepare the medium (MSBIS) for inducing and selecting stable transgenic cell lines.

   (a) MSBIS medium contains MS inorganic salts and B_5 vitamins. Make the basic medium with MS inorganic salts and B_5 vitamins.

   (b) Add 0.5 mg/L 2,4-D and 0.1 mg/L ZT into the medium.

   (c) Add 3% sucrose into the MSBIS medium.

   (d) Adjust the pH value to 5.8 using 1 N KOH or 1 N HCl.

   (e) Add 0.7 g/L agar to the MSBC medium.

   (f) Autoclave under 121°C for 15 min.

   (g) Make antibiotics kanamycin and carbenicillin stock solution and sterilize by filtering through a ProX 0.22 μm membrane filter.

   (h) Add the sterilized antibiotics kanamycin and carbenicillin to the cooled medium (~50–60°C) after autoclaving with final concentration of 50 mg/L for kanamycin (see Note 14) and 400 mg/L for carbenicillin (see Note 15).

   (i) Aliquot medium to 175 mL flasks. Each flask contains 50 mL medium.

   (j) Allow the sterilized medium to be solidified in room temperature.

2. Induction, selection, and maintenance of callus.

   (a) After 48 h of coculture, the cotyledon disks or hypocotyls segments are transferred to MSBIS medium for inducing callus resistant to antibiotic kanamycin (see Note 16).

   (b) Culture them at 28 ± 2°C under a 14–16-h photoperiod with a light intensity of approximately 2,000 lux provided by cool white fluorescent lamps.

   (c) After 1 week of culture, callus is observed.

   (d) After 4 weeks of culture, the callus is transferred to fresh medium.

   (e) Regular subculture to fresh medium with 0.1 mg/L ZT and 0.1 mg/L IAA every 4 weeks.

</td></tr>
<tr><td valign="top">

3.3. Recovery of Plants from Transgenic Cells

</td><td valign="top">

Plant regeneration is an important step for obtaining transgenic cotton, in which inducing embryogenic callus is a critical step. Comparing with other plant species, it is difficult to obtain embryogenic callus from cotton tissue culture. Usually, embryogenic callus will appear after three to four times of subculture on MSBIS medium with 0.1 mg/L ZT and 0.1 mg/L IAA.

1. Prepare the medium (MSBE) for the induction and proliferation of embryogenic callus.

</td></tr>
</table>

(a) MSBE medium contains MS inorganic salts and B5 vitamins. Make the basic medium with MS inorganic salts and B5 vitamins.

(b) Add 0.1 mg/L 2,4-D and 0.1 mg/L ZT into the medium.

(c) Add 3% sucrose into the MSBE medium.

(d) Adjust the pH value to 5.8 using 1 N KOH or 1 N HCl.

(e) Add 0.7 g/L agar to the MSBC medium.

(f) Autoclave under 121°C for 15 min.

(g) Add the sterilized antibiotics kanamycin and carbenicillin to the cooled medium (~50–60°C) after autoclaving with final concentration of 150 mg/L for kanamycin (see Note 17).

(h) Aliquot medium to 175 mL flasks. Each flask contains 50 mL medium.

(i) Allow the sterilized medium to be solidified in room temperature.

2. Prepare the medium (MSBEF) for the formation and development of somatic embryos.

   The only different between MSBEF and MSBE medium is that 0.1 mg/L ZT and 2 g/L charcoal are added into the basic MSB medium instead of 0.1 mg/L ZT and 0.1 mg/L 2,4-D. The method for making MSBEF medium is same as preparing MSBE medium.

3. Prepare the medium (MSBP) for recovery of plant from somatic embryos.

   The only difference between MSBP and MSBE medium is that 0.1 mg/L ZT and 0.1 mg/L IAA are added into the basic MSB medium instead of 0.1 mg/L ZT and 0.1 mg/L 2,4-D. The method for making MSBP medium is same as preparing MSBE medium.

4. Transfer callus from Subheading 3.2, step 2e, to the MSBE medium.

5. Select and subculture the callus with high embryogenesis to fresh MSBE medium after 3 weeks (see Note 18).

6. Repeat step 5 if necessary.

7. Transfer embryogenic callus with high embryogenesis into MSBEF medium.

8. Subculture to fresh medium every 3 weeks.

9. Select somatic embryos at the stage of torpedo-shaped embryos or later into the MSBO medium.

10. Recovered plantlets are developed in about 2–3 weeks.

3.4. Detection and Expression Analysis of Transgenic Plants

3.4.1. Isolation of Total RNA

Young cotton leaves are collected from the regenerated plants or the plants from the Greenhouse. Following the manufacturer's protocol, total RNAs are isolated using mirVana™ miRNA Isolation Kit (Ambion, Austin, TX) (see Note 19).

1. Collect cotton leaves and immediately drop the samples into liquid nitrogen.

2. Grind the tissue sample into a fine powder in a prechilled mortar.

3. Quickly transfer the fine powder into a 1.5 mL microcentrifuge tube on ice.

4. Add 300 μL of Lysis/Binding buffer on ice.

5. Sonic for 15–20 s using an Ultrasonic Convertor on ice.

6. Add 30 μL (1/10 the volume of Lysis/Binding buffer) miRNA Homogenate Additive to the homogenate and mix well by vortexing.

7. Keep the mixture on ice for 10 min.

8. Add 300 μL (the same volume equal to the Lysis/Binding buffer before miRNA Homogenate Additive addition) Acid-Phenol/Chloroform to each tube (see Note 20).

9. Mix gently and thoroughly by inverting or vortexing the tube.

10. Centrifuge at $10,000 \times g$ at room temperature for 5 min (see Note 21).

11. Remove the aqueous upper phase to a new 1.5 mL tube and record the volume of transferred aqueous phase (see Note 22).

12. Add 100% ethanol (1.25 volumes of the aqueous phase or 375 μL if only 300 μL aqueous phase transferred) at room temperature.

13. Mix well but gently by inverting or vortexing.

14. Pass the mixture through a Filter Cartridge that sits on a collection tube by centrifuge or vacuum (see Note 23).

15. Add 700 μL of miRNA Washing Solution 1 to wash the filter cartridge.

16. Add 500 μL of miRNA Wash Solution 2/3 to continuously wash the filter cartridge.

17. Repeat step 16 with a second aliquot of equal volume of miRNA Wash Solution 2/3.

18. Dry the cartridges to remove all residual fluid from the filter.

19. Transfer the filter cartridge to a new collection tube.

20. Elute RNAs from the filter cartridges by applying 50 μL of preheated (95°C) nuclease-free water by centrifuge or vacuum (see Note 24).

21. Mix the recovered RNAs by gently flicking the tube.

22. Measure the quality and quantity of the total RNAs using a NanoDrop ND-100 (NanoDrop Technologies, Wilmington, DE).

23. Store RNA samples in a –80°C freezer until further use.

3.4.2. RT-PCR

1. Use 1 μg of total RNA per 15-μL RT reaction. Calculate how much volume of RNA sample is needed.

2. Calculate how much nuclease-free water is needed to make up a total of 15 μL RT reaction.

3. Make one RT reaction for each sample as per the following volume for each component (see Note 25):

Component	Volume/15 μL reaction
Nuclease-free water	??
RNase inhibitor, 20 U/μL	0.19
100 mM dNTPs (with dTTP)	0.15
10× Reverse transcription buffer	1.50
Poly T primer	1.00
MultiScribe™ reverse transcriptase, 50 U/μL	1.00
RNA sample	??
Total	15

Note: the volume of nuclease-free water is calculated based on the volume of RNA sample and other components

4. Mix the reaction gently. Then, briefly centrifuge to bring solution down to the bottom of the tube (2,000 ×*g* for 10 s).

5. Incubate the tube on ice for 5 min and keep on ice until you are ready to load the thermal cycler.

6. Perform reverse transcription:

Temperature (°C)	Time (min)
16	30
42	30
85	5
4	Hold

7. Store the RT products at –20°C or below.

3.4.3. qRT-PCR

1. Prepare one qRT reaction as per the following volume for each component:

Component	Volume/20 µL reaction
Nuclease-free water	6
2×PCR SYBE Green Master mix	10
Product from RT-PCR reaction	2
Forward and reverse primers	2
Total	20

2. Perform qRT-PCR

	Enzyme activation	PCR	
		Cycle (40 cycles)	
Step		**Denature**	**Anneal/extend**
Time	10 min	15 s	60 s
Temp (°C)	95	95	60

3. After running, store the qRT-PCR products at –20°C or below.

4. Notes

1. The major difficulty for cotton transformation using *Agrobacterium*-mediated method is the genotype-dependence. Currently, only a limited number of upland cultivars can be used for gene transformation because of their capability for obtaining regenerated plants via somatic embryogenesis.

2. Up to date, there are several *Agrobacterium* strains successfully employed to obtaining transgenic cotton. The most used strains are LBA 4404 and EHA105. Although each strain works, different strains affect the transformation rate in a different way. Two studies show that stain LBA 4404 is significantly better than stain EHA 105 (19) or C58C3 (26).

3. Cotton is rich in pigments and polyphenolic compounds. Compared to the old mature leaves, it is easier to obtain high yield and quality of RNA from young leaves.

4. Bleach can damage the seeds, and cause low generation rate and later damaged leaves. During the rising process, bleach should be completely removed until no bleach odor remains. In most cases, three times of rinse is enough; however, if not, more rinse should be performed.

5. There are many housekeep genes used in cotton expression profile analysis. Based on previous studies, we found that UBQ7, GAPDH, and EF1A8 are better reference genes in cotton leaves, particularly under stress condition.

6. Seed should be placed in a right direction; the bottom of seeds (roots) should face downside and it is better to insert into the medium about 2–3 mm. This way will help the seed germination and later development.

7. Antibiotics are unstable at high temperature. Before adding them into medium, you should allow the medium to cool down to about 70°C.

8. Before adding *Agrobacterium* in, the medium should be cooled down to room temperature; high temperature will kill the bacterium.

9. After 24 h of culture at 28°C, the *Agrobacterium* will reach to the exponential phase of growth; the value of A600 will be around 1.6–1.9.

10. Acetosyringone is a phenolic natural product, which is isolated from a variety of plant resources, particularly, in relation to wounding and other physiologic changes; it is well known that acetosyringone functions as signaling molecule, induces the expression of Agrobacterium vir genes, and then initiates the transformation process (27–30). Several studies show that adding acetosyringone into the coculture medium or pre-culture of *Agrobacterium* significantly enhances *Agrobacterium*-mediated transformation (31–34). Recent reports also demonstrate that acetosyringone enhances *Agrobacterium*-mediated transformation of cotton (19, 26, 35).

11. Antibiotics cannot be added in this step. If added, it will damage the wounded cells and significantly reduce or even eliminate the transformation.

12. Many plant tissues can be used for transformation using *Agrobacterium*. However, different explants may affect the transformation efficiency; cotyledon and hypocotyls are two tissues often used for cotton transformation.

13. Coculture can be carried out at any temperature between 21 and 28°C. However, several studies show that low temperature may enhance the *Agrobacterium*-mediated transformation efficiency in several plant species (36–38), including cotton (19).

14. The sensitivity varies from tissue to tissue to kanamycin exposure. According to a study, non-embryogenic callus is most sensitive to kanamycin, followed by cotyledon disk, hypocotyls segment, and then embryogenic callus (39). At the early stage of inducing transformed cells, 50 mg/L kanamycin is enough for inhibiting most cell growth. However, for selecting transgenic

embryogenic callus, the concentration of kanamycin should be increased to at least 150 mg/L (39).

15. Carbenicillin is used to kill the *Agrobacterium*, but it does not affect cotton cells at the used concentrations. *Agrobacterium* can continuously grow on the medium without carbenicillin and affect cotton callus induction, thus carbenicillin must be added into the medium.

16. Since then on, antibiotic kanamycin needs to be added into the medium. Without kanamycin, both transformed and non-transformed cells will survive and be divided, which will cause so many false positives in the regeneration process.

17. There are two major reasons for using high concentration (150 mg/L) of antibiotic kanamycin instead of 50 mg/L used in callus induction: (a) cotton embryogenic callus is not sensitive as non-embryogenic callus or other explants, high concentrations of kanamycin inhibits more the growth of non-embryogenic callus than that of embryogenic callus, such it will become easy to screen embyogenic callus.

18. The induction of embryogenic callus is the most critical step for obtaining transgenic cotton. In some cases, it may be a long process to obtain embryogenic callus and it depends on many factors, including genotype, medium, and combination of plant hormones.

19. mirVana™ miRNA Isolation Kit can be used to isolate total RNAs, including small RNAs. Other traditional RNA isolation methods can also be used in this step.

20. There are two layers for the solution of Acid-Phenol/Chloroform. The upper aqueous layer is water and the down layer is phenol/chloroform. Please make sure to draw from the bottom phase of the bottle. Otherwise, it will affect the participation of RNA and significantly reduce RNA yield.

21. The purpose of centrifuge is to separate the aqueous phase from the organic phase. If the interphase is not clear after the centrifugation, repeat with a second-round centrifuge.

22. This step should be done very carefully as not to disturb the lower phase. Otherwise, phenol/chloroform may contaminate the aqueous part, which will result in low quality and yield of RNAs. The volume of the transferred aqueous phase should be carefully measured and recorded because adding ethanol volume is based on this volume. If you mess up the interphase, you should centrifuge again. To increase the quality of RNAs, it is also better to recentrifuge after transferring the aqueous phase to a new tube.

23. The Filter Cartridge only can hold 700 μL solution. If more than 700 μL mixture is needed to filter through, repeat it until all solutions are filtered.

24. Adding the water on the center of the filter cartridge, preheating the water to 95°C, and also resting there for at least 1 min will increase the yield.

25. Thaw all components on ice. Always add the RNA samples as the last one.

References

1. Birch RG (1997) Plant transformation: problems and strategies for practical application. Annu Rev Plant Physiol Plant Mol Biol 48:297–326

2. Gelvin SB (2003) *Agobacterium*-mediated plant transformation: the biology behind the "gene-Jockeying" tool. Microbiol Mol Biol Rev 67:16–37

3. Firoozabady E, Deboer DL, Merlo DJ, Halk EL, Amerson LN, Rashka KE, Murray EE (1987) Transformation of cotton (*Gossypium hirsutum* L.) by *Agrobacterium tumefaciens* and regeneration of transgenic plants. Plant Mol Biol 10:105–116

4. Umbeck P, Johnson G, Barton K, Swain W (1987) Genetically transformed cotton (*Gossypium hirsutum* L.) plants. Bio-Technology 5:263–266

5. Asad S, Mukhtar Z, Nazir F, Hashmi JA, Mansoor S, Zafar Y, Arshad M (2008) Silicon carbide whisker-mediated embryogenic callus transformation of cotton (*Gossypium hirsutum* L.) and regeneration of salt tolerant plants. Mol Biotechnol 40:161–169

6. Chen TZ, Wu SJ, Zhao J, Guo WZ, Zhang TZ (2010) Pistil drip following pollination: a simple in planta Agrobacterium-mediated transformation in cotton. Biotechnol Lett 32:547–555

7. Hashmi JA, Zafar Y, Arshad M, Mansoor S, Asad S (2011) Engineering cotton (*Gossypium hirsutum* L.) for resistance to cotton leaf curl disease using viral truncated AC1 DNA sequences. Virus Genes 42:286–296

8. Katageri IS, Vamadevaiah HM, Udikeri SS, Khadi BM, Kumar PA (2007) Genetic transformation of an elite Indian genotype of cotton (*Gossypium hirsutum* L.) for insect resistance. Curr Sci 93:1843–1847

9. Kim HJ, Murai N, Fang DD, Triplett BA (2009) Functional analysis of *Gossypium hirsutum* cellulose synthase catalytic subunit 4 promoter in transgenic Arabidopsis and cotton tissues. Plant Sci 180:323–332

10. Li FF, Wu SJ, Chen TZ, Zhang J, Wang HH, Guo WZ, Zhang TZ (2009) Agrobacterium-mediated co-transformation of multiple genes in upland cotton. Plant Cell Tissue Organ Cult 97:225–235

11. Liu JF, Zhao CY, Ma J, Zhang GY, Li MG, Yan GJ, Wang XF, Ma ZY (2009) Agrobacterium-mediated transformation of cotton (*Gossypium hirsutum* L.) with a fungal phytase gene improves phosphorus acquisition. Euphytica 181:31–40

12. Nandeshwar SB, Moghe S, Chakrabarty PK, Deshattiwar MK, Kranthi K, Anandkumar P, Mayee CD, Khadi BM (2009) Agrobacterium-mediated transformation of cry1Ac gene into shoot-tip meristem of diploid cotton Gossypium arboreum cv. RG8 and regeneration of transgenic plants. Plant Mol Biol Rep 27:549–557

13. Wu JH, Luo XL, Zhang XR, Shi YJ, Tian YC (2011) Development of insect-resistant transgenic cotton with chimeric TVip3A* accumulating in chloroplasts. Transgenic Res 20:963–973

14. Wu SJ, Wang HH, Li FF, Chen TZ, Zhang J, Jiang YJ, Ding YZ, Guo WZ, Zhang TZ (2008) Enhanced *Agrobacterium*-mediated transformation of embryogenic calli of upland cotton via efficient selection and timely subculture of somatic embryos. Plant Mol Biol Rep 26:174–185

15. Zhang J, Cai L, Cheng JQ, Mao HZ, Fan XP, Meng ZH, Chan KM, Zhang HJ, Qi JF, Ji LH, Hong Y (2008) Transgene integration and organization in cotton (*Gossypium hirsutum* L.) genome. Transgenic Res 17:293–306

16. Ikram Ul H (2004) Agrobacterium-mediated transformation of cotton (*Gossypium hirsutum* L.) via vacuum infiltration. Plant Mol Biol Rep 22:279–288

17. Leelavathi S, Sunnichan VG, Kumria R, Vijaykanth GP, Bhatnagar RK, Reddy VS (2004) A simple and rapid Agrobacterium-mediated transformation protocol for cotton (*Gossypium hirsutum* L.): embryogenic calli as a source to generate large numbers of transgenic plants. Plant Cell Rep 22:465–470

18. Satyavathi VV, Prasad V, Lakshmi BG, Sita GL (2002) High efficiency transformation protocol for three Indian cotton varieties via Agrobacterium tumefaciens. Plant Sci 162:215–223

19. Sunilkumar G, Rathore KS (2001) Transgenic cotton: factors influencing Agrobacterium-mediated transformation and regeneration. Mol Breed 8:37–52

20. Tohidfar M, Mohammadi M, Ghareyazie B (2005) Agrobacterium-mediated transformation of cotton (*Gossypium hirsutum*) using a heterologous bean chitinase gene. Plant Cell Tissue Organ Cult 83:83–96

21. Yuceer SU, Koc NK (2006) Agrobacterium-mediated transformation and regeneration of cotton plants. Russian J Plant Physiol 53:413–417

22. Zhao FY, Li YF, Xu PL (2006) Agrobacterium-mediated transformation of cotton (*Gossypium hirsutum* L. cv. Zhongmian 35) using glyphosate as a selectable marker. Biotechnol Lett 28:1199–1207

23. Zhu SW, Gao P, Sun JS, Wang HH, Luo XM, Jiao MY, Wang ZY, Xia GX (2006) Genetic transformation of green-colored cotton. Vitro Cell Dev Biol Plant 42:439–444

24. Murashige T, Skoog F (1962) A fdvised medium for rapid growth and bioassays with tobacco tissue cultures. Physiol Plant 15:473–497

25. Gamborg OL, Miller RA, Ojima K (1968) Nutrient requirements of suspension cultures of soybean root cells. Exp Cell Res 50:151–158

26. Jin SX, Zhang XL, Liang SG, Nie YC, Guo XP, Huang C (2005) Factors affecting transformation efficiency of embryogenic callus of upland cotton (*Gossypium hirsutum*) with *Agrobacterium tumefaciens*. Plant Cell Tissue Organ Cult 81:229–237

27. Joubert P, Beaupere D, Lelievre P, Wadouachi A, Sangwan RS, Sangwan-Norreel BS (2002) Effects of phenolic compounds on Agrobacterium vir genes and gene transfer induction – a plausible molecular mechanism of phenol binding protein activation. Plant Sci 162:733–743

28. Lai E-M, Shih H-W, Wen S-R, Cheng M-W, Hwang H-H, Chiu S-H (2006) Proteomic analysis of *Agrobacterium tumefaciens* response to the vir gene inducer acetosyringone Proteomics 6:4130–4136

29. Nair GR, Lai X, Wise AA, Rhee BW, Jacobs M, Binns AN (2011) The integrity of the periplasmic domain of the VirA sensor kinase is critical for optimal coordination of the virulence signal response in *Agrobacterium tumefaciens*. J Bacteriol 193:1436–1448

30. Stachel SE, Messens E, Vanmontagu M, Zambryski P (1985) Identification of the signal molecules produced by wounded plant cells that activate T-DNA transfer in *Agrobacterium tumefaciens*. Nature 318:624–629

31. Aggarwal D, Kumar A, Reddy MS (2011) *Agrobacterium tumefaciens* mediated genetic transformation of selected elite clones of *Eucalyptus tereticornis*. Acta Physiol Plant 33:1603–1611

32. Bhuiyan MSU, Min SR, Jeong WJ, Sultana S, Choi KS, Lim YP, Song WY, Lee Y, Liu JR (2011) An improved method for *Agrobacterium*-mediated genetic transformation from cotyledon explants of *Brassica juncea*. Plant Biotechnol 28:17–23

33. Mehrotra M, Sanyal I, Amla DV (2011) High-efficiency *Agrobacterium*-mediated transformation of chickpea (*Cicer arietinum* L.) and regeneration of insect-resistant transgenic plants. Plant Cell Rep 30:1603–1616

34. Rashid H, Chaudhry Z, Khan MH (2011) Effect of explant plant source and acetosyringone concentration on transformation efficiency of wheat cultivars. Afr J Biotechnol 10:8737–8740

35. Wu S-J, Wang H-H, Li F-F, Chen T-Z, Zhang J, Jiang Y-J, Ding Y, Guo W-Z, Zhang T-Z (2008) Enhanced Agrobacterium-mediated transformation of embryogenic calli of upland cotton via efficient selection and timely subculture of somatic embryos. Plant Mol Biol Rep 26:174–185

36. Dillen W, De Clercq J, Kapila J, Zambre M, Van Montagu M, Angenon G (1997) The effect of temperature on *Agrobacterium tumefaciens*-mediated gene transfer to plants. Plant J 12:1459–1463

37. Uranbey S, Sevimay CS, Kaya MD, Ipek A, Sancak C, Basalma D, Er C, Ozcan S (2005) Influence of different co-cultivation temperatures, periods and media on *Agrobacterium tumefaciens*-mediated gene transfer. Biol Plant 49:53–57

38. Wang B, Liu L, Wang X, Yang J, Sun Z, Zhang N, Gao S, Xing X, Peng D (2009) Transgenic ramie [*Boehmeria nivea* (L.) Gaud.]: factors affecting the efficiency of *Agrobacterium tumefaciens*-mediated transformation and regeneration. Plant Cell Rep 28:1319–1327

39. Zhang B-H, Liu F, Liu Z-H, Wang H-M, Yao C-B (2001) Effects of kanamycin on tissue culture and somatic embryogenesis in cotton. Plant Growth Regul 33:137–149

Chapter 4

Biolistic Transformation of Cotton Zygotic Embryo Meristem

Kanniah Rajasekaran

Abstract

Biolistic transformation of cotton (*Gossypium hirsutum* L.) meristems, isolated from mature seed, is detailed in this report. A commercially available, helium-driven biolistic device (Bio-Rad PDS1000/He) was used to bombard gold particles coated with a marker gene (*uidA* or "β-glucuronidase") into the shoot meristem. The penetration of gold particles was dependent on bombardment parameters and it was mostly one to two cell layers deep. Stable transformation of epidermal L1 layer was consistently observed in approximately 5% of the seedlings. Germ line transformation was observed in up to 0.71% of bombarded meristems by several laboratories. Using this method identification of germ line transformation is laborious and time-consuming. However, the protocol described here represents a simple and efficient method for generating germ line transformation events. In addition, this procedure offers a quick method to evaluate gene constructs in cotton tissues (embryos, cotyledons, leaf), especially fibers which originate as single cells from the maternal epidermis layer.

1. Introduction

It is a well-known fact that cotton regeneration in vitro is a tedious, time-consuming procedure and is generally only applicable to limited, obsolete genotypes (1–4). One approach to overcome these difficulties involves the bombardment of apical meristems from which transformed shoots and plants can be derived (5–7). This method bypasses regeneration procedures and the associated problems, thereby reducing the time required to produce transformed plants for subsequent evaluation (8). Of several devices that have been reported in the literature (9), only one is commercially available from Bio-Rad (PDS 1000/He) and this device is widely used in many laboratories around the world. A modified version of this device was used to transform Brazilian cotton varieties to obtain germ line transformation from bombarded meristems (10, 11).

Baohong Zhang (ed.), *Transgenic Cotton: Methods and Protocols*, Methods in Molecular Biology, vol. 958,
DOI 10.1007/978-1-62703-212-4_4, © Springer Science+Business Media New York 2013

Using another proprietary Accell® Technology, McCabe and Martinell (5), Keller et al. (6) successfully demonstrated biolistic transformation of cotton embryo meristems.

We examined the shoot apex of germinating cottonseeds using scanning electron microscopy and evaluated conditions for transforming the shoot meristems with the biolistic device from Bio-Rad (PDS 1000/He). The transformation method described in this report should prove useful for routine testing of gene constructs directed toward the improvement of agronomic as well as cotton fiber quality traits.

2. Materials

2.1. DNA Constructs

A typical construct should include a visual marker gene (β-glucuronidase, GUS)—for histochemical or fluorescence detection, green fluorescent protein (GFP) or luciferase, and your favorite gene (YFG). The pUC19-based construct containing a chimeric GUS (*uidA*) gene with a promoter consisting of a duplicated Cauliflower Mosaic Virus 35S enhancer element and the 5′ leader sequence from Tobacco Mosaic Virus (12) has been used for developing the meristem transformation procedure presented here.

2.2. Seed Materials

Use acid-delinted and/or fungicide-treated cottonseed (see Note 1). The procedure that is presented here uses the Acala variety—Maxxa (California Planting Cotton Seed Distributors Inc., Shafter, CA). Cottonseed freshly harvested or stored for less than 2 years at room temperature is preferable. Delint cottonseed to remove fuzz fibers by carefully stirring the seed with concentrated sulfuric acid for 2–3 min in a fume hood. Decant the acid; carefully neutralize the residual acid with copious amounts of sodium bicarbonate and water. Rinse several times before air-drying and storage in a cool ($21 \pm 3°C$) and dry environment.

2.3. Stock Solutions and Materials

Prepare all stock solutions and dilutions using ultrapure water (prepared by purifying deionized water to attain a resistivity of 18.2 MΩ.cm at 25°C). Prepare and store all stock solutions in the freezer or the refrigerator as indicated below.

1. 6-(γ,γ-dimethylallylamino) purine (2iP; Sigma-Aldrich)—1.0 mg/ml stock—dissolve in a small volume of 0.1 N KOH and dilute to volume; add before autoclaving medium. Store in siliconized glassware for up to 3 months at 7°C or longer at –20°C.

2. 6-benzyl aminopurine (BA; Sigma-Aldrich)—1.0 mg/ml stock—dissolve in a small volume of 0.1 N KOH and dilute to volume; add before autoclaving medium. Store in siliconized glassware for up to 3 months at 7°C or longer at –20°C.

3. Indoleacetic acid (IAA; Sigma-Aldrich)—1.0 mg/ml stock in a small volume of 0.1 N KOH and dilute to volume; always use a freshly prepared solution; filter-sterilize using a membrane filter (0.22 μm) before adding to the medium.

4. α-naphthaleneacetic acid (NAA; Sigma-Aldrich)—1.0 mg/ml stock in a small volume of 0.1 N KOH and dilute to volume; add before autoclaving medium. Store at 7°C for up to 6 months.

5. GUS Substrate solution: 5 mM potassium ferricyanide; 5 mM potassium ferrocyanide; 0.3% X-glucuronide; 0.1 M sodium phosphate, pH 7.0; 0.06% (v/v) Triton X-100. Store at 7°C.

6. GUS Extraction Buffer: 50 mM $NaPO_4$, pH 7.0; 10 mM β-mercaptoethanol; 10 mM Na_2EDTA; 0.1% Sodium Lauryl Sarcosine; 0.1% Triton X-100. Store at 7°C.

7. GUS Assay Buffer: 1 mM 4-methylumbelliferyl-beta-D-glucuronide (MUG; Bio-World, Dublin, OH) in GUS Extraction Buffer, 20% methanol. Store at 7°C.

8. 100% ethanol (stored at 4°C with Dri-rite).

9. 2.5 M $CaCl_2$, filter-sterilize (0.22 μm) and store at 4–7°C.

10. 0.1 M Spermidine, free-base (Sigma-Aldrich), filter-sterilize (0.22 μm) and store frozen at –20°C.

11. Probe ultrasonicator (Vibra Cell, Danbury, CT).

12. Microcentrifuge (location inside a refrigerator preferred but not essential).

13. Spectrophotometer or a plate reader.

2.4. Biolistic Devices

1. There are several biolistic devices available for use (9); the only commercially available delivery system is from Bio-Rad (PDS 1000/He) that employs high-pressure helium to propel micro-carriers coated with DNA into target cells. This improved system is based on the earlier design by Klein et al. (13). The biolistic device should be positioned in a laminar flow cabinet to avoid contamination and should be connected to a vacuum pump and an ultrahigh-purity compressed helium gas cylinder.

2. Supplies for biolistic bombardment:
    (a) 1.0 μm or 1.6 μm Au particles (Bio-Rad, Hercules, CA)—(see Note 2).
    (b) Macrocarrier holders (Bio-Rad).
    (c) Macrocarriers (Bio-Rad).

2.5. Tissue Culture Media

1. Seed Germination medium (SG): Murashige and Skoog (MS) (14) or White's medium as modified by Singh and Krikorian (15) (see Note 3). Add BA or 2iP (5 mg/L) to promote uniform germination and cotyledon unfolding.

2. Surface sterilant: Sodium hypochlorite solution (dilute to obtain 1% available chlorine; Clorox Co., Oakland, CA, USA) containing 0.1% (v/v) Tween-20 (Polyethylene glycol sorbitan monolaurate; Sigma-Aldrich) as a dispersing agent.

3. All plant tissue culture media are solidified with 6–8 g/L agar-agar (Caisson Labs, Sugar City, ID, USA) and adjusted to a final pH of 5.8. When gellan gum from several commercial sources (e.g., Phytogel from Sigma-Aldrich; Gel-Gro from ICN Biochemicals, Cleveland, OH, USA; or Gelrite from Merck, Rahway, NJ, Kelco Division, USA) is used as the solidifying agent at 2 g/L, add an additional source of bivalent cation (e.g., 0.75 g/L $MgCl_2$) to aid in gelling.

3. Methods

3.1. Isolation of Zygotic Embryo Axes and Preparation for Bombardment

1. Surface sterilize cottonseed first in 70% ethanol for 2–3 min and then in 20% (v/v) Clorox (1% available chlorine) plus a surfactant (Tween 20, 0.01% v/v) for 20 min. All operations should be handled in a laminar flow cabinet under sterile conditions.

2. Rinse the seeds thoroughly in sterile water and then soak overnight at room temperature (21°C) in sterile water containing 5 mg/L 2iP or BA. Cytokinin treatment results in uniform protrusion of radicle through the micropylar ends.

3. Isolate zygotic embryos from imbibed seeds showing radicle extrusion after 24 h of cytokinin treatment by gently squeezing on chalazal end of the seed coat without touching the embryo. The zygotic embryo pops out cleanly and is ready for use (Fig. 1a). Remove cotyledon pieces if they are still attached to the embryo (see Note 4).

4. Keep all the isolated embryos moist either in a drop of sterile water or on moist filter papers.

5. Using a stereomicroscope, remove the first leaf primorium that covers the meristem dome (Fig. 1b). This can be accomplished by gently dissecting with the sharp end of a sterile disposable needle (first leaf primordium has been removed to expose the meristem dome in Fig. 1c) (see Note 5).

6. Arrange axes in 3 cm dia. circles in a concentric manner on a solidified agar medium (MS or modified White), usually 25–30 axes per plate, so that the exposed apical shoot meristem domes are facing the flight path of the microcarriers (gold particles).

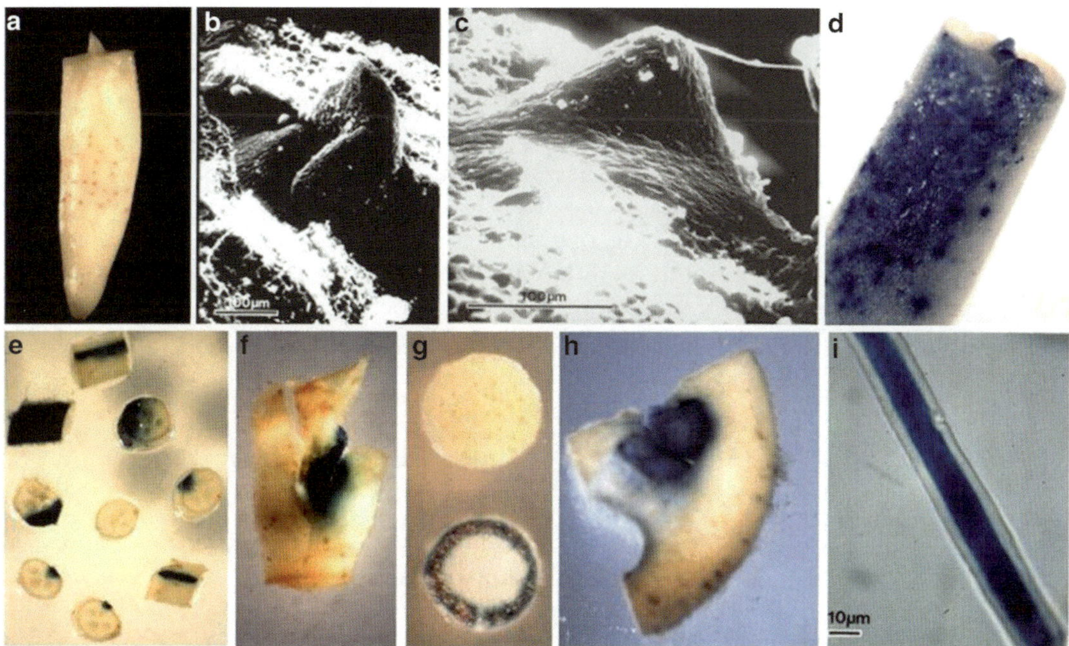

Fig. 1. (**a**) A cotton zygotic embryo axis isolated after soaking the seed overnight in water containing 5 mg/L 2iP. (**b**) Scanning electron micrograph of a 1-day-old cotton shoot meristem. Note that the meristem dome is completely covered by two leaf primordia. (**c**) Same as in 2. The first leaf primordium has been removed (foreground) to expose the meristem dome. The second leaf primordium is seen in the background. (**d**) A zygotic embryo axis showing ample transient expression of GUS gene on the meristem and hypocotyl area. Stained 24 h after bombardment with 1.0 μm gold particles coated with the plasmid DNA containing the GUS gene. (**e**) Stem sections from a 2-week-old R_0 seedling showing stable GUS expression in L1 and L2 layers. (**f**) Stable expression of GUS gene in an axillary bud from a 3-week-old R_0 seedling. (**g**) Stable expression of GUS gene in a leaf disk (bottom) compared to a non-transformed control disk (top) from an R_0 seedling. (**h**) Cross section of an ovary (collected on the day of anthesis) showing GUS expression in fiber initials on the outer integument of the ovule. (**i**) An immature fiber (1 day post anthesis) from an R_0 plant showing GUS expression in the lumen.

3.2. Gold Particle (Microcarrier) Preparation

1. Place 15 mg Au in a sterile 1.5 ml microcentrifuge tube.
2. Add 500 μl of 100% EtOH.
3. Ultrasonicate for 15 s in a water bath and let sit for 30 min to settle the particles.
4. Centrifuge at 960 rcf for 1 min. Remove the supernatant and discard.
5. Add 1 ml sterile ddH$_2$O (4°C). Flick. Allow particles to settle.
6. Centrifuge at 2,700 rcf for 15 s. Remove the supernatant and discard.
7. Repeat steps 5 and 6.
8. Repeat steps 5 and 6. (Rinsing 3× total.)
9. Suspend pellet in 250 μl sterile ddH$_2$O. You now have a 5× Au concentration.

10. Vortex the 5× microcentrifuge tube.

11. Aliquot 50 µl of the suspension to five 0.5 ml microcentrifuge tubes. Vortex to resuspend in between each tube.

12. Parafilm tubes and place in freezer. Each tube is now a 1× concentration—has 3 mg Au in 50 µl of H_2O.

3.3. DNA Preparation

1. Add 10 µg of DNA to 50 µl of gold—keep on ice. Volume amount of DNA is dependent on the concentration of the DNA solution. Flick tube—between each step it is very important to flick the tube to keep the gold particles suspended in solution.

2. Add 50 µl of 2.5 M $CaCl_2$. Flick tube.

3. Add 20 µl of 0.1 M spermidine—Flick tube; keep on ice.

4. Let sit for 5 min at room temperature.

5. Spin in microcentrifuge. Let speed go up to 900 rcf and stop, for about 10 s. Remove the supernatant.

6. Add 300 µl of 95% EtOH. Flick tube.

7. Vortex to resuspend about 30–60 s.

8. Spin in microcentrifuge. Let speed go up to 960 rcf and stop, for about 10 s. Remove the supernatant.

9. Add 65 µl of absolute EtOH. Flick tube.

10. Parafilm tubes and put on ice until ready to use.

3.4. Macrocarrier Preparation

1. For every tube of DNA, use 6 macrocarriers.

2. Place 6 macrocarriers and 6 macrocarrier holding rings in 95% EtOH. Rings should be flat side up. Let sit in the EtOH; this can be done before DNA prep. Do not stack up the macrocarriers, as they can stick together.

3. Remove the macrocarriers and rings from the EtOH. Dip the macrocarriers through the EtOH to remove any particles and shake off slightly to remove excess EtOH. Place in large Petri dish to dry off, rings should be placed flat side up, and lean the macrocarriers up against the rings.

4. After the ring and macrocarrier have dried, flip the ring over and place the macrocarrier inside. Use the CA plug to push the macrocarrier past a little lip in the ring—push and twist.

5. Flick the DNA microcarrier tubes and then vortex for about 30 s.

6. Pulse-ultrasonicate the tubes briefly (2 s). You need to ultrasonicate after every two rings.

7. Spread the microcarrier solution in a thin film over the macrocarrier, where the ring opening is. Dispense 5–6 µl for each of the six macrocarriers, and then go back and spread the rest of

the solution left in the tube—usually about 2 µl the second time. You do this to make sure that you have enough to cover all 6 macrocarriers first, and then you can add the rest. On average, use about 5–8 µl of the DNA/Au solution (at least 1 mg/ml) per bombardment.

3.5. Bombardment of Zygotic Embryo Meristems

1. Bombard the meristems with the plasmid DNA of choice (see Subheading 3.2 to 3.4). The embryos should be bombarded only once. Multiple bombardments of meristems usually results in higher rates of mortality.

2. Bombardment conditions—Conditions for bombardment that optimized the level of transient GUS expression included using 1,550 psi rupture disks and setting the distance between the stopping screen and the meristems to 7.5 cm and the macrocarrier flying distance to 10 mm. The desired vacuum in the sample chamber is 28″ Hg during the bombardment (see Note 6).

3. After bombardment, plate the axes for germination in baking trays (30 cm × 12 cm) containing agar-solidified MS (14) or White's (15) medium. If necessary, incorporate a broad-spectrum fungicide (e.g., benomyl 0.05% w/v) to prevent possible fungal contamination. Up to 300 axes can be planted in one baking tray.

4. Place the trays in a growth cabinet or a culture room (28°C, 16/8-h photoperiod, 60 µE/m^2 s) to allow growth of the seedlings.

5. Seedlings were then transferred to soil within a week after bombardment or they can be maintained on agar trays for 2 more weeks, after which only positive plants can be transferred to soil.

6. Perform visual selection of transgenic expression as below (step 14 or 15).

7. GUS histochemical assay (16, 17): Place plant tissue in a test tube containing GUS reaction buffer [GUS substrate solution with 20% (v/v) methanol] (see Note 7) and incubate at 37°C for 30 min to 24 h. GUS-positive tissues will be staining blue (Fig. 1d–i) (see Notes 8 and 9).

8. GUS fluorogenic assay: Samples (<1 g) are transferred to a mortar and ground in 1 ml of GUS Extraction Buffer. The homogenate is transferred to an Eppendorf tube and centrifuged (2,700 rcf) for 10 min. Fifty µl of the supernatant is mixed with 450 µl of GUS Assay Buffer (1 mM MUG in Extraction Buffer, 20% methanol) to start the reaction. Incubate the samples at 37°C, remove 10 µl aliquots at 1-h time periods, and dilute with 1 ml 0.2 M Na_2CO_3 (carbonate stop buffer). Measure fluorescence of 4-MU with excitation at $\lambda = 365$ nm

and emission at $\lambda = 455$ nm. Avoid excessive light exposure of tubes because 4-MU is light sensitive (see Note 9).

9. Selection for transformed meristems is not possible due to chimeric gene expression by biolistic bombardment (Fig. 1e–h). Such an analysis of genomic DNA by PCR is not advisable at this stage since a positive result is not indicative of germ line transformation. Demonstration of gene expression by Western blotting should also be carried out with caution since a positive result is not indicative of germ line transformation. However, it is possible to obtain germ line transformation if stable expression is obtained in the corpus region of the meristems that comprises L2 and L3 layers from which parts of the vascular cylinder and the reproductive cells of anthers and ovules arise (18–20). Reported germ line transformation efficiency from bombarded shoot meristems is very low—from 0.001 to 0.71% (10, 11, 21). Transformation in L2 and L3 layers, which gives rise to stable germ line, is very sporadic and requires careful screening, identification, and processing by selective pruning (see Note 10). It is essential to document germ line transformation by demonstration of the inheritance of transgenes in seed progeny (see Note 11).

10. For the reasons explained above, the most common type of transformation by biolistic bombardment of meristems is in L1 layer. However, this type of expression is useful to evaluate gene expression in cotton plants as well as in cotton fibers, which originate as single cells from the maternal epidermis of ovules and are of L1 layer origin (Fig. 1i) (see Note 10).

4. Notes

1. Commercially available cottonseed is often treated with fungicide. Remove the fungicide coating with several rinses in water.

2. We have also used gold particles (0.6–2.5 μm) from other suppliers with equal success—from Alfa Products (Danvers, MA) or InBio Gold (Eltham, VIC, Australia).

3. Prepackaged media formulations are suitable for growing cotton embryo axes (e.g., M5519 or W0876 from Sigma-Aldrich).

4. It is possible to prepare up to 300 axes from soaked seeds in 1 h (=10 plates).

5. The isolated zygotic embryos (Fig. 1a) are very slippery to handle; hence long forceps with serrated tips will be useful. We observed that meristems with the first leaf primordia removed were subject to transformation four or five times more than the meristem tips with the leaf primordia intact.

6. The following conditions can be adjusted in the PDS 1000/ He device—distance between stopping screen and the target plate, psi rated rupture disk, size of gold particles, and chamber vacuum. Gold particles above 1.6 μm size are not preferred as they cause significant damage to meristem dome.

7. The addition of methanol to the GUS reaction has been shown to increase the activity of the bacterial enzyme and inhibit endogenous GUS-like activities in plant tissues (17).

8. X-gluc substrate penetration into cotton leaf disks is rather slow and the blue staining occurs only along the cut-ends (Fig. 1g). Staining for a longer period is necessary for uniform staining. Fluorometric detection of GUS is more sensitive than histochemical staining.

9. To visualize the expression of GUS gene, destructive sampling is needed for both histochemical and fluorogenic assays. GFP gene (GFP from the jellyfish *Aequorea victoria*) or luciferase (from the firefly *Photinus pyralis*) will be useful to identify transformed meristems without destructive sampling.

10. Cotton shoot buds or meristems of elite cotton varieties are not amenable to rapid multiplication in vitro. Recent reports on multiple shoot regeneration rate of up to four shoots per explants (22–24) are encouraging but they do not reach the high multiplication rate obtained from other species (e.g., soybean or peanuts) where biolistic bombardment has been applied very successfully (25, 26).

11. Due to the high possibility of obtaining chimeric transformants from bombarded meristems it is essential to confirm stable integration and inheritance of transgenes by molecular and genetic analyses.

---

5. Conclusions

A step-by-step procedure on transformation of cotton zygotic embryo meristem by biolistic bombardment is provided in this chapter. This transformation method is very appealing because it is applicable to all cotton genotypes without the need to work with regenerable, obsolete lines only, a prerequisite for the currently employed *Agrobacterium*-mediated transformation. However, biolistic bombardment method is laborious and time consuming and reproducibility of effecting germ line transformation is very low. Improvements in physical delivery of transgene into deep layers of the meristem (L2 and L3 layers) explants will be highly beneficial to improve the efficiency of transformation. Visual reporter genes such as GFP or luciferase that do not require destructive sampling will also be useful to improve the efficiency of this method. Lastly, increasing the multiplication rate of cotton shoot meristems will also be helpful in cleaning the chimeric expression by selection as well as in rapid propagation of selected transgenic events.

References

1. Trolinder NT, Goodin JR (1987) Somatic embryogenesis and plant regeneration in cotton (*Gossypium hirsutum* L.). Plant Cell Rep 6:231–234
2. Rajasekaran K, Grula JW, Hudspeth RL, Pofelis S, Anderson DM (1996) Herbicide-resistant Acala and Coker cottons transformed with a native gene encoding mutant forms of acetohydroxyacid synthase. Mol Breed 2:307–319
3. Wilkins TA, Rajasekaran K, Anderson DM (2000) Cotton biotechnology. Crit Rev Plant Sci 19:511–550
4. Rajasekaran K (2004) *Agrobacterium*-mediated genetic transformation of cotton. In: Curtis IS (ed) Transgenic crops of the world – essential protocols. Kluwer Academic Publishers, Dordrecht, The Netherlands, pp 243–254
5. McCabe DE, Martinell BJ (1993) Transformation of elite cotton cultivars via particle bombardment of meristems. BioTechnology 11:596–598
6. Keller G, Spatola L, McCabe D, Martinell B, Swain W, John ME (1997) Transgenic cotton resistant to herbicide bialaphos. Trangenic Res 6:385–392
7. Chlan CA, Lin J, Cary JW, Cleveland TE (1995) A procedure for biolistic transformation and regeneration of transgenic cotton from meristematic tissue. Plant Mol Biol Rep 13:31–37
8. Christou P (1992) Genetic transformation of crop plants using microprojectile bombardment. Plant J 2:275–281
9. Christou P (1996) Particle bombardment for genetic engineering of plants. R.G. Landes Company and Academic Press, Austin, TX, p 199
10. Aragão FJL, Vianna GR, Carvalheira SBRC, Rech EL (2005) Germ line genetic transformation in cotton (*Gossypium hirsutum* L.) by selection of transgenic meristematic cells with a herbicide molecule. Plant Sci 168:1227–1233
11. Rech EL, Vianna GR, Aragao FJL (2008) High-efficiency transformation by biolistics of soybean, common bean and cotton transgenic plants. Nat Protoc 3:410–418
12. Skuzeski JM, Nichols LM, Gesteland RF (1990) Analysis of leaky viral translation codons in vivo by transient expression of improved β-glucuronidase vectors. Plant Mol Biol 15:65–79
13. Klein TM, Gradziel T, Fromm ME, Sanford JC (1988) Factors influencing gene delivery into *Zea mays* cells by high-velocity microprojectiles. BioTechnology 6:559–563
14. Murashige T, Skoog F (1962) A revised medium for rapid growth and bioassays with tobacco tissue cultures. Physiol Plant 15:473–497
15. Singh M, Krikorian AD (1981) White's standard nutrient solution. Ann Bot 47:133–139
16. Jefferson RA (1987) Assaying chimeric genes in plants: the GUS gene fusion system. Plant Mol Biol Rep 5:387–405
17. Kosugi S, Ohashi Y, Nakajima K, Arai Y (1990) An improved assay for β-glucuronidase in transformed cells: methanol almost completely suppresses a putative endogenous β-glucuronidase activity. Plant Sci 70:133–140
18. Satina S, Blakeslee AF, Avery AG (1940) Demonstration of the three germ layers in the shoot apex of *Datura* by means of induced polyploidy in periclinal chimeras. Am J Bot 27:895–905
19. Tilney-Bassett RAE (1986) Plant chimeras. Edward Arnold, London, p 199
20. Sussex I (1989) Developmental programming of the shoot meristem. Cell 56:225–229
21. John ME (1997) Cotton crop improvement through genetic engineering. Crit Rev Biotech 17:185–208
22. Agrawal DC, Banerjee AK, Kolala RR, Dhage AB, Kulkarni AV, Nalawade SM, Hazra S, Krishnamurthy KV (1997) In vitro induction of multiple shoots and plant regeneration in cotton (*Gossypium hirsutum* L.). Plant Cell Rep 16:647–652
23. Hemphill JK, Maier CGA, Chapman KD (1998) Rapid in-vitro plant regeneration of cotton (*Gossypium hirsutum* L.). Plant Cell Rep 17:273–278

24. Bazargani MM, Tabatabaei BES, Omidi M (2011) Multiple shoot regeneration of cotton (*Gossypium hirsutum* L.) via shoot apex culture system. African J Biotechnol 10:2005–2011

25. McCabe DE, Swain WF, Martinell BJ, Christou P (1988) Stable transformation of soybean (*Glycine max*) by particle acceleration. BioTechnology 6:923–926

26. Brar GS, Cohen BA, Vick CL, Johnson GW (1994) Recovery of transgenic peanut (*Arachis hypogaea* L.) plants from elite cultivars utilizing Accell® technology. Plant J 5:745–753

Chapter 5

Biolistic Transformation of Cotton Embryogenic Cell Suspension Cultures

Kanniah Rajasekaran

Abstract

Genetic transformation of cotton (*Gossypium hirsutum* L.) is highly dependent on the ability to regenerate fertile plants from transgenic cells through somatic embryogenesis. Induction of embryogenic cell cultures is genotype dependant. However, once embryogenic cell cultures are available, they can be effectively used for transformation by *Agrobacterium* or biolistic bombardment methods. Here I describe a detailed procedure to transform cotton embryogenic cell suspension cultures by biolistic bombardment. A commercially available, helium-driven biolistic device (Bio-Rad PDS1000/He) was used to bombard gold particles coated with plasmid DNA (for visual identification of transformed cells and/or selection) into embryogenic cells. Stable transformation at a high frequency (up to 4% of the transiently expressing cells) is possible. Regeneration of fertile transgenic plants from embryogenic cells takes only about 2 months. Another advantage of the embryogenic cell suspension cultures is that they are amenable for cryopreservation and long-term storage. It is highly preferable to transform commercial varieties of choice than obsolete varieties to avoid the genetic drag due to backcrossing.

---

1. Introduction

Recent success in cotton biotechnology is largely due to the successful combination of somatic embryogenesis procedures and the ability to transform with *Agrobacterium* vectors to transfer the gene(s) of interest (1). As a result, transgenic cotton varieties resistant to insect pests and invasive weedy species are being grown in 160 M ha of cotton-growing countries around the world (2). Nearly 90% of cotton being grown in the United States belongs to transgenic varieties with insect resistance and/ or herbicide-tolerance traits (2). The success rate of somatic embryogenesis is highly genotype dependent (3–7). Obsolete varieties such as Coker are the easiest and most of the commercial varieties (e.g., Acala, DeltaPine, Pima, and stripper varieties)

Baohong Zhang (ed.), *Transgenic Cotton: Methods and Protocols*, Methods in Molecular Biology, vol. 958,
DOI 10.1007/978-1-62703-212-4_5, © Springer Science+Business Media New York 2013

are recalcitrant. Coker varieties produce embryogenic callus in about 2–5 months; on the other hand, commercial varieties take an average of 10 months. In our experience with diverse species of cotton and more than 100 cotton upland varieties, only once we were able to obtain embryogenic callus from a Coker variety within a month after culture. Once embryogenic callus is obtained from a variety of choices, it can be used for quick transformation with *Agrobacterium* followed by successful regeneration of transgenic plants (8–11). In addition, embryogenic cell suspension cultures can be initiated from those cultures for high-frequency stable transformation by biolistic bombardment (12, 13). An added advantage of embryogenic cell cultures is that it can be successfully cryopreserved for several years (14). Transforming an advanced variety of choice is extremely beneficial to avoid time-consuming backcrosses and dilution of important genetic traits of the target variety (e.g., fiber quality) due to undesired linkages through successive backcrosses with an obsolete, transgenic variety such as Coker. A simple biolistic transformation procedure is presented here with embryogenic cell cultures as the starting material.

2. Materials

2.1. Embryogenic Cell Suspension Cultures

Freshly initiated embryogenic cell suspension cultures—of an Acala variety B1654 (13)—is used here. Cell suspension cultures of var. Coker 315 were used to demonstrate typical growth curve (Fig. 1). Always use 3–4 days after subculture (see Note 1).

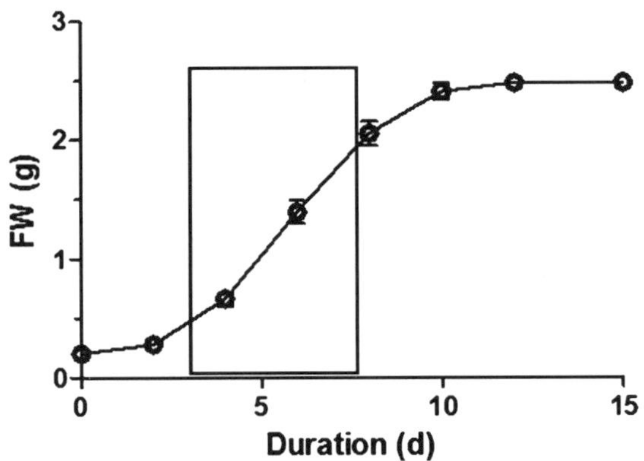

Fig. 1. A typical growth curve for embryogenic cell suspension cultures of var. Coker 315 (mean of three replicates). This is very similar to the growth curve reported for Acala var. B1654 (13). The preferred growth phase during days 3–7 after subculture for biolistic bombardment is indicated in the *box*.

2.2. DNA Constructs

A typical construct should include a visual marker gene (β-glucuronidase, GUS—for histochemical or fluorescence detection, green fluorescent protein [GFP] or luciferase), antibiotic or herbicide selectable marker genes, and your favorite gene (YFG) for agronomic or fiber quality trait. The pUC19-based construct containing a chimeric GUS (*uidA*) gene driven by a duplicated 35S CaMV promoter was used for transformation of cotton embryogenic cell cultures presented here. The plasmid also contained the NPT II gene for selection for resistance to the antibiotic G418 (13).

2.3. Stock Solutions and Materials

Prepare all stock solutions and dilutions using ultrapure water (prepared by purifying deionized water to attain a resistivity of 18.2 MΩ.cm at 25°C). Prepare and store all stock solutions in the freezer or the refrigerator as indicated below. High-purity plant cell culture-tested chemicals were used when possible.

1. α-naphthaleneacetic acid (NAA; Sigma-Aldrich)—1.0 mg/mL stock in a small volume of 0.1 N KOH and dilute to volume; add before autoclaving medium. Store at 7°C for up to 6 months.

2. GUS Substrate solution: 5 mM potassium ferricyanide; 5 mM potassium ferrocyanide; 0.3% X-glucuronide; 0.1 M sodium phosphate, pH 7.0; 0.06% (v/v) Triton X-100. Store at 7°C.

3. GUS Extraction Buffer: 50 mM $NaPO_4$, pH 7.0; 10 mM β-mercaptoethanol; 10 mM Na_2EDTA; 0.1% Sodium Lauryl Sarcosine; 0.1% Triton X-100. Store at 7°C.

4. GUS Assay Buffer: 1 mM 4-methylumbelliferyl-beta-D-glucuronide (MUG; Bio-World, Dublin, OH) in GUS Extraction Buffer, 20% methanol. Store at 7°C.

5. Gyratory shaker (Innova 44 or 4000 shakers, New Brunswick Scientific, Edison, NJ).

6. 500 mL side-arm flask and porcelain filters for vacuum filtration (presterilized).

7. Vacuum pump.

8. 120 μm and 400 mesh Nylon screens (Tetco Inc., Kansas City, MO) trimmed to fit funnels and Petri dishes (presterilized).

9. Hemocytometer for counting cells.

10. Tally counter.

11. 100% ethanol (stored at 4°C with Dri-rite).

12. 2.5 M $CaCl_2$, filter-sterilize (0.22 μm) and store at 4–7°C.

13. 0.1 M Spermidine, free-base (Sigma-Aldrich), filter-sterilize (0.22 μm) and store frozen at –20°C.

14. Probe ultrasonicator (Vibra Cell, Danbury, CT).

15. Microcentrifuge (location inside a refrigerator preferred but not essential).

16. Spectrophotometer or a plate reader.

2.4. Biolistic Devices

1. There are several biolistic devices available for use (15); the only commercially available delivery system is from Bio-Rad (PDS 1000/He) that employs high-pressure helium to propel microcarriers coated with DNA into target cells. This improved system is based on the earlier design by Klein et al. (16). The biolistic device should be positioned in a laminar flow cabinet to avoid contamination and should be connected to a vacuum pump and an ultrahigh-purity compressed helium gas cylinder.

2. Supplies for biolistic bombardment:

    (a) 1.0 μm Au particles (Bio-Rad, Hercules, CA)—(see Note 2).

    (b) Macrocarrier holders (Bio-Rad).

    (c) Macrocarriers (Bio-Rad).

2.5. Tissue Culture Media

1. Agar (6–8 g/L agar-agar; Caisson Labs, Sugar City, ID, USA)-solidified Beasley and Ting (BT) medium (17) supplemented with 500 mg/L casein hydrolysate (acid or enzyme hydrolyzed; Sigma-Aldrich) (BTCH).

2. Liquid Murashige and Skoog (MS) (18) medium containing 100 mg^{-1} myo-inositol and 2 mg/L NAA (MSNAA). This medium can be autoclaved or filter-sterilized (using 22 μm Millipore or Gelman membrane filters) and stored for 4–6 weeks at 7°C.

3. Methods

3.1. Multiplication of Embryogenic Callus

1. Identify and isolate embryogenic callus (Fig. 2a) from seedling cotyledon, hypocotyl, leaf, or stem explants (see Note 3).

2. Embryogenic callus obtained from cotton callus cultures (regardless of the starting medium; see Note 3) can be aseptically multiplied on agar-solidified BTCH.

3. Quality of the cell cultures should be checked under a compound microscope for the type of cells. Pure embryogenic callus (comprising small, isodiametric, cytoplasmic-rich cells, measuring 30–60 μm) devoid of long, vacuolated non-embryogenic cells should be selectively subcultured every 2–3 weeks (see Note 4).

3.2. Initiation of Embryogenic Cell Suspension Cultures

1. Collect 1,000 mg of actively growing, homogeneous embryogenic callus cultures. Globular or later stage somatic embryos, if present, should be avoided.

2. Using a sterile 18 gauge needle or a scalpel (No. 11 sterile, surgical blade, both from Bectin Dickinson AcuteCare, Franklin

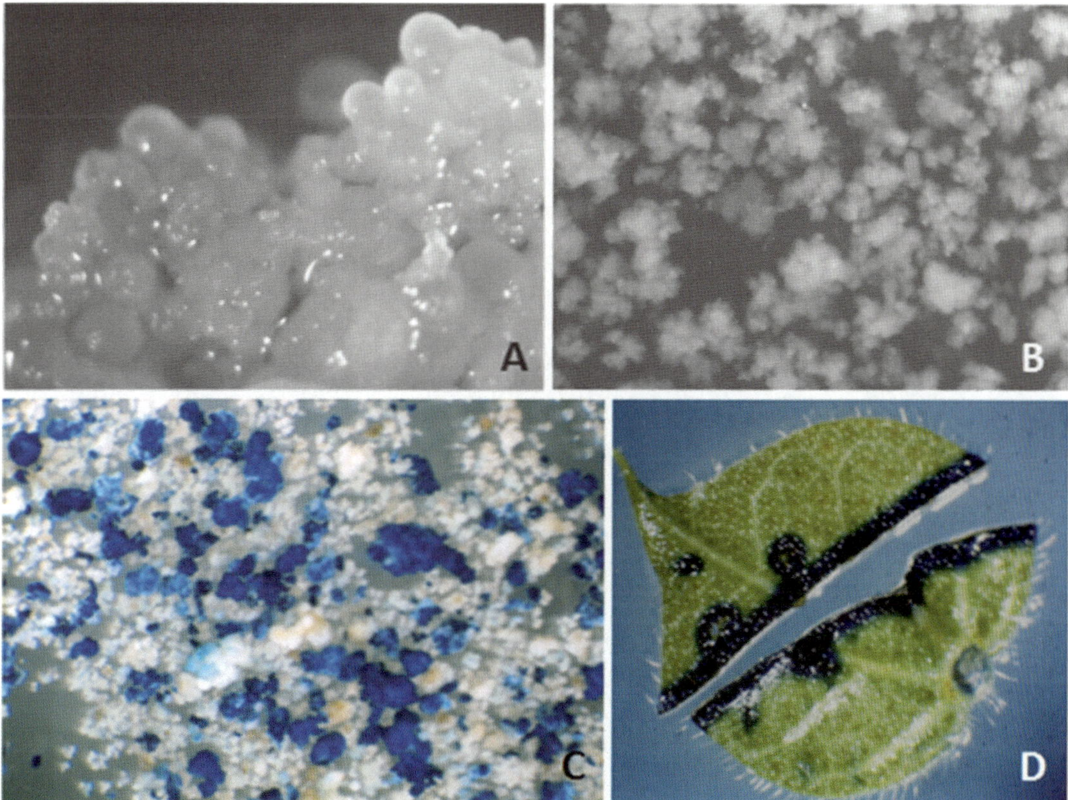

Fig. 2. (**a**) A highly homogeneous embryogenic callus of Acala cotton var. B1654 obtained from cotton seedling explants. (**b**) A freshly prepared embryogenic cell suspension culture of var. B1654. On average, the cell aggregates contain anywhere from 4 to 25 cells. (**c**) Stable expression of GUS gene in bombarded embryogenic cell suspension cultures showing deep blue color. Non-transformed cells appear white due to bleaching in ethanol. (**d**) A young leaf from a transgenic cotton plantlet showing cut expression along the cut ends. Photographed 4 h after treatment with the substrate.

Lakes, NJ) gently tease the embryogenic callus clumps into smaller aggregates (see Note 5).

3. Transfer the embryogenic callus to 250 mL Erlenmeyer flasks containing 40 mL liquid MSNAA (see Note 6).

4. Repeat the steps 1–3 to start several flasks of suspension cultures.

5. Place the suspension culture flasks on gyratory shakers (120 strokes per minute) maintained at $26 \pm 2°C$, with a light irradiance of 40 W/m^2 (provided by cool white fluorescent lamps).

3.3. Maintenance/ Multiplication of Embryogenic Cell Suspension Cultures

1. After 1 week of suspension culture initiation, filter the suspension culture to remove large clumps. Resuspend the cells in MSNAA liquid medium in a 50 mL disposable tubes and remove floating, non-embryogenic or dead cells (see Note 4). Using a wide-mouth pipet transfer 400–500 mg cells to a new flask containing 40 mL MSNAA.

2. Repeat step 1 every week. In about 3–4 weeks suspension culture will be proliferating and will be doubling in fresh weight every 2–3 days depending on the variety and the homogeneity of embryogenic cells (Fig. 2b). Growth of cell suspension cultures from most of the upland cotton varieties will be similar to the growth curve presented in Fig. 1 (for var. Coker-315) with slight variation in doubling time.

3. Cell cultures from day 3 to day 6 after subculture are ideally suitable for biolistic bombardment (Fig. 1).

3.4. Preparation of Cell Cultures for Biolistic Bombardment

1. Using a vacuum filtration device, collect actively growing cell cultures (from Subheading 3.3) onto a moist filter paper (Whatman No.1; 3.5 cm dia.) placed on top of a 120 µm nylon screen. Cells should be spread as a thin layer onto the filter paper at approximately 1×10^6 cells.

2. Place the filter paper containing cells on an agar medium (MSNAA preferred). Keep the cells just moist with the medium.

3. Place a 400 mesh nylon screen (Tetco Inc.) on top of the cells to serve as a baffle during bombardment.

4. Repeat steps 1–3 to prepare additional target of cell culture plates.

3.5. Gold Particle (Microcarrier) Preparation

1. Place 15 mg Au in a sterile 1.5 mL microcentrifuge tube.

2. Add 500 µl of 100% EtOH.

3. Ultrasonicate for 15 s in a water bath and let sit for 30 min to settle the particles.

4. Centrifuge at 960 rcf for 1 min. Remove the supernatant and discard.

5. Add 1 mL sterile ddH_2O (4°C). Flick. Allow particles to settle.

6. Centrifuge at 2,700 rcf for 15 s. Remove the supernatant and discard.

7. Repeat steps 5 and 6.

8. Repeat steps 5 and 6. (Rinsing 3× total.)

9. Suspend pellet in 250 µl sterile ddH_2O. You now have a 5× Au concentration.

10. Vortex 5× microcentrifuge tube.

11. Aliquot 50 µl of the suspension to five 0.5 mL microcentrifuge tubes. Vortex to resuspend in between each tube.

12. Parafilm tubes and place in freezer. Each tube is now a 1× concentration—has 3 mg Au in 50 µl of H_2O.

3.6. DNA Preparation

1. Add 10 µg of DNA to 50 µl of gold—keep on ice. Volume amount of DNA is dependent on the concentration of the DNA solution. Flick tube—between each step it is very important to flick the tube to keep the gold particles suspended in solution.

2. Add 50 µl of 2.5 M $CaCl_2$. Flick tube.

3. Add 20 µl of 0.1 M spermidine—Flick tube; keep on ice.

4. Let sit for 5 min at room temperature.

5. Spin in microcentrifuge. Let speed go up to 900 rcf and stop, for about 10 s. Remove the supernatant.

6. Add 300 µl of 95% EtOH. Flick tube.

7. Vortex to resuspend about 30–60 s.

8. Spin in microcentrifuge. Let speed go up to 960 rcf and stop, for about 10 s. Remove the supernatant.

9. Add 65 µl of absolute EtOH. Flick tube.

10. Parafilm tubes and put on ice until ready to use.

3.7. Macrocarrier Preparation

1. For every tube of DNA, use 6 macrocarriers.

2. Place 6 macrocarriers and 6 macrocarrier holding rings in 95% EtOH. Rings should be flat side up. Let sit in the EtOH; this can be done before DNA prep. Do not stack up the macrocarriers, as they can stick together.

3. Remove the macrocarriers and rings from the EtOH. Dip the macrocarriers through the EtOH to remove any particles and shake off slightly to remove excess EtOH. Place in large Petri dish to dry off, rings should be placed flat side up, and lean the macrocarriers up against the rings.

4. After the ring and macrocarrier have dried, flip the ring over and place the macrocarrier inside. Use the CA plug to push the macrocarrier past a little lip in the ring—push and twist.

5. Flick the DNA microcarrier tubes and then vortex for about 30 s.

6. Pulse-ultrasonicate the tubes briefly (2 s). You need to ultrasonicate after every two rings.

7. Spread the microcarrier solution in a thin film over the macrocarrier, where the ring opening is. Dispense 5–6 µl for each of the 6 macrocarriers, and then go back and spread the rest of the solution left in the tube—usually about 2 µl the second time. You do this to make sure that you have enough to cover all 6 macrocarriers first, and then you can add the rest. On average, use about 5–8 µl of the DNA/Au solution (at least 1 mg/mL) per bombardment.

3.8. Bombardment of Embryogenic Cell Cultures

1. Bombard the cell suspension cultures with the plasmid DNA of choice (see Subheading 3.5 to 3.7). The suspension cultures can be bombarded multiple times during the growth to increase the transformation frequency ((13); see Note 7).

2. Bombardment conditions—Conditions for bombardment that optimized the level of transient GUS expression included using 1,450 psi rupture disks and setting the distance between the stopping screen and the Petri plate containing the embryogenic cells to 7.5 cm and the macrocarrier flying distance to 10 mm. The desired vacuum in the sample chamber is 28″ Hg (95 kPa) during bombardment (see Note 8).

3. After bombardment, transfer the cells to MSNAA in shake-cultures.

4. Include antibiotics or herbicides in liquid medium for selection of transformed cells (12, 13). Usually the cells respond to selection better after 5–7 days of growth without selection pressure (see Note 9). Gradual increase in selection pressure (for example, increasing the concentration of the antibiotic G418 from 10 μg to 50 μg/mL over a period of 4–5 weeks) has been shown to increase the recovery of transgenic events (13) (see Note 10).

5. The number of independent transformation events can be determined by plating a known number of cells after biolistic bombardment at a low density (2×10^4 cells per plate) on solidified MSNAA containing the selection agent. This method of counting helps to avoid overestimating the number of independent transformation events due to clonal multiplication of transformed cells in liquid cultures.

6. Assay for stable transformation using visual marker genes after 4 weeks of selection (Fig. 2c; see Note 11).

7. Transfer embryogenic cell clumps from selection media to agar-solidified MSNAA or BTCH media to develop embryos and plantlets. Selection pressure at this stage slows down the growth and development of somatic embryos and plantlets and hence it is avoided. Somatic embryos take about 4–6 weeks to develop into four to five leafed plantlets in Magenta boxes. Use vented lids to avoid vitrification of leaves due to high humidity (see Note 12).

8. Test leaf segments from each plant for the presence of the visual marker genes (e.g., GUS) as in step 9 or 10 (Fig. 2d).

9. GUS histochemical assay (19, 20): Place plant tissue in a test tube containing GUS reaction buffer [GUS substrate solution with 20% (v/v) methanol] (see Note 7) and incubate at 37°C for 30 min to 24 h. GUS-positive cell aggregates will be staining blue (Fig. 2c, d) (see Notes 13 and 14).

10. GUS fluorogenic assay: Cell aggregate samples (< 1 g) are transferred to a mortar and ground in 1 mL of GUS Extraction Buffer. The homogenate is transferred to an Eppendorf tube and centrifuged (2,700 rcf) for 10 min. Fifty µl of the supernatant is mixed with 450 µl of GUS Assay Buffer (1 mM MUG in Extraction Buffer, 20% methanol) to start the reaction. Incubate the samples at 37°C, remove 10 µl aliquots at 1-h time periods, and dilute with 1 mL 0.2 M Na_2CO_3 (carbonate stop buffer). Measure fluorescence of 4-MU with excitation at $\lambda = 365$ nm and emission at $\lambda = 455$ nm. Avoid excessive light exposure of tubes because 4-MU is light sensitive (see Note 14).

11. Transfer transgenic plants to soil in 10 cm diameter pots—use a soil mix that provides good drainage (e.g., perlite:peat:sand at 4:4:2). Cover plants with plastic cups or plastic bags (Ziploc bags, SC Johnson Co., Racine, WI, USA) for about 5 days before gradual acclimatization to ambient greenhouse temperature.

12. Keep the top soil in pots dry and water sparingly. Too much moisture in soil often results in poor root growth and fungal growth around the stem. Fertilize the plants once a week with nutrient solution (e.g., Hoagland's nutrient solution). Commercially available slow-release nutrient pellets (e.g., Osmocote 14-14-14 from Scotts Co., Marysville, OH, USA) can be used instead. Plants flower in about 3–4 months after transfer to greenhouse.

13. Transfer the putatively transformed plants to a BL-2 greenhouse. Perform molecular analyses to confirm integration of the introduced gene and transmission to the progeny by, for example, PCR, Southern blot analyses of R_0 and R_1 plants followed by Northern and Western blot analyses to confirm their expression at the RNA and protein levels, respectively.

4. Notes

1. Freshly initiated (less than 6 months) embryogenic cell suspension cultures are preferable for use in transformation experiments. Care should be taken to avoid using long-term cultures (more than 6 months), which are known to accumulate somatic mutations and cytogenetic abnormalities leading to undesirable morphological and fertility-related problems in regenerated cotton plants (13).

2. Quality grade gold particles from other suppliers can also be used with equal success—from Alfa Products (Danvers, MA) or InBio Gold (Eltham, VIC, Australia).

3. Embryogenic callus initiation medium from seedling explants according to several published reports includes, for example, Rangan and Rajasekaran (4)—kinetin (1 mg/L) plus IAA or NAA (2 mg/L); Firoozabady et al. (21)—2iP (5 mg/L) plus NAA (0.1 mg/L); Trolinder and Goodin (22)—0.5 mg/L kinetin and 0.1 mg/L 2,4-D; Davidonis and Hamilton (23)—1.0 mg/L kinetin and 2.0 mg/L NAA; and Gawel et al. (24)—kinetin 1.0 mg/L and 4.0 mg/L NAA.

4. Non-embryogenic cells are produced in embryogenic cell cultures due to quantal mitosis; hence it is best to check the quality of the suspension cultures every 4–6 weeks. Non-embryogenic cells should be removed during subcultures.

5. Excessive manipulation of embryogenic callus will result in the production of phenolics in the medium. If this is the case, remove the spent medium and replace with fresh medium until the problem goes away.

6. Use about 1,000 mg FW of embryogenic callus in 40 mL medium. If higher amount of callus is used, make sure that it does not hinder movement of cells during shake-cultures.

7. Suspension cell cultures can be bombarded four to five times in a week. The cells are returned to shake-culture after every bombardment (13).

8. The following conditions can be adjusted in the PDS 1000/He device—distance between stopping screen and the target plate, psi rated rupture disk, size of gold particles, and chamber vacuum. The conditions provided in Subheading 3 are adequate to obtain nuclear transformation.

9. It is highly beneficial to culture cells without any selection pressure for 5–7 days. This provides enough time for integration of the introduced DNA into nuclei.

10. Selection with antibiotics hygromycin (12), kanamycin, or G418 (13) or with herbicides sulfonylureas or imidazolinones (5) have been applied successfully in cotton cell cultures.

11. To visualize the expression of GUS gene, destructive sampling is needed for both histochemical and fluorogenic assays. GFP gene (GFP from the jellyfish *Aequorea victoria*) or luciferase (from the firefly *Photinus pyralis*) will be useful to identify transformed cell aggregates without destructive sampling. Accelerated regeneration from such positive cells is preferable without subjecting them through selection procedures.

12. Poor germination of somatic embryos is another bottleneck in cotton regeneration. Germination percentage can vary from 10 to 70% depending on the genotype and the quality of somatic embryos. Dehydration of somatic embryos and subsequent germination in a low-salt medium in vented (0.22 μm)

jars (Sigma-Aldrich) are often recommended and practiced in several laboratories to improve germination and subsequent acclimatization in soil.

13. X-gluc substrate penetration into cotton leaf disks is rather slow and the blue staining occurs only along the cut-ends (Fig. 2d). Staining for a longer period is necessary for uniform staining. Fluorometric detection of GUS is more sensitive than histochemical staining.

14. The addition of methanol to the GUS reaction has been shown to increase the activity of the bacterial enzyme and inhibit endogenous GUS-like activities in plant tissues (20).

5. Conclusions

A procedure for biolistic bombardment of cotton embryogenic cell suspension cultures is provided in this chapter. This transformation procedure with the embryogenic cell cultures as the starting material saves time and labor in successful production of transgenic events from readily regenerable cell cultures. The frequency of transformation is also high using this method. Additional advantages of using embryogenic cell cultures of choice varieties are the following: (1) it is not necessary to depend on obsolete, easy-to-regenerate varieties which invariably contaminate the genetic makeup, thereby reducing agronomic or fiber quality traits of the target variety due to unavoidable genetic linkages in backcrosses, and (2) they are amenable for long-term storage after cryopreservation.

Disclaimer: Product names and trademarks are mentioned to report factually on available data; however, the USDA neither guarantees nor warrants the standard of the product, and the use of the name by USDA does not imply the approval of the product to the exclusion of others that may also be suitable.

References

1. Wilkins TA, Rajasekaran K, Anderson DM (2000) Cotton biotechnology. Crit Rev Plant Sci 19:511–550

2. James C (2011) Annual biotech crop report by International Service for the Acquisition of Agri-biotech Applications (ISAAA). Available at http://www.isaaa.org/

3. Trolinder NL, Xhixian C (1989) Genotype specificity of the somatic embryogenesis response in cotton. Plant Cell Rep 8:133–136

4. Rangan TS, Rajasekaran K (1996) Regeneration of cotton plants in suspension culture. US Patent 5,583,036

5. Rajasekaran K, Grula JW, Hudspeth RL, Pofelis S, Anderson DM (1996) Herbicide-resistant Acala and Coker cottons transformed with a native gene encoding mutant forms of acetohydroxyacid synthase. Mol Breed 2:307–319

6. Zhang B-H, Liu F, Yao C-B (2000) Plant regeneration via somatic embryogenesis in cotton. Plant Cell Tissue Organ Cult 60:89–94

7. Rajasekaran K (2004) *Agrobacterium*-mediated genetic transformation of cotton. In: Curtis IS (ed) Transgenic crops of the world essential protocols. Kluwer Academic Publishers, Dordrecht, The Netherlands, pp 243–254

8. Leelavathi S, Sunnichan VG, Kumria R, Vijaykanth GP, Bhatnagar RK, Reddy VS (2004) A simple and rapid *Agrobacterium*-mediated transformation protocol for cotton (*Gossypium hirsutum* L.): embryogenic calli as a source to generate large numbers of transgenic plants. Plant Cell Rep 22:465–470

9. Wu J, Zhang X, Nie Y, Luo X (2005) High-efficiency transformation of *Gossypium hirsutum* embryogenic calli mediated by *Agrobacterium tumefaciens* and regeneration of insect-resistant plants. Plant Breed 124:142–146

10. Wu S-J, Wang H-H, Li F-F, Chen T-Z, Zhang J, Jiang Y-J, Ding Y, Guo W-Z, Zhang T-Z (2008) Enhanced *Agrobacterium*-mediated transformation of embryogenic calli of upland cotton via efficient selection and timely subculture of somatic embryos. Plant Mol Biol Rep 26:174–185

11. Obembe OO, Khan T, Popoola JO (2011) Use of somatic embryogenesis as a vehicle for cotton transformation. J Med Plant Res 5:4009–4020

12. Finer JJ, McMullen MD (1990) Transformation of cotton (*Gossypium hirsutum* L.) via particle bombardment. Plant Cell Rep 8:586–589

13. Rajasekaran K, Hudspeth RL, Cary JW, Anderson DM, Cleveland TE (2000) High-frequency stable transformation of cotton (*Gossypium hirsutum* L.) by particle bombardment of embryogenic cell suspension cultures. Plant Cell Rep 19:539–545

14. Rajasekaran K (1996) Regeneration of plants from cryopreserved embryogenic cell suspension and callus cultures of cotton (*Gossypium hirsutum* L.). Plant Cell Rep 15:859–864

15. Christou P (1996) Particle bombardment for genetic engineering of plants. R. G. Landes Company and Academic Press, Austin, TX, p 199

16. Klein TM, Gradziel T, Fromm ME, Sanford JC (1988) Factors influencing gene delivery into *Zea mays* cells by high-velocity microprojectiles. BioTechnology 6:559–563

17. Beasley CA, Ting IP (1973) The effects of plant growth substances on in vitro development from fertilized cotton ovules. Am J Bot 60:130–139

18. Murashige T, Skoog F (1962) A revised medium for rapid growth and bioassays with tobacco tissue cultures. Physiol Plant 15:473–497

19. Jefferson RA (1987) Assaying chimeric genes in plants: the GUS gene fusion system. Plant Mol Biol Rep 5:387–405

20. Kosugi S, Ohashi Y, Nakajima K, Arai Y (1990) An improved assay for β-glucuronidase in transformed cells: methanol almost completely suppresses a putative endogenous β-glucuronidase activity. Plant Sci 70:133–140

21. Firoozabady E, DeBoer DL (1993) Plant regeneration via somatic embryogenesis in many cultivars of cotton (*Gossypium hirsutum* L.). Vitro Cell Dev Biol Plant 29:166–173

22. Trolinder NL, Goodin JR (1987) Somatic embryogenesis and plant regeneration in cotton (*Gossypium hirsutum* L). Plant Cell Rep 6:231–234

23. Davidonis GH, Hamilton RH (1983) Plant regeneration from callus tissue of *Gossypium hirsutum* L. Plant Sci Lett 32:89–93

24. Gawel NJ, Rao AP, Robacker CD (1986) Somatic embryogenesis from leaf and petiole callus cultures of *Gossypium hirsutum* L. Plant Cell Rep 5:457–459

Chapter 6

Cotton Transformation via Pollen Tube Pathway

Min Wang, Baohong Zhang, and Qinglian Wang

Abstract

Although many gene transfer methods have been employed for successfully obtaining transgenic cotton, the major constraint in cotton improvement is the limitation of genotype because the majority of transgenic methods require plant regeneration from a single transformed cell which is limited by cotton tissue culture. Comparing with other plant species, it is difficult to induce plant regeneration from cotton; currently, only a limited number of cotton cultivars can be cultured for obtaining regenerated plants. Thus, development of a simple and genotype-independent genetic transformation method is particularly important for cotton community. In this chapter, we present a simple, cost-efficient, and genotype-independent cotton transformation method—pollen tube pathway-mediated transformation. This method uses pollen tube pathway to deliver transgene into cotton embryo sacs and then insert foreign genes into cotton genome. There are three major steps for pollen tube pathway-mediated genetic transformation, which include injection of foreign genes into pollen tube, integration of foreign genes into plant genome, and selection of transgenic plants.

1. Introduction

Traditional transformation methods, such as *Agrobacterium*-mediated gene transformation, have made significant progress on cotton improvement, particularly on insect resistance and herbicide tolerance (1–3). Transgenic *Bt* cotton and herbicide-tolerant cotton have been widely adopted around the world. However, the majority of those methods are based on cotton tissue culture, in which it requires to obtain regenerated plants from a single transformed cell. Compared with other plant species, it is difficult to induce somatic embryogenesis and plant regeneration in cotton; currently, only a limited number of cotton cultivars can be induced to obtain somatic embryogenesis and plant regeneration. Thus, developing a genotype-independent gene transformation method for cotton is of great interest in the past two decades.

Baohong Zhang (ed.), *Transgenic Cotton: Methods and Protocols*, Methods in Molecular Biology, vol. 958,
DOI 10.1007/978-1-62703-212-4_6, © Springer Science+Business Media New York 2013

Pollen tube pathway-mediated transformation is a transformation technique for transferring foreign genes/DNA into embryo sacs and then inserting into plant genome after self-pollination. During pollination, a pollen falls on the stigma, and then germinates and form a pollen tube and grows through the style tissues until it reaches the ovary. During this process, a long distance of pollen tube pathway is formed, which can be used to deliver foreign DNAs into embryo sacs and transform the eggs and zygotes. This technology was first developed and applied to cotton genetic transformation in 1978 (4); at that time, Zhou and colleagues injected exogenous genomic DNAs into cotton embryo sacs via the pollen tube pathway and then obtained a great number of transformants (4); from these transformants, they had selected many cotton germplasms with fine characteristics (5). Huang and colleagues used GFP as a reporter gene to investigate the pathway of inducing foreign genes into cotton through the pollen tube; they found that GFP was expressed in cotton embryo and in the transformed seedling, which provides direct and convinced evidence in cytology and molecular biology for the pollen tube pathway-mediated transformation method (6).

Pollen tube pathway-mediated transformation is a tissue culture (or genotype)-independent transformation method and also it does not require expensive instrument and everyone can perform this technique. Thus, since it was developed, this technique has been widely adopted by scientists, particularly in China, and employed to transform many different crop species using different resources of DNA and genes (5). These important crops include cotton (4–6), watermelon (7, 8), soybean (9–11), wheat (12–15), papaya (16), and corn (7, 17).

There are three major steps for pollen tube pathway-mediated genetic transformation, which include injection of foreign genes into pollen tube, integration of foreign genes into plant genome, and selection of transgenic plants. In this chapter, we step by step introduce the pollen tube pathway-mediated transformation protocol for cotton.

2. Materials

2.1. Microinjection of Foreign Genes into Cotton Pollen Tube

1. Seeds of cotton (*Gossypium hirsute* L.) cultivar CRI 12 that is bred by the scientists of Cotton Research Institute (CRI) of the Chinese Academy of Agricultural Sciences (CAAS). CRI 12 is a fine cotton non-transgenic cultivar with many great characteristics, including high yield and high quality.

2. Plasmid DNA, such as pCNL56, containing the selected marker gene *npt*II, which codes for the aminoglycoside

3'-phosphotransferase (denoted *aph*(3')-II or NPTII) enzyme (see Note 1). NPTII enzyme inactivates by phosphorylation a range of aminoglycoside antibiotics, including kanamycin. Thus, transgenic cotton containing *npt*II gene will offer tolerance to antibiotic kanamycin.

3. Gibberellic Acid (GA).

4. Label.

5. 50 μL microinjector.

6. Sterilized deionized water (ddH_2O).

7. Field suitable for growing cotton.

8. Pesticides and herbicides potentially used for pest, disease, and weed control.

9. Fertilizers for normal agricultural practices.

2.2. Identification, Selection, and Expression Analysis of Transgenic Plants

There are many methods that can be used to identify transgenic plants. These methods include bioassay (such as field and lab bioassay) and molecular techniques (such as Northern blotting, Sothern blotting, PCR). Here, we only present a rapid method using qRT-PCR for both identification and expression analysis of transgenic plants.

2.2.1. Isolation of Total RNA

mirVana™ miRNA Isolation Kit is used to isolate total RNAs from cotton leaves.

1. Leaves from tested plants.

2. Liquid Nitrogen.

3. *mir*Vana™ miRNA Isolation kit (Ambion, Austin, TX).

   (a) Lysis/Binding Buffer: Store at 4°C.

   (b) miRNA Homogenate Additive: Store at 4°C.

   (c) miRNA Wash Solution 1.

   (d) Wash Solution 2/3.

   (e) Collection Tubes.

   (f) Filter Cartridges.

   (g) Acid-Phenol:Chloroform. Store at 4°C.

   (h) Nuclease Free Water.

4. 100% RNase-free ethanol.

5. RNase-free water.

2.2.2. RT-PCR

1. TaqMan® MicroRNA Reverse Transcription Kit (Applied Biosystems, Foster City, CA). Store at –20°C.

   (a) 10× RT Buffer.

   (b) dNTP mix with dTTP (100 mM).

    (c) RNase Inhibitor (20 U/μL).

    (d) Multiscribe™ RT enzyme (50 U/μL).

2. Nuclease-Free water.

3. Poly T RT Primers.

2.2.3. qRT-PCR

1. PCR master mix with SYBR green.

2. PCR Primers for nptII gene (Forward primer: TCATCGACT GTGGCCGGCTG; reversed primer: AAGCGGTCAGCCCA TTCGCC).

3. Reference gene GAPDH (Forward primer: TGATGCCAAG GCTGGAATTGCTT; reversed primer: GADPH is GTGTC GGATCAAGTCGATAACACGG).

4. Nuclease-Free water.

3. Methods

3.1. Microinjection of Foreign Genes into Cotton Pollen Tube

There are three major methods for pollen tube pathway-mediated transformation. They are microinjection, direct drop of DNA to stigma, and culture foreign genes with pollen and pollination with the pollens. Among these three methods, microinjection produces more stable transgenic plants with high transformation rate. Here, we only describe the protocol for microinjection.

1. Cotton cultivar CRI 12 is sown by conventional methods at a rate of approximately 10 seed/m of row on a 1.5 m row spacing. Final plant density is about 3–5 plants/m^2 (see Note 2).

2. Standard cultivation practices and insect control are used in an attempt to optimize yields.

3. Prepare plasmid DNA to a final concentration of 0.1 μg/mL.

4. During flowering season, select the flowers that will bloom next day (see Note 3).

5. Tie up the upper part (corolla) of the flower bud with a cotton thread (see Note 4).

6. In the next morning around 8–10 am after the blooming, gently remove petal and style (see Note 5).

7. Clean the microinjector.

8. Take 10 μL plasmid DNA.

9. Puncture vertically with the microinjector into the ovary about 5 mm deep from the top of the ovary; then, draw back about 2 mm; finally, slowly inject 5–10 μL plasmid DNA into the ovary. Remove the microinjector carefully from the ovary (see Notes 6 and 7).

10. Drop GA on the pedicel to prevent the young cotton bolls from shedding.

11. Label the cotton bolls with data.

12. Remove the rest of the branch to keep all the nutrients transferred to the cotton boll.

13. During harvest season, harvest the cotton bolls individually.

14. Take seeds out of each boll.

15. Keep the seeds dry after they are harvested.

3.2. Identification, Selection, and Expression Analysis of Transgenic Plants (see Note 8)

3.2.1. Isolation of Total RNA

1. Young leaves are collected from the tested cotton plants and immediately frozen in liquid nitrogen (see Note 9).

2. Total RNAs are extracted using the mirVana miRNA Isolation Kit (Ambion, Austin, TX) according to the manufacturer's protocol. The detailed methods for extracting total RNAs can be found in the kits or can be found in Chapter 3.

3. The total RNA is quantified and the quality is assessed by using a Nanodrop ND-1000 (Nanodrop Technologies, Wilmington, DE).

4. RNA samples are stored in a $-80°C$ freezer until further use and analysis.

3.2.2. RT-PCR

1. Calculate how much volume (μL) of RNA samples containing 500 ng total RNAs.

2. Calculate how much volume (μL) of water is needed for preparing a 15 mL of RT reaction.

3. Prepare reverse transcription reaction. Add the following items in order in a 200 μL sterilized microcentrifuge tube: ddH_2O, 1.5 μL 10× Reverse Transcription Buffer, 0.19 μL RNase Inhibitor, 1.0 μL Poly T primer, 0.15 μL 100 mM dNTPs, and 1.0 μL reverse transcriptase, and 500 ng RNA samples.

4. Gently mix the reaction and centrifuge down all the solution.

5. Incubate on ice for 5 min.

6. Perform the reverse transcription as follows: $16°C$ for 30 min, followed by $42°C$ for 30 min and $85°C$ for 5 min.

7. Store the RT products at $-20°C$ or below.

3.2.3. qRT-PCR

1. Prepare PCR reaction. The total volume for each reaction is 20 μL, which contains 10 μL 2× SYBE green PCR master mixture, 2 μL RT product, 2 μL forward and reverse primers, and 6 μL nuclease-free water.

2. Add items in order into a 96-well RT-PCR plates or RT-PCR tube: ddH_2O, PCR mixture, primers, and RT products.

3. Gently mix the reaction and centrifuge down all the solution.

4. Perform the RT-PCR as follows: temperature program: 95°C for 10 min followed by 40 cycles of 95°C for 15 s and 60°C for 60 s.

5. Manually set up the threshold value (see Note 10).

6. Determine which plants are transgenic plants (see Note 11).

7. Calculate the expression level of each transgenic plant compared to the reference gene.

4. Notes

1. Pollen tube pathway-mediated transformation method can be used to transfer any type of DNA into plants, which can be but not limited to total genomic DNA, partial genomic DNA, plasmid DNA, or a single gene.

2. Comparing with traditional cotton culture, the plant density of cotton should be low for increasing the pollen tube pathway-mediated gene transformation. High density of cotton usually result in high percentage fall down of the flower bud and young cotton bolls.

3. It is very critical for pollen tube pathway-mediated transformation to select the right flower for transformation. It is better to remove the first several flowers before starting transformation process. Always select the flower close to the stem.

4. The major purpose of this operation is to keep the self-pollination and to avoid the potential effect by cross-pollination.

5. Gently and carefully remove petal and style to reduce the damage to ovary. Otherwise, it will result in falling down of the cotton bolls.

6. This is another very critical step for pollen tube pathway-mediated transformation. Usually, the right hand holds the microinjector while the left hand holds the young cotton boll. All operation should be done gently and carefully to minimize any possible damage to the ovary.

7. Cannot inject too much volume of DNA. If injected too much, it will damage the ovary and result in cotton ovary to stop growth and shed.

8. The transformation rate of pollen tube pathway-mediated method is low. For majority cases, it is only 0.5–1.0%. Thus, the majority seeds will be non-transgenic seeds; to reduce the cost, it is better to do field bioassay first and then use molecular technology to test those potential transgenic plants.

9. Because of the interference of gossypol, pigment, and other components, it is difficult to get high yield of RNAs with high

quality from cotton. Using young leaves is a little bit easy for isolating RNAs because of fewer amounts of those components. If RNAs are not extracted immediately, the leaf samples can be stored at −80°C.

10. The threshold should be set in the exponential phase. If running multiple plates, each plate should have the same threshold value. Otherwise, it is hard to compare the data from different plates.

11. If the Ct value is higher than 40, it can be considered that that plants do not contain the transformed gene, suggesting that the plant is not a transgenic plant. Otherwise, if a Ct value is less than 40, it is likely to be a transgenic plant.

References

1. Zhang BH, Feng R (2000) Cotton resistance to insect and transgenic cotton. China Agricultutal Science and Technology Press, Beijing

2. Zhang BH, Liu F, Yao CB, Wang KB (2000) Recent progress in cotton biotechnology and genetic engineering in China. Curr Sci 79:37–44

3. Showalter, A. M., Heuberger, S., Tabashnik, B. E., and Carriere, Y. (2009) A primer for using transgenic insecticidal cotton in developing countries, *Journal of Insect Science 9*.

4. Zhou G, Weng J, Zheng Y, Huang J, Qian S, Liu G (1983) Introduction of exogenous DNA into cotton embryos. Methods Enzymol 101:433–481

5. Ni WC, Guo SD, Jia SR (2000) Cotton transformation with the pollen tube pathway. Rev China Agricult Sci Technol 2:27–32

6. Huang GC, Dong YM, Sun JS (1999) Introduction of exogenous DNA into cotton via the pollen-tube pathway with GFP as a reporter. Chinese Sci Bull 44:698–701

7. Yang A, Su Q, An L, Liu J, Wu W, Qiu Z (2009) Detection of vector- and selectable marker-free transgenic maize with a linear GFP cassette transformation via the pollen-tube pathway. J Biotechnol 139:1–5

8. Hao J, Niu Y, Yang B, Gao F, Zhang L, Wang J, Hasi A (2011) Transformation of a marker-free and vector-free antisense ACC oxidase gene cassette into melon via the pollen-tube pathway. Biotechnol Lett 33:55–61

9. Hu CY, Wang LZ (1999) In planta soybean transformation technologies developed in China: procedure, confirmation and field performance. In Vitro Cell Dev Biol Plant 35:417–420

10. Shou HX, Palmer RG, Wang K (2002) Irreproducibility of the soybean pollen-tube pathway transformation procedure. Plant Mol Biol Rep 20:325–334

11. Yang S, Li G, Li M, Wang J (2011) Transgenic soybean with low phytate content constructed by *Agrobacterium transformation* and pollen-tube pathway. Euphytica 177:375–382

12. Martin N, Forgeois P, Picard E (1992) Investigations on transforming *Triticum aestivum* via pollen tube pathway. Agronomie 12:537–544

13. Qiu Z, Su Q, An L-J (2008) Application of FITC tracing in the optimization of wheat transformation via pollen-tube pathway. Xibei Zhiwu Xuebao 28:611–616

14. Yin J, Yu G-R, Ren J-P, Li L, Song L (2004) Transforming anti-TrxS gene into wheat by means of pollen tube pathway and ovary injection. Xibei Zhiwu Xuebao 24:776–780

15. Zeng JZ, Wang DJ, Wu YQ, Zhang J, Zhou WJ, Zhu XP, Xu NZ (1994) Transgenic wheat obtained with pollen tube pathway method. Sci China B Chem 37:319–325

16. Wei J-Y, Liu D-B, Chen Y-Y, Cai Q-F, Zhou P (2008) Transformation of PRSV-CP dsRNA gene into papaya by pollen-tube pathway technique. Xibei Zhiwu Xuebao 28:2159–2163

17. Zhang YS, Yin XY, Yang AF, Li GS, Zhang JR (2005) Stability of inheritance of transgenes in maize (*Zea mays* L.) lines produced using different transformation methods. Euphytica 144:11–22

Silicon Carbide Whisker-Mediated Transformation of Cotton (*Gossypium hirsutum* L.)

Muhammad Arshad, Yusuf Zafar, and Shaheen Asad

Abstract

Plant transformation methods are invaluable biotechnological tools to generate specific and targeted genetic variation for performance improvement of crop plants. Genetic information is created by proper modification during gene cloning flanked by proper regulatory sequences and delivered to plants via different plant transformation techniques. Due to being a multipurpose plant, cotton has been subjected to different genetic transformation methods to provide the breeders with an opportunity to develop alien traits or improve the endogenous gene performance that are very difficult or impossible to develop through conventional breeding methods. Here we describe the novel physical way of cotton transformation with different genes by using embryogeneic calli as continuous source of explants.

1. Introduction

After 1970s, green revolution has led to fair improvement in crop production largely explored by natural genetic variation through classical breeding methods. But the era from end of last century and start of twenty-first century has witnessed a tremendous boom in farm crops by "Gene Revolution" with the adoption of transgenic crops developed by artificial delivery of specific genes for particular traits. After successful demonstration of transgenic crop adoption in modern agricultural systems (1) and availability of rich source of useful genes, there is a pressing need to expedite the delivery of specific genes in crop plants to meet the challenges faced in the era of "Third Generation" of transgenic plants. The production of transgenic plants in any species by any methods entails two essential and equally important steps. The first step involves transfer and stable integration of transgene into the host/recipient plant genome. The second step comprises the recovery of a fertile transgenic plant from the tissues subjected to stable

Baohong Zhang (ed.), *Transgenic Cotton: Methods and Protocols*, Methods in Molecular Biology, vol. 958,
DOI 10.1007/978-1-62703-212-4_7, © Springer Science+Business Media New York 2013

transformation. Cotton transformation was first published in 1987. Since then, despite early success, there are scarce reports on transgenic cotton during last two decades. The reason of this reality is the extreme difficulty in the production of fertile transgenic cotton plants from inefficient transformation methods which require high degree of tissue culture and media preparation skills (2). Cotton has certain inherent tissue culture difficulties like very long culture time and limited number of cultivars that can be cultured for regeneration as observed at the start of culture 38 years ago making cotton transformation a challenging process in the era of "Evergreen Revolution" till to date. Possibility of transforming embryogenic calli was a major development towards reducing the time required to produce plants from different transgenic events (3–5). Major impediments in the transformation of plants with genes are limitations found in existing transformation protocols like *Agrobacterium* spps. which are only applicable to dicots. *Agrobacterium*-mediated cotton transformation is sensitive with high risk of bacterial contamination during the course of subculturing, with lengthy tissue culture time periods and requiring sophisticated culturing, storage, media, and sterile conditions. Moreover monocots are not prone to *Agrobacterium*-mediated transformation and largely rely on biolistic methods which are highly expensive, with very low transformation efficiency associated with the problem of high copy number insertions of transgenes. Due to economic importance of cotton in developed and developing countries, there is an urgent need to accelerate the application of biotechnological tools to address the problems associated with this crop in applied and fundamental research. Cotton transformation requires a simple and robust gene delivery mechanism and methods. A protocol for the production of transgenic cotton via silicon carbide-mediated embryogenic transformation was developed in our Lab. which is equally valid for monocots as well (4, 6, 8, 10). Silicon carbide whiskers are elongated microneedles which facilitate transgene delivery into target cells by abrasion/perforation during mixing/vortexing (7, 11). This is a newly applied cotton transformation protocol which is easy and elegant in simplicity, efficiency, and throughput.

2. Materials

Prepare all stock solutions in ultrapure and autoclaved water. Use hormone stocks not old more than 2 weeks. Prepare and store all reagents at room temperature (unless indicated otherwise). Use laminar airflow for subculturing and shifting. Diligently follow all waste disposal regulations when disposing potentially health hazard waste materials.

2.1. Gene Constructs and Selection Marker

1. pGreen0029 (GUS) Kanamycin for *E. coli*.

2. pRG229 (AVP1) spectinomycin (12) for *E. coli* (see Note 1). pGreen0029 (GUS) and pRG229 are plant transformation vectors. Kanamycin/spectinomycin (stock 100 mg/mL); weigh Kanamycin-sulfate/spectinomycin 1,000 mg/10 mL, add double-distilled water up to 10 mL volume, filter sterilize using Millipore filter 0.22 μm, aliquot in 1.5 mL Eppendorf, and store at –20°C.

2.2. E. coli Cultures

1. DH$_5$α, *E. coli* competent cells (Invitrogen) (see Note 2).
   Pick a single colony from a freshly grown plate of *E. coli* DH$_5$α and inoculate into 100 mL LB liquid medium in 250 mL autoclaved flask using sterile toothpick followed by incubation at 37°C for 10 h with vigorous shaking on orbital shaker. Reinoculate 5 mL of the overnight grown culture into a 1 L flask containing 200 mL of the same LB medium and incubate at 37°C on orbital shaker until O.D$_{600}$ of cells is 0.5–1.0 (10^8 cells/mL). Transfer grown cell culture aseptically to an ice-cold 50 mL propylene tube and kept cool on ice for 10 min, then centrifuge at 4,000×g for 10 min in a benchtop Sorvall RT6000 refrigerated centrifuge at 4°C to pellet cells and decant supernatant. Resuspend cells in 50 mL of sterile cold d$_2$H$_2$O. Again centrifuge at 4,000×g in the same centrifuge tube at 4°C for 10 min, decant supernatant, and resuspend pelleted cells in 25 mL of sterile cold d$_2$H$_2$O. After another wash, resuspend pelleted cells in 10 mL sterile cold d$_2$H$_2$O containing filter-sterilized cold 10% glycerol. Repeat for another wash. Finally resuspend cells in 1–1.5 mL filter-sterilized cold 10% glycerol, aliquot in 50 μl competent cells, and store at –70°C.

2. Luria Bertani (LB): Add tryptone 10 g, Yeast extract 5 g, NaCl 5 g, pH 7.0, Bacto agar 15 g (see Note 3). Dissolve one by one all LB ingredients keeping initial low volume, adjust pH with 1 N KOH, and make up the volume to 1 L at the end. Again check pH after making up volume and adjust if necessary. Add agar after adjusting pH and autoclave.

3. Electroporator.
   LB solid plates containing Spectinomycin, Phytotechnology Lab. (100 mg/L) and Kanamycin Phytotechnology Lab. (50 mg/L) for growing DH$_5$α (see Note 4). Take 2 μl (1 μg) of each clone for electro transformation by electro cell manipulator 600 (BTX San Diego, California). Keep electroporation cuvettes of 1 mm gap placed on ice. Thaw frozen vials of electro-competent cells of DH$_5$α on ice. Mix 1 μg DNA of the recombinant plasmid with 50 μl of electro-competent cells in the electroporation cuvettes on ice. Set conditions for electroporation as recommended by the manufacturer: Choose

mode 2.5 kV, set resistance R R5 (129 Ω), set charging voltage 1.44 kV. Mix electro-competent cells with DNA followed by transfer to electroporation cuvette. Give pulse and add 1 mL of liquid LB medium immediately, mix gently, transfer to a 1.5 mL Eppendorf tube, and incubate at 37°C for 1 h with vigorous shaking. Spread 200 µl of transformed culture on petri plates containing solid medium supplemented with 50 µg of kanamycin per mL so that only transformed cells should multiply. Seal plates with parafilm and keep at 37°C for 10–12 h. Pick colonies and culture in 5 mL liquid LB medium in a 50 mL tube containing 50 µg of kanamycin per mL. Keep culture tubes at 37°C on shaker growth room in the dark for 10–12 h with vigorous shaking. Confirm transformants via PCR using primers specific to the AVP1 and GUS genes.

4. Microbalance.

5. Spectrophotometer.

6. pH meter.

7. Orbital shaker.

2.3. Harvest of Plasmids

1. Miniprep (alkaline lysis solutions)
   Mini prep solutions: Solution I (Suspension buffer): Tris–HCl (pH 7.4–7.6) 50 mM, EDTA 1 mM, RNAse 100 µg/mL. Solution II (Denaturation soln.) NaOH 0.2 N, SDS 1% and Solution III (Neutralization soln.) Potassium acetate 3 M, Glacial acetic acid 11.5 mL/100 mL (pH 4.8–5.0). Store at room temperature (see Note 5).

2. 70% ethanol Fluka.

3. Isopropanol Fluka (chilled).

4. Ultrapure water.

5. Centrifuge.

6. Agarose.

7. Gel Electrophoresis apparatus.

2.4. 1% Agarose Gel

1. Prepare 50× TAE buffer For 50× Tris–acetate–EDTA buffer (TAE): Weigh Tris base 242 g, Glacial acetic acid, 57.1 mL, 0.5 M EDTA (pH 8.0) 100 mL. Make up the final volume with distilled water to 1 L (see Note 6).

2. Prepare 1% agarose gel
   Weigh 1 g agarose and transfer to 100 mL 1× TAE. Boil for 3–4 min in microwave oven until agarose is dissolved completely (see Note 7).

3. Ethidium bromide 2 µg/mL (see Note 8).

2.5. Cotton Embryogenic Calli

1. Cottonseeds (*Gossypium hirsute L.* cv. Coker-312 seeds delinted with conc.H_2SO_4 @1 mL/10 g (100 seeds) (13).

2. 0.1% SDS and 0.1% $HgCl_2$ (see Note 9).

3. MS0 medium; weigh 4.43 g MSSalt (Sigma Aldrich) (14), 30 g glucose, B5 vitamins (15) and make 1-L medium, adjust pH at 5.7–5.8. B5 vitamins: Gamborg vitamins: Nicotinic Acid 0.1 g, thiamine HCL 1 g, Pyridoxine HCL 0.1 g, and myo-inositol 10 g. Dissolve one by one in double-distilled deionized water and store at 4°C (see Note 10).

4. MS0 petri plates 25–28 mL/petri plate (25×90 mm) Autoclaving conditions: psi 20, 20 min, 120°C. Cool medium to 50–55°C, add antibiotic (kanamycin) 25 μg/mL, shake well, and pour into petri plate (25–28 mL) in laminar airflow (see Note 11).

5. Cotton tissue culture growth room maintained at 28°C with 16/8-h photoperiod.

6. Germinated seeds in glass jar (sigma) and shifted to growth rooms with 16/8-h photoperiod with 1,600–2,000 lux.

7. Hypocotyls of 3–5 mm length.

8. Culture hypocotyls on MSB1 CCi medium: Weigh 4.43 g of Ms Salt, 30 g glucose, 1.60 g $MgCl_2$, B5 vitamins and transfer to 1-L dH_2O. Add hormones: kinetin 0.5 mg and 2,4-D 0.1 mg. Adjust pH at 5.7–5.8 and add 2.5 g of phytagel (Sigma). Add plant growth hormones before pH adjustment. Dissolve Kinetin in 0.1 N NaOH (200 μL), then make up volume with autoclaved ultrapure water and 2, 4-D in 50% ethanol. Wrap in Al-foil and store at 4°C. Avoid using over 2-week-old hormone stocks (see Note 12).

9. Two-month-old calli (1 g) were transferred on MSB2 (16) MSK medium (Fig. 2a): Weigh 4.43 g of Ms Salt, 30 g glucose, 1.90 g KNO_3, 1.60 g $MgCl_2 \cdot 6H_2O$, B5 vitamins and transfer to 1 L dH_2O. Adjust pH at 5.7–5.8 and autoclave (see Note 13).

10. Embryogenic callus cultures (16) (Fig. 1).

2.6. Silicon Carbide Whisker Components

1. Silicon carbide whiskers (Alfa Aesar Avocado Organic Cat # 38787) 250 mg and transfer to a 1.5 mL Eppendorf. Wrap in double layer of aluminum foil and shift to a glass jar (see Note 14).

2. Transfer 250 mg of silicon carbide whiskers to 5 mL of MSB2 medium in a 50 mL falcon tube (Sanaka).

3. Add 5 μg of plasmid DNA (see Note 15).

4. 2 g of embryogenic calli (Fig. 2b) (see Note 16) (16).

5. Vortex genie II.

6. Whatman 3 filter paper for semidrying.

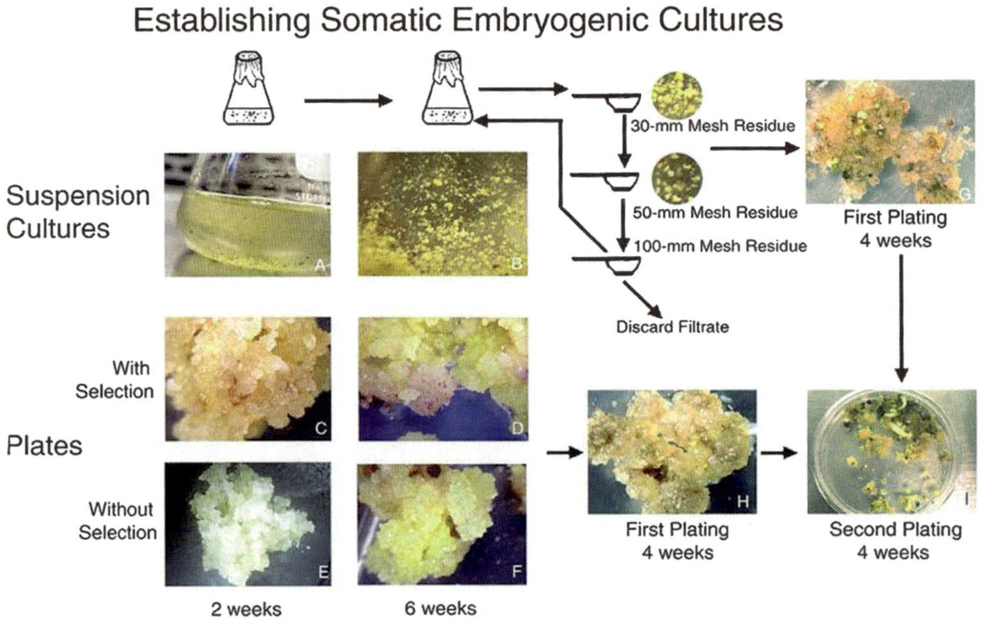

Fig. 1. Friable, creamy-colored callus is generally subdivided into three parts for transfer to hormone-free MSK media (see Note 13). (**a**, **b**) shows liquid medium for establishing embryogenic suspension cultures. (**c**–**f**) shows solid medium with or without selection. Growing suspension cultures appear greenish in color (**a**), and depending on the cell-line, embryogenic suspension cultures (**b**) are generally established in ~3–4 weeks. Selection is added at one-half strength once the cultures are well established. Large cell clumps or organized masses appear in the cultures in 2–4 weeks, with well-formed somatic embryos occurring in 4–6 weeks (b). Callus transferred to plates used primarily as backup in case of contamination of suspension cultures also become embryogenic (**d**, **f**). A small accumulation of anthocyanin (*red pigmentation*) in cultures under selection is an indicator of somatic embryogenesis (**c**, **d**). Callus grown without selection appears light green in color (**e**, **f**). After 6–8 weeks, embryogenic suspension cultures are size fractionated by sieving through a series of 30-, 50-, and 100-mm mesh screens to obtain somatic embryos of similar and uniform size and developmental stage for synchronizing growth for plating onto semisolid medium. The 100-mm mesh residue is used to maintain embryogenic suspension cultures. Embryogenic material can be harvested from cultures on plates as well. On the "first" plating, somatic embryos are present in low abundance (**g**, **h**). The number, size, and quality of somatic embryos increase after the "second" plating (reproduced from ref. (16) with permission from springer link).

2.7. Selection and Expression of Transgenes

1. Transfer putatively transformed calli to MSB2 medium.

2. Recover putative transgenic calli colonies on kanamycin 25 mg/L (Fig. 2d, e).

3. GUS substrate (17) (Fig. 2c, d, f).

4. Regenerate transgenic plants km-resistant calli (Fig. 2g–i).

5. Transfer embryos on S-medium for root/shoot development (18).

6. Hardening of transgenic plants (Fig. 2i).

Fig. 2. Tissue culture steps in whisker-mediated cotton callus transformation: (**a**) nonembryogenic callus, (**b**) embryogenic callus, (**c**) transient GUS expression in whisker-treated embryo after 72 h. (**d**) GUS expression in kanamycin-resistant callus colonies after 6 weeks. (**e**) Development of km-resistant callus colonies for AVP1 after 6 weeks. (**f**) GUS expression in stable colony; *Right control*, left transgenic callus. (**g**) Transgenic embryogenic callus with regenerated embryos. (**h**) Regenerated transgenic plant, (**i**) population of To regenerated transgenic cotton plants. (**j**) To regenerated plant in soil pot. (**k**) Km-wiping test of transgenic (lower) and control (upper) plant cotyledonary leaves. (**l**) Salt treatment of control (left 2) and AVP1 transgenic T1 cotton plants (*Right*) (reproduced from ref. (4) with permission from springerlink).

2.8. Molecular Analysis of Transgenic Tissues

1. Transgenic tissues 0.5–1.0 g.

2. Genomic DNA: 2%CTAB (19).

   Prepare 2% CTAB (w/v), 100 mM Tris–HCl (pH 8.0), 20 mM EDTA (pH 8.0), 1.4 M NaCl, 1% polyvinylpyrrolidone

(PVP). Tris–HCl means that concentrated HCl (12 N) can be used at first to narrow the gap from the starting pH to the required pH. From then on it would be better to use a series of HCl (e.g., 6N and 1N) with lower ionic strengths to avoid a sudden drop in pH below the required pH and make up the final volume (see Note 17).

3. Pestle mortar.

4. β-Mercaptoethanol (strong reducing agent) is mixed in 2%CTAB short before the start of DNA isolation (see Note 18).

5. 70% ethanol.

6. Isopropanol.

7. PCR thermal cycler (Eppendorf).

8. PCR reagents (Paq DNA polymerase, Paq buffer, dNTPS, Primers (F&R) (Fig. 3).

9. Nylon membrane.

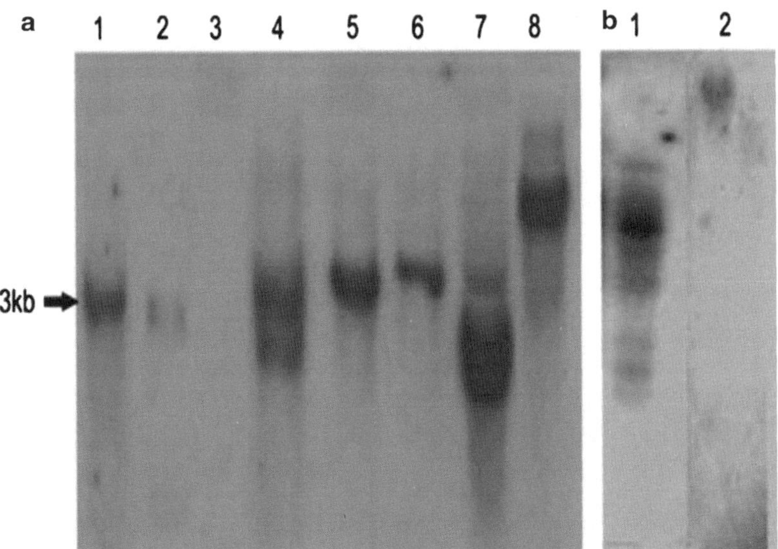

Fig. 3. Southern blot analysis of nuclear DNA isolated from silicon carbide-mediated kana-mycin-resistant and nontreated control calli probed with the *AVP1* gene. (**a**) Total genomic DNA (approx. 15 μg) was digested with *Hind*III and electrophoresed on 0.8% agarose gel followed by transfer to nylon membrane. Lanes 1 (1×-5 pg) and 2 (5×-25 pg) are one copy and five copies of 3.0 kb AVP1 containing fragment obtained by *Hind*III digestion of pRG229, lanes 3 (nontreated control), 4, 5, 6, 7, and 8 represent integration of *AVP1* gene into nuclear DNA either in one or more copies. (**b**) shows transgene integration in uncut genomic DNA (*Lane 2*) and in multiple copies in genomic DNA digested with *Pst*1 (*lane 1*). The blots were probed with a ^{32}P-labelled 2.2 kb AVP1 gene (reproduced from ref. (4) with permission from springerlink).

10. Southern; (20) Depurination (0.25 M HCl). Take 28 mL of 37% HCl and make up volume 1 L with ultrapure water. Wear coat, gloves, and mask while handling HCl to prevent any health hazard (see Note 19).

11. Denaturation Solution: 20 g of NaOH, 58.4 g of NaCl/L of ultrapure water. NaOH is caustic, so be careful with it. Denature gel for 30 min. Neutralizing Solution: 78.8 g Tris-base, 87.6 g NaCl in 800 mL of ultrapure water. Adjust the pH to 7.4. Add ultrapure water to 1 L (see Note 20).

12. 20×SSC: 175.3 g NaCl, 88.2 g of trisodium citrate (citric acid); add ultrapure water up to 1 L (see Note 21).

13. Blotting papers.

14. 1% agarose gel.

2.9. Salt Treatment

1. Screen T_1 transgenic cotton plants by km wiping (15 µg/leaf) (Fig. 2k).

2. NaCl solution (50, 100, 150, 200 mM).

3. Progeny of transgenic plants.

4. Control plants (non-transformed).

5. Hoagland nutrient solution.

3. Methods

Carry out all procedures at room temperature unless otherwise specified.

3.1. Miniprep and 1% Gel Electrophoresis

1. Grow *E. coli* culture for 10–12 h at 190 rpm at 37°C until OD 0.75×10^8 cells/mL is attained.

2. Take 1 mL *E. coli* grown culture and harvest cells by centrifuging at $4,000 \times g$ for 1 min.

3. Decant supernatant and dissolve pellet in soln I, II, and III (see Note 5).

4. Spin for 10 min and collect supernatant (~600 µl).

5. Add 0.6 vol of isopropanol to supernatant.

6. Incubate at –20°C for 30 min.

7. Spin for 15 min at $10,000 \times g$.

8. Discard supernatant and wash with chilled 70% ethanol.

9. Add 20 µl ultrapure water.

10. Load on 1% gel and measure OD on spectrophotometer.

11. Adjust concentration to 1 µg/µl (see Note 22).

3.2. Embryogenic Calli Transformation

1. Weigh 2 g embryogenic calli and transfer to 5 mL MSB2 medium containing 250 mg of silicon carbide whiskers.

2. Vortex for 60 s on vortex genie.
   Semidry calli and transfer to MSB2 medium.

3. Filter the calli suspension by using a Whatman 3 filter paper.

4. Incubate at 28°C for 48 h.

5. Transfer calli to MSB2 medium containing 25 mg/L kanamycin.

6. Recover embryos and transfer to S-medium (19) for root/shoot development.

3.3. Selection and Analysis of Transgenic Calli/Plants

1. Transfer transformed calli to MSB2 medium containing 25 mg/L kanamycin.

2. Subculture calli after every 18–20 days (see Note 23).

3. Transgenic calli start to emerge in the form of small clusters (Fig. 2e).

4. A part of emerged calli were used in histochemical assays of GUS (Fig. 2d) (see Note 18).

3.4. Southern Analysis of Transgenic Cotton Plants

1. Isolate total genomic DNA
   Isolate total genomic DNA by 2% CTAB by Doyle and Doyle (18). Take 20 mL 2× cetyltriethylmethylammonium bromide (CTAB) and 100 μl 2β-mercaptoethanol in a 50 mL falcon tube. Keep the tube in water-bath at 65°C for 30 min. Take 0.3 g calli in mortar and pestle, and add liquid nitrogen. Grind the sample to a fine powder. Pour the hot 2× CTAB in ground powder suspension and incubate at 65°C for 30 min with occasional swirling. Add equal volume of chloroform:isoamyl alcohol (24:1) and mix gently. Centrifuge at $6,000 \times g$ for 10 min at room temp. Take the upper phase in a 50 mL tube and add 2.5 volume of absolute ethanol. Centrifuge at $6,000 \times g$ for 5 min. Decant the supernatant and wash the pellet with 5 mL 70% ethanol. Dissolve the pellet in 300 μl distilled autoclaved water, transfer to an eppendorf tube, and spin for 5–10 s. Take the supernatant in another tube and run 2 μl on 1% agarose gel to check the concentration and quality of nuclear DNA (see Note 24).

2. Restriction of genomic DNA with *Hind*III
   Restrict total genomic DNA with two enzymes to know the copy number of delivered DNA molecules. *Hind*III restriction yielded expression cassette and intensity of hybridization signals is indication copy number while kpn1 restriction gives unique DNA fragment size peculiar number of integrations (Fig. 2a, b) (see Note 25).

3. Transfer restricted and electrophoresed DNA to Nylon membrane
   Transfer resolved DNA from gel to Nylon membrane in 10×SSC transfer buffer for 12–14 h. Dry and bake (50°C) DNA blot for 15 min (see Note 26).

    4. Cross-linking

        Then cross-link for 90 s on Cl-1,000 Ultraviolet Cross linker wrap DNA blots in cling film for subsequent storage (room temperature) or hybridization studies (see Note 27).

    5. Pre-hybridization and probing (20).

    6. Washing and developing X-ray film (Fig. 3).

3.5. Salt Treatment of Transgenic Cotton Plants

1. Evaluate transgenic plants containing AVP1 for improved salt tolerance (4).

4. Notes

1. Confirm recombinant pGreen0029 (GUS) and pRG229 (AVP1) plant transformation vectors by PCR using gene-specific primers and restriction analysis.

2. Grow good-quality *E. coli* cultures. Inspect cultures for cell lysis; if it happens regrow culture, reduce volume in the flask, and harvest culture at exponential growth phase with foaming.

3. Prepare LB medium and divide into 25, 50, 100, and 200 mL in flasks and do autoclaving. These media bottles can be stored at room temperature for 2 weeks.

4. Place cuvettes in ice for 30 min. First mix well *E. coli* competent cells and plasmid DNA and then add in the cuvette. Ensure that cells/DNA mixture is in contact with electrode plates in the cuvette and clean/dry cuvette from outside. Add LB(L) immediately after electric shock. If sparking occurs then change/prepare fresh competent cells.

5. Miniprep solutions can be stored for extended periods. Add RNase in solution I short before use and it can be stored at 4°C for a week.

6. Use 1× TAE buffer; use fresh gel and running buffer after every DNA run.

7. Before pouring melted 1% gel, make sure that agarose is uniformly dissolved; if agarose is undissolved, reheat in microwave oven for another 30 s. Maintain weight before and after heating agarose gel. Use dH_2O for maintaining weight.

8. Wear gloves while handling ethidium bromide as it is potentially carcinogenic. Dispose off used gloves and gels in a safe way.

9. $HgCl_2$ and SDS are highly toxic and has high risk of SDS ingestion with air. Wear gloves and aspirator while handling.

10. B5 vitamins are perishable media components and require storage at 4°C. Check for contamination before adding and use fresh for optimal results.

11. Autoclaving conditions: psi 20, 20 min, and 120°C. Cool medium to 50–55°C, add antibiotic (kanamycin) 25 μg/mL, shake well, and pour into petri plate (25–28 mL) in laminar airflow.

12. Plant growth hormones are perishable and remain no more viable after 2 weeks. Some time precipitation and contamination may appear, prepare fresh stocks in case of precipitation and contamination.

13. Transfer fresh, granular calli to embryogenic medium MSB2. Visit callus culture daily, shift calli to fresh medium in case of contamination, and subculture every 2 weeks. Select granular, parrot green calli portions with tiny embryos during subculturing.

14. Wear gloves and mask to handle silicon carbide whisker to avoid any lung hazards in case of inhalation, double layer whisker carrying tubes.

15. Use good quality (no smear/degradation on ethidium bromide gel) and concentrations of RNA free plasmid (1 μg/μ). Use spectrophotometer and gel electrophoresis to confirm quality and concentration of plasmid.

16. Use fast-growing, embryogenic calli which are loose and granular in texture with light green. Maintain good embryogenic cultures by selecting fast-growing calli containing immature embryos.

17. Shake well and warm in microwave oven for 3 min. It is a strong detergent and use gloves during handling.

18. β-Mercaptoethanol has a strong and persistent smell. Use fume hood for adding into the CTAB.

19. Wear lab coat, gloves, and mask and use fume hood to avoid the spread of acid fumes in the working environment. Use freshly prepared depurination solution.

20. Reuse denaturation solution and store in the dark at room temperature.

21. 20×SSC is a strong salt which precipitates after sometime and heat before using it.

22. Measure plasmid concentration after miniprep and adjust at 1 μg/mL. In case of low concentration increase concentration by vacuum dryer.

23. Keep transformed calli on selection medium containing 25 mg/L of each calli cluster separated. Subculture within 20–25 days and keep cultures contamination free.

24. Take healthy calli for nuclear DNA isolation and grind calli to fine powder. Use 5 mL CTAB for 0.5 g callus. Total genomic DNA bands should be intact and RNA free. In case of DNA smear on 1% agarose gel which indicates the degradation of DNA samples, reisolate the genomic DNA.

25. Avoid enzyme overconcentration and incubation of restriction reactions for extended periods as they lead to degradation of genomic DNA and may give nonspecific restriction of DNA samples. Check restriction by loading 0.5 µl restricted DNA on 1% gel. In case of partial restriction, add fresh enzyme and ensure complete restriction.

26. Depurinate 1% DNA gel in 0.2N HCl solution, during depurination loading DNA dye color changes from blue to yellow orange. After DNA transfer to hybond membrane, restain gel with ethidium bromide (1 µg-mL) for 20 min, and view under UV light. Absence of any DNA illumination indicates complete DNA transfer to hybond membrane. Short circuit in DNA transfer assembly or trapping of air bubbles in between nylon membrane and 1% gel are the main reasons of inefficient DNA transfer. Remover the air bubbles by rolling pencil on the membrane and avoid short circuiting by cutting Hybond Nylon membrane exactly of the same size as of DNA gel.

27. Dry blots completely before cross-linking. Cross-link blots twice for optimal results.

Acknowledgments

This work was supported by PIBS and PSF, Pakistan. Technical help by Miss. Shahnaz Akhtar (media preparations), Mrs. Kalsoom Akhtar, and Mr. karim Bukhsh (plant tissue culture) is gratefully acknowledged.

References

1. Hasnsen G, Wright MS (1999) Recent advances in the transformation of plants. Trends Plant Sci 4(6):226–231

2. Rathore KS, Sunilkumar G, Campbell LM (2010) Cotton: methods in molecular biology, edited by: Kan Wang C, vol 343. *Agrobacterium* protocols, 2/e, vol 1. Humana Press Inc., Totowa Nj

3. Wu J, Zhang X, Nie Y, Luo X (2005) High-efficiency transformation of *Gossypium hirsutum* embryogenic calli mediated by Agrobacterium tumefaciens and regeneration of insect resistant plants. Plant Breed 124:142–146

4. Asad S, Mukhtar Z, Arshad M (2008) Silicon carbide mediated embryogenic callus transformation of cotton and regeneration of salt tolerant plants. Molecular Biotechnol 40:161–169

5. Duncan DR (2010) Biotechnology in agriculture and forestry, vol 65. pp 65–77. doi: 10.1007/978-3-642-04796-1-4

6. Petolino JF, Hopkins NL, Kosegi BD, Skokut M (2000) Whisker-mediated transformation of embryogenic callus of maize. Plant Cell Rep 19:781–786

7. Mizuno K, Takahashi W, Ohyama T, Shimada T, Tanaka O (2004) Improvement of the aluminum borate-whisker medtaed method of DNA delivery into rice callus. Plant Prod SCi 7(1):45–49

8. Terakawa T, Hisakazu H, Masanori Y (2005) Efficient whiskermediated gene transformation in a combination with supersonic treatment. Breed Sci 55:456–458

9. Khalafalla M, El-Sheny HA, Rahman SM, Teraishi M, Hasegawa H, Terakawa T, Ishimoto M (2006) Efficient production of transgenic soybean (*Glycine max* [L] Merrill) plants mediated via whisker-supersonic (WSS) method. Afr J Biotechnol 5(18):1594–1599

10. Matsushita M, Otani M, Wakita M, Tanaki O, Shimida T (1999) Transgenic plant regeneration through silicon carbide mediated transformation of rice (*Oryza sativa* L.). Breed Sci 49:21–26

11. Asaad S, Arshad M (2011) Silicon carbide mediated-plant transformation. Properties and application of silicon carbide. ISBN 978-953-307-2012. Intech April, 2011

12. Gaxiola RA, Li J, Undurraga S, Dang LM, Allen GJ, Alper SL, Fink GR (2001) Drought- and salt-tolerant plants results from overexpression of the AVP1 H$^+$-pump. Proc Natl Acad Sci USA 98:11444–11449

13. Hashmi JA, Arshad M, Asad S (2011) Engineering cotton against CLCuD using truncated AC1 sequences. Virus Genes 42(2): 286–296

14. Murashige T, Skoog F (1962) A revised medium for rapid growth and bioassays with tobacco tissues cultures. Physiol Plantarum 15: 473–479

15. Gamborg OL, Miller RA, Ojima K (1968) Nutrient requirements of suspension culture of soybean roots cells. Exp Cell Res 50:150–158

16. Wilkins TA, Mishra R, Trolinder NL (2004) Agrobacterium mediated transformation regeneration of cotton. Food Agri Environ 2(1):179–187

17. Jefferson RA, Kavanagh TA, Bevan MW (1987) GUS fusions: b-glucuronidase as a sensitive and versatile gene fusion in higher plants. EMBO J 6:3901–3907

18. Doyle JJ, Doyle JL (1990) A rapid DNA isolation procedure for small quantities of fresh leaf tissue. Phytochem Bull 19:11–15

19. Stewart J, Mc D, Hsu CL (1977) In ovulo embryo culture and seedling development of cotton (*Gossypium hirsutum* L.). Planta 137: 113–117

20. Sambrook J, Fritsch EF, Maniatis T (1989) Molecular cloning: a laboratory manual. Cold Spring Harbor Laboratory Press, Cold Spring Harbor, NY

Part III

Detection

Chapter 8

Investigating Transgene Integration and Organization in Cotton (*Gossypium hirsutum* L.) Genome

Jun Zhang and Yan Hong

Abstract

In this chapter, we present detailed experimental procedures for investigating integration patterns of transgenes in cotton genome. We use conventional PCR and genomic Southern blot hybridization to characterize integration of T-DNA components and vector backbone fragments. For multiple copy insertions into the same site (complex loci), transgene/transgene junctions (including canonical and truncated T-DNA and transgene involved vector backbone sequences) are characterized by PCR and sequencing. Inverse PCR (see Note 1) and sequencing is used to characterize transgene/cotton genome junctions. Distribution of T-DNA insertion in cotton genome is evaluated by analysis of transgene flanking sequences. The pre-insertion sites can also be cloned and sequenced (based on the flanking sequences) for survey of genomic structure changes brought by transgene integration by comparing a pre-insertion site with corresponding transgene/plant junctions.

1. Introduction

Microprojectile bombardment and Agrobacterium-mediated T-DNA transfer are widely used for the production of transgenic plants useful for basic plant research such as random mutagenesis (1, 2) and functional genomics studies (3) and biotechnology. Since the structure of a transgene locus can have a major influence on the level and stability of transgene expression, comprehensive molecular characterization of transgene integration and organization in cotton genome is an essential step of developing transgenic cotton for both commercial and basic research aimed at functional genomics. Most of our limited knowledge on transgene integration and organization (e.g., the information on locus structure, frequency of T-DNA integration sites in target plant genome, and distribution of transgene loci) comes from studies on model plants like Arabidopsis (4–8), tobacco (9), and rice (10–16). However,

Baohong Zhang (ed.), *Transgenic Cotton: Methods and Protocols*, Methods in Molecular Biology, vol. 958,
DOI 10.1007/978-1-62703-212-4_8, © Springer Science+Business Media New York 2013

the patterns of which the T-DNA integrates into the host genome of cotton are still poorly understood although transgenic cotton is ranked the most successful commercial GMO crop of the world (17). PCR or DNA gel blot analysis is generally used to prove the existence of transgenes and evaluate transgene copy number (18). Tail-PCR and inverse PCR and other genome walking methods have also been used to isolate and clone the T-DNA flanking sequences (19–22). In this chapter, we present the experimental procedures of PCR and sequencing, genomic Southern blot hybridization, and inverse PCR and sequencing for detailed characterization of transgene integration and organization in 139T0 transgenic cotton plants generated by *Agrobacterium tumefaciens* AGL1-mediated transformation by a binary plasmid pPZP-GFP.

2. Materials

2.1. Binary Vector Construct and Transgenic Cotton

2.1.1. Binary Vector Construct

To construct the binary plasmid pPZP-GFP (Fig. 1) used in this study, a 1,863 bp fragment containing the 35S promoter and GFP2 gene was cut from pGFP2 vector (23) with two restriction enzymes PstI/EcoRI (New England Biolabs, Ipswich, MA) and inserted into the binary vector pPZP111 (24) to obtain the plasmid pPZP-GFP (Fig. 1) that was later transformed into *Agrobacterium tumefaciens* strain AGL1 (17).

2.1.2. Transformation

Cotyledon and hypocotyl of upland cotton (*Gossypium hirsutum* L.) variety Coker 312 is used as the explant for generating transgenic cotton lines (17, 25). 50 mg/L of Kanamycin is included in somatic embryo induction media and GFP fluorescence is also used for positive selection of transformants. T0 cotton plants were maintained in a greenhouse under natural lighting with day length of about 12 h and temperature ranged from 25 to 34°C. They were allowed to self-pollinate in greenhouse.

Numbering of nucleotide base pairs starts from the R border to L border (Fig. 1). Single-arrowed PCR primers are used for

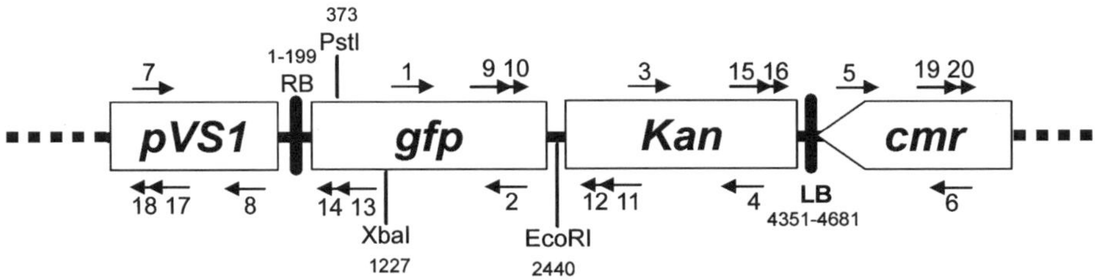

Fig. 1 Illustration of the binary vector pPZP-GFP (17).

analysis of the presence of different transgenic elements and generating probes for genomic Southern blot hybridization (with primers 1 + 2, 3 + 4, and 5 + 6 for gfp, Kan and cmr respectively). Double-arrowed primers are used for nested PCR cloning of junction sequences. Restriction enzymes PstI, XbaI, and EcoRI all have unique cut site in the binary vector (17).

2.2. Primers for PCR and Inverse PCR

Table 1 lists all the PCR primers used in this protocol with location, orientation, annealing temperature, and sequence information; primers 1–8 are used as pairs to characterize if the target gene (*gfp*) and selection marker gene (*Kan*) and vector backbone fragment outside the two T-DNA borders integrated into the cotton

Table 1
Primers and nested primers for this chapter (17)

Primer ID	Position and orientation on construct	Target or usage	Primer sequence (5′-3′)	Annealing temperature (°C)
1	F1251	*gfp*	ATGGGTAAGGGAGAAGAACTTTTCACTGG	65
2	R1792		TGGTCTGCTAGTTGAACGCTTCCATC	
3	F3354	*Kan*	CTATTCGGCTATGACTGGGCACAACA	63
4	R4031		GTAAAGCACGAGGAAGCGGTCAGC	
5	F4804	*Cmr*	CAGTGGCGGTTTTCATGGCTTCTTGT	65
6	R5321		AGAAACTGCCGGAAATCGTCGTGGTA	
7	F10141	*pVS1*	CGGGTGCGGTCGATGATTAGGGAACG	60
8	R10729		TGATCCAACCCCTCCGCTGCTAT	
9	F2017	1st	AGATTGAATCCTGTTGCCGGTCTTG	62
10	F2026	2nd	CCTGTTGCCGGTCTTGCGATGATTAT	65
11	R2349	1st	CATTAGGCACCCCAGGCTTTACACTT	62
12	R2340	2nd	CCCCAGGCTTTACACTTTATGCTTCC	64
13	R428	1st	CATCTGTGGGTTAGCATTCTTTCTGA	62
14	R398	2nd	GAAAAGGCTAATCTGGGGACCTGCAG	66
15	F4059	1st	ATCGCCTTCTATCGCCTTCTTGACG	60
16	F4062	2nd	GCCTTCTATCGCCTTCTTGACGAGT	62
17	R10310	1st	AGCACCCGCGACCTACTGGACAT	60
18	R10299	2nd	GACCTACTGGACATTGCCGAGCGC	64
19	F5536	1st	ATTCAGGTTCATCATGCCGTTTGTG	60
20	F5608	2nd	GATGAGTGGCAGGGCGGGGCGTAAT	65

genome. The nested primers numbered from 9 to 20 are used to isolate the flanking sequences from any situations of transgene integration by inverse PCR. The transgenic cotton DNA samples are digested by EcoRI thoroughly and circularized by self-ligation using T4 ligase to be used as the templates of inverse PCR. R428, R398/F2017, F2026 and F4059, F4062/R2349, and R2340 are used to amplify the canonical T-DNA/plant DNA junctions at the two border end regions while R10310, R10299/F2017, F2026 and F5536, F5608/R2349, and R2340 are used to amplify the junctions formed by the integrated vector backbone and plant DNA. All the 2nd primers, i.e., the nested primers, are also used as sequencing primers in directly sequencing.

2.3. DNA Preparation

2.3.1. Instruments

1. Centrifuge 4-15C/Centrifuge 4K15C (Sigma).
2. Sterile polypropylene tube (50 ml and 15 ml) and 1.5 ml Eppendorf tubes.
3. Pipettes and suitable sterile tips.
4. Liquid nitrogen.
5. Mortar and pestle.

2.3.2. Extraction Buffer (see Note 2)

0.35 M glucose, 1 M Tris HCl (pH 8.0), 5.0 mM Na-EDTA (pH 8.0), 2% Polyvinylpyrrolidone (PVP) 40.
1% Beta-mercaptoethanol (β-ME, added just before use).

2.3.3. Lysis Buffer

0.1 M Tris HCl (pH 8.0), 1.4 M NaCl, 20 mM Na EDTA (pH 8.0), 2% Cetyltrimethyl Ammonium Bromide (CTAB), 2% PVP 40,000 MW, 1% β-ME (added just before use).

2.4. Conventional PCR and Inverse PCR

1. Primer pairs (Table 1 and see Note 3).
2. 10× Amplification buffer (Qiagen).
3. Taq polymerase (Qiagen).
4. dNTPs (10 mM, containing 2.5 mM of the dATP, dCTP, dGTP, and dTTP).
5. PTC100 Thermal Cycler (MJ Research).
6. Qiagen Gel Extraction Kit (Qiagen, Chatsworth, CA).
7. pGEM-T-easy vector (Promega, Madison, WI).
8. ABI 3730XL DNA sequencer (Applied Biosystems, Foster City, CA).
9. Restriction enzyme EcoRI (see Note 4) (New England Biolabs, Ipswich, MA).
10. T4 Ligase (New England Biolabs, Ipswich, MA).
11. Phenol:chloroform (1:1, v/v).
12. Phenol.

13. Sodium acetate (3 M, pH 7.0).

14. Pipette and suitable sterile tips.

2.5. Southern Blot Hybridization Analysis

1. Submerge Agarose electrophoresis set.

2. Nylon membrane (Hybond N+, Amersham Biosciences).

3. Whatman 3 M paper and paper towels.

4. Hybridization oven and hybridization tubes.

5. Autoradiograph cassette and intensifying screens.

6. Roche DIG Nuclear Detection System (Roche, Indianapolis, IN, USA).

7. X-ray film (Kodak BioMax).

8. 65°C shaking water bath.

9. 20× SSC: 3 M NaCl, 0.3 M Citrate, pH 7.4.

10. 10%SDS.

3. Methods

3.1. Genomic DNA Preparation for Genomic Southern Blot and PCR Analysis

DNA for Southern blot analysis and PCR is isolated using CTAB method followed by Paterson et al. (26) with modification.

Before Experiment preheat Nuclei Lysis Buffer to 65°C and add 1% β-ME before use.

Wear safety glasses throughout the procedure.

1. Grind 10–20 g leaf tissue (either fresh or stored frozen at −80°C after submergence in N2) with mortar and pestle in liquid nitrogen, transfer the fine powder into a 50 ml falcon tube, and store at −20°C.

2. Add 20 ml of ice-cold Extraction Buffer, mix well, and keep on ice for at least 30 min before centrifugation at $2,600 \times g$ (3,500 rpm, rotor 11150, Sigma 4K15C) for 20 min at 4°C; remove supernatant.

3. Add 10 ml Lysis Buffer and vortex to resuspend pellet; incubate in 65°C water bath for at least 30 min. Mix the tubes periodically by stirring or rocking the tubes every 10 min.

4. Add 12 ml of chloroform:isoamyl alcohol (24:1) and mix gently by inverting tubes, until an emulsion forms (performed in hood).

5. Centrifuge at $2,600 \times g$ for 20 min at 15°C.

6. Transfer upper phase to a new 50 ml tube.

7. Add 0.6 volume of cold isopropanol; mix gently; and let set at −20°C for more than 1 h. Centrifuge at $3,400 \times g$ for 10 min (15°C). Decant supernatant, and let tubes air-dry.

8. Add 5 ml of TE buffer (pH 8.0) to dissolve pellet at 65°C for 30 min (see Note 5).

9. Transfer the dissolved genomic DNA into 15 ml falcon tubes and add RNaseA to a concentration of 20 µg/ml, mix gently, and incubate at 37°C for 15 min.

10. Add equal volume of chloroform:isoamyl alcohol (24:1) to each tube, and mix thoroughly by gently inverting tubes.

11. Centrifuge at 3,400×g (4000 rpm, rotor 11150, Sigma 4K15C) for 15 min (15°C). Carefully transfer the supernatant into a new 15 ml falcon tube.

12. Add 1/10 volume of 3 M Na Acetate (pH 7.0) and equal volume of cold isopropanol into the tubes. Mix thoroughly by carefully inverting the tubes.

13. Pick out the fiber-like DNA using sterile tips and transfer them into 1.5 ml sterile Eppendorf tubes. Or centrifuge at 3,400×g for 10 min at 15°C and carefully transfer the pellet into a 1.5 ml sterile Eppendorf tube.

14. Pour in 70% ethanol to wash the DNA pellet two times (see Note 6).

15. Air-dry the pellet and dissolve the pellet with 200–400 µL TE buffer (pH 8.0).

16. Quantify DNA by measuring OD_{260}.

17. Quality check of DNA by electrophoresis on 0.8% agarose gel.

DNA is stable for several days when stored at 4°C in Buffer TE. For long-term storage, freezing at –20°C is recommended.

3.2. Conventional PCR and Inverse PCR

3.2.1. Conventional PCR

Primer pairs are used to amplify the gene of interest, the selection marker gene inside T-DNA (1/2 for *gfp* and 3/4 for *npt*II), and vector backbone sequences outside RB and LB (5/6 and 7/8) (see Table 1) by the following procedures:

1. 20–50 ng of DNA is used as templates for the conventional PCR with 0.5 µM of each of the specific primers above and 1 U of Taq polymerase (Qiagen) in a 25 µL reaction volume.

2. The PCR reactions are initiated by heating the samples for 3 min at 95°C, followed by 35 cycles performed at 95°C for 20 s, optimal annealing temperatures to special primer pairs for 30 s, and suitable durations of extension according to the fragment size amplified at 72°C. Additional 10 min of extension at 72°C is needed after the cycles.

3. PCR products are separated by 1% agarose gel electrophoresis.

4. PCR fragments are excised from gel and purified from agarose gel by using Qiagen Gel Extraction Kit (Qiagen, Chatsworth, CA, USA) (see Note 7).

5. All fragments purified are used for direct sequencing and/or cloned into the vector pGEM-T or T-easy vector (Promega USA) for sequencing.

6. Sequencing is carried out with the nested primers using a Perkin-Elmer ABI377XL DNA sequencer.

3.2.2. Inverse PCR

Inverse PCR is conducted based on the method reported by Thomas et al. (27). 100 U of restriction enzyme EcoRI is used to digest 2–5 µg of DNA from a transgenic T0 plant in 100 µL of reaction volume and incubation at 37°C overnight.

1. The digestion products are then purified by phenol extraction (see Note 8).

2. Circularize the digestion products by T4 Ligase (New England) at low temperature (typically 12°C) overnight (see Note 9).

3. The inverse PCR is then initiated using products of ligation as templates with the 1st-round primers at relatively low annealing temperatures (see Table 1).

4. The first-round PCR procedure was as follows: 95°C for 3 min followed by 35–40 cycles of 95°C for 20 s, 55°C for 30 s, and 72°C for 2.5 min and end after 10 min of additional extension at 72°C.

5. The 2nd-round (nested) PCR is carried out using the diluted products of the 1st-round PCR with the nested primers in Table 1.

6. The PCR reaction is similar to the 1st-round PCR besides the relatively higher annealing temperature used.

7. The products of 2nd-round PCR are fractionated by 1% agarose gel.

8. PCR fragments are purified from agarose gel by using Qiagen Gel Extraction Kit (Qiagen, Chatsworth, CA, USA).

9. All fragments purified are used for direct sequencing and/or cloned into the vector pGEM-T or T-easy vector (Promega USA) for sequencing.

10. Sequencing is carried out with the nested primers using a Perkin-Elmer ABI377XL DNA sequencer.

3.2.3. Sequence Analysis

The flanking sequences in cotton genome and the transgene junction sequences are determined by BLASTN DNA homology searches against the NCBI GenBank (http://www.ncbi.nlm.nih.gov/BLAST) database and alignment with the construct pPZP-GFP sequences. An E value of e^{-10} is adopted as the cutoff threshold for positive homology.

3.2.4. Characterization on Pre-insertion Sites in Cotton Genome

Primers located on either side of one insertion can be designed based on the transgene flanking sequences. They can be used to amplify the pre-insertion sites from non-transgenic (wild type) cotton genomic DNA. The amplified products are either directly sequenced with the primers per se or cloned into pGEM T- or T-easy vectors and sequenced by the sequencing primers in the vector. The sequences are then compared with the insertion sites to see the changes caused by the transgene insertion.

3.3. Southern Blot Analyses

Southern analyses are conducted using Roche DIG Nuclear Detection System (Roche, Basal, Switzerland) followed by the protocols from the manufacturer's recommendations. It has been suggested that multiple-copy T-DNAs can either integrate into one single site as repeats to form the so-called complex T-DNA locus (28) or inserted into independent sites. An enzyme cutting inside the T-DNA (XbaI) will not be able to distinguish the two situations; we digested the same genomic DNAs with an enzyme (*Bam*HI) with no cut site on the T-DNA to help determine the number of insertion loci. All probes/Restriction Enzyme used for characterization copy numbers, vector backbone integration, and number of insertion sites are illustrated in Fig. 1.

3.3.1. Digestion of Genomic DNA

1. The total reaction system contains:

10× buffer	20 μL
DNA	10–20 μg
Restriction enzyme	150 u

2. Add distilled water to total volume of 200 μL and digest overnight.
3. Add 2.5 volume of ethanol to precipitate the digested genomic DNA (see Note 10).
4. Dissolve the digested DNA with 40 μL of distilled water.

3.3.2. Agarose Gel Electrophoresis

Run a gel (0.8% agarose) of the genomic DNA samples that have been digested with restriction enzymes.

3.3.3. Southern Transfer to Nylon Membrane in NaOH

1. After staining and taking photograph, rinse the gel in distilled water for 10–20 min.
2. Soak the gel in a tray filled with 500 ml of 250 mM HCl for 15 min with gentle agitation.
3. Pure off HCl and rinse the gel with distilled water twice.
4. Denature the gel by soaking the gel in denaturing solution (0.4 N NaOH) for 30 min (15 min × 2) with agitation.
5. Blot transfer DNAs to Hybond N + membrane in 0.1 NaOH overnight (see Fig. 2 for an illustration).

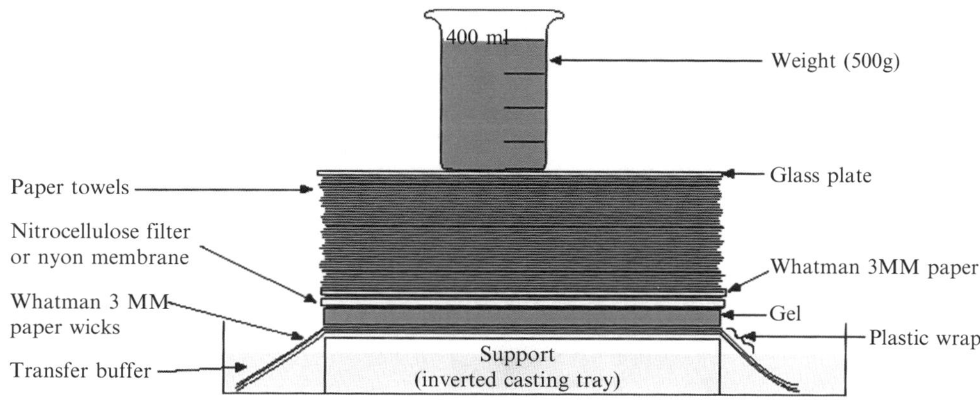

Fig. 2. Illustration of the setup for blot transfer DNA to membrane.

3.3.4. Probe Labeling

Probes are prepared and labeled with Digoxigenin-11-dUTP by PCR amplification of the plasmid pPZP-GFP (see Fig. 1 and Note 11).

3.3.5. Hybridization

1. Remove the nylon membrane from the filter paper and wash with 2×SSC and then dry with paper towel.

2. Cross-link the DNA and membrane with UV crosslinker (Stratagene).

3. Wet the membrane with 2×SSC and carefully place into a hybridization tube.

4. Add 9 ml of hybridization solution (Dig Easy Hyb) to pre-hybridize in 65°C hybridization oven for at least 2 h.

5. Discard the pre-hybridization solution and add 9 ml of hybridization solution containing labeled probe (about 100 ng per membrane). Hybridize at 65°C for up to 4 h.

3.3.6. Membrane Washing

1. Drain the solution from the hybridization tube.

2. For room-temperature wash, add in W1. Mix it up and discard W1 solution.

3. Repeat twice.

4. Add W2. Leave tube rotating in the hybridization oven at 55°C for 30 min.

5. Repeat once.

6. Add W3. Leave the tube rotating in oven at 55°C for 30 min.

7. Repeat once.

8. Take out the membrane and blot dry on paper towel.

     W1 = 2×SSC, 0.1% SDS

     W2 = 0.5×SSC, 0.1% SDS kept at 65°C

     W3 = 0.1×SSC, 0.1% SDS kept at 65°C

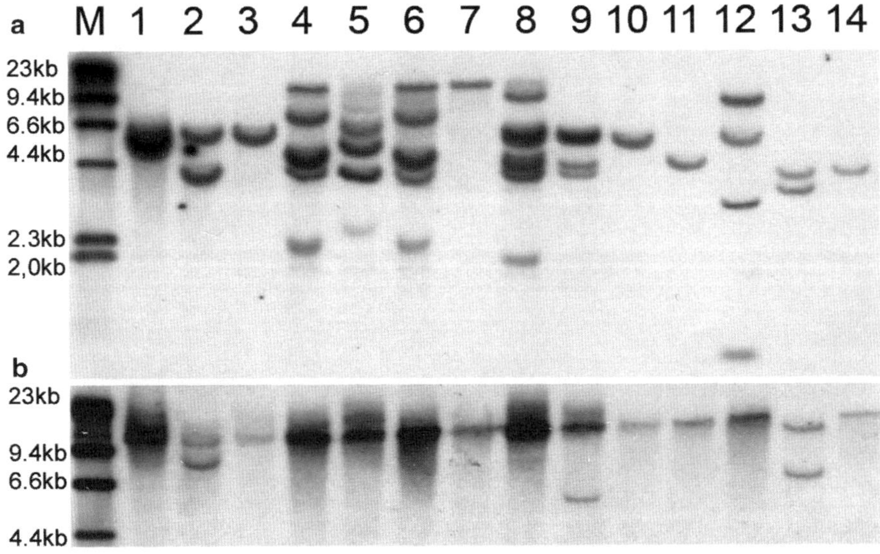

Fig. 3. Genomic Southern blot analysis to evaluate the number of transgene copy number insertion sites of T0 plant (17). M is DNA ladder; all blots are probed with *gfp* and the DNA samples of the upper panel (**a**) are digested with XbaI while the down panel (**b**) with BamHI.

3.3.7. Signal Detection

The hybridized digoxigenin-labeled probes are incubated with anti-digoxigenin antibody conjugated with AP. Incubation with ready-to-use CDP-Star (see Note 12) will generate fluorescent signals that can be detected with X-ray film. All the procedures for signal detection are followed by the Procedures for Nonradioactive Labeling and Detection of DIG Application Manual (Roche, Basal, Switzerland).

Examples of genome Southern blot hybridization are given in Fig. 3. The number of bands in the upper panel (panel A) represents the copy number of the transgene *GFP* while the single band in panel B indicated that the transgene are integrated into a unique locus of the host genome. For example, the transgenic lines detected in lanes 1, 3, 7, 10, 11, and 14 should be single transgene copy integrated into single locus and lanes 2, 9, and 13 represented lines with different copies of transgene integrated into different loci, while lanes 4, 5, 6, 8, and 12 represented the transgenic lines with multiple copies of transgene integrated into a single locus. And most of the multiple copies of the transgene should be complex loci. This can be verified by the results of PCR to amplify the junctions of the tandem repeated transgene.

4. Notes

1. Inverse PCR is used to amplify and clone unknown DNA that flanks one end of a known DNA sequence. The technique involves digestion by a restriction enzyme of a preparation of DNA containing the known sequence and its flanking region. The individual restriction fragments are converted into circles by intramolecular ligation, and the circularized DNA is then used as a template in the PCR. The unknown sequence is amplified by two primers that bind specifically to the known sequence and point in opposite directions. The product of the amplification reaction is a linear DNA fragment containing a single site for the restriction enzyme originally used to digest the DNA.

2. Prepare all solutions using sterile ultrapure water (prepared by purifying deionized water to attain a sensitivity of 18 M Ω cm at 25°C) and analytical grade reagents. Prepare and store all reagents at room temperature (unless indicated otherwise). Diligently follow all waste disposal regulations when disposing waste materials.

3. Primer pairs 1/2, 3/4, 5/6, and 7/8 are used to amplify the T-DNA (gfp and Kan) and vector backbone fragments outside right and left border, respectively (Table 1). The nested primers numbered from 9 to 20 are used for inverse PCR amplification of transgene flanking sequences. The primers listed in Table 1 are also combined and used for amplifying the transgene repeat junctions based on their locations in the construct. Primers located at the ends of the T-DNA can also amplify the junctions of repeats formed by "canonical T-DNA units" and primers at the expected ends of integrated backbone fragments are designed to investigate the possible junction structures formed by the T-DNA/vector backbone and/or vector backbone each other. Some other primers inside the T-DNA are also used to check for transgene integration involving T-DNA deletions. All the 2nd (nested) primers are also used as sequencing primers in directly sequencing.

4. We also used XbaI to cut the transgenic T0 DNA samples for inverse PCR experiment using corresponding primer combinations to increase the possibilities of the flanking sequence cloning.

5. Make sure that DNA is fully dissolved in TE before proceeding to the next step.

6. If white mash is not seen, repeat steps 11–14; then transfer pellet to an Eppendorf tube.

7. PCR products generated by primer pairs 1/2, 3/4, 5/6, and 7/8 (see Table 1) which are used to amplify the T-DNA (*gfp* and *Kan* here) and vector backbone fragments outside right and left border are not to be recollected from the gel and sequenced. Steps 4–7 are necessary for the PCR products of the transgene repeat junctions to investigate the different junction structures of the multiple copies that formed complex transgenic loci.

8. Purifying digestion products by phenol extraction will increase ligation efficiency in the next step. And phenol extraction also inactivates restriction enzymes.

9. Inverse PCR requires a circular DNA as template. Steps 1–3 of this protocol describe how such templates can be generated. Ligation of the digested fragments at low temperature for a long time will increase circlization efficiency of DNA fragments from a plant species with a big genome size such as cotton.

10. Purify digestion products by phenol extraction, if necessary.

11. The probes used for the transgenes (including gfp, nptII, and cmr) are illustrated in Fig. 1. The PCR conditions depend on the probe sizes and the annealing temperature of the primer pairs.

12. CDP-Star Reagent is a 1,2-dioxetane compound utilized in chemiluminescent detection assays. CDP-Star Reagent produces a light signal when it is activated by alkaline phosphatase, accumulates in its dephosphorylated form, and decomposes at a constant rate for up to several days. When assayed on nylon membrane at 1:100 in the buffer provided, CDP-Star signals reach a maximum within 15 min and decay slowly over 3 days. Typical film exposure times range from 15 s to 15 min (see DIG Application Manual, Roche, Basal, Switzerland).

References

1. Newell CA (2000) Plant transformation technology. Developments and applications. Mol Biotechnol 16:53–65

2. Topping JF, Wei W, Clarke MC, Muskett P, Lindsey K (1995) Agrobacterium-mediated transformation of Arabidopsis thaliana. Application in T-DNA tagging. Methods Mol Biol 49:63–76

3. Simone M, McCullen CA, Stahl LE, Binns AN (2001) The carboxy-terminus of VirE2 from Agrobacterium tumefaciens is required for its transport to host cells by the virB-encoded type IV transport system. Mol Microbiol 41:1283–1293

4. Ichikawa T, Nakazawa M, Kawashima M, Muto S, Gohda K, Suzuki K, Ishikawa A, Kobayashi H, Yoshizumi T, Tsumoto Y, Tsuhara Y, Iizumi H, Goto Y, Matsui M (2003) Sequence database of 1172 T-DNA insertion sites in Arabidopsis activation-tagging lines that showed phenotypes in T1 generation. Plant J 36:421–429

5. Windels P, De Buck S, Van Bockstaele E, De Loose M, Depicker A (2003) T-DNA integration in Arabidopsis chromosomes. Presence and origin of filler DNA sequences. Plant Physiol 133:2061–2068

6. Alonso JM, Stepanova AN (2003) T-DNA mutagenesis in Arabidopsis. Methods Mol Biol 236:177–188

7. Alvarez R, Toribio M, Cortizo M, Ordas Fernandez RJ (2006) Cork oak trees (Quercus suber L.). Methods Mol Biol 344:113–123

8. Gelvin SB, Kim SI (2007) Effect of chromatin upon Agrobacterium T-DNA integration and transgene expression. Biochim Biophys Acta 1769:410–421

9. Chilton MD, Que Q (2003) Targeted integration of T-DNA into the tobacco genome at double-stranded breaks: new insights on the mechanism of T-DNA integration. Plant Physiol 133:956–965

10. Sha Y, Li S, Pei Z, Luo L, Tian Y, He C (2004) Generation and flanking sequence analysis of a rice T-DNA tagged population. Theor Appl Genet 108:306–314

11. Wang J, Li L, Wan X, An L, Zhang J (2004) Distribution of T-DNA carrying a Ds element on rice chromosomes. Sci China 47:322–331

12. Sallaud C, Gay C, Larmande P, Bes M, Piffanelli P, Piegu B, Droc G, Regad F, Bourgeois E, Meynard D, Perin C, Sabau X, Ghesquiere A, Glaszmann JC, Delseny M, Guiderdoni E (2004) High throughput T-DNA insertion mutagenesis in rice: a first step towards in silico reverse genetics. Plant J 39:450–464

13. Afolabi AS, Worland B, Snape JW, Vain P (2004) A large-scale study of rice plants transformed with different T-DNAs provides new insights into locus composition and T-DNA linkage configurations. Theor Appl Genet 109:815–826

14. Jeong DH, An S, Park S, Kang HG, Park GG, Kim SR, Sim J, Kim YO, Kim MK, Kim SR, Kim J, Shin M, Jung M, An G (2006) Generation of a flanking sequence-tag database for activation-tagging lines in japonica rice. Plant J 45:123–132

15. Hsing YI, Chern CG, Fan MJ, Lu PC, Chen KT, Lo SF, Sun PK, Ho SL, Lee KW, Wang YC, Huang WL, Ko SS, Chen S, Chen JL, Chung CI, Lin YC, Hour AL, Wang YW, Chang YC, Tsai MW, Lin YS, Chen YC, Yen HM, Li CP, Wey CK, Tseng CS, Lai MH, Huang SC, Chen LJ, Yu SM (2007) A rice gene activation/knockout mutant resource for high throughput functional genomics. Plant Mol Biol 63:351–364

16. Himmelbach A, Zierold U, Hensel G, Riechen J, Douchkov D, Schweizer P, Kumlehn J (2007) A set of modular binary vectors for transformation of cereals. Plant Physiol 145:1192–1200

17. Zhang J, Cai L, Cheng J, Mao H, Fan X, Meng Z, Chan KM, Zhang H, Qi J, Ji L, Hong Y (2008) Transgene integration and organization in cotton (*Gossypium hirsutum* L.) genome. Transgenic Res 17:293–306

18. Kohli A, Twyman RM, Abranches R, Wegel E, Stoger E, Christou P (2003) Transgene integration, organization and interaction in plants. Plant Mol Biol 52:247–258

19. Liu YG, Mitsukawa N, Oosumi T, Whittier RF (1995) Efficient isolation and mapping of Arabidopsis thaliana T-DNA insert junctions by thermal asymmetric interlaced PCR. Plant J 8:457–463

20. Chen S, Jin W, Wang M, Zhang F, Zhou J, Jia Q, Wu Y, Liu F, Wu P (2003) Distribution and characterization of over 1000T-DNA tags in rice genome. Plant J 36:105–113

21. Thole V, Worland B, Wright J, Bevan MW, Vain P (2010) Distribution and characterization of more than 1000T-DNA tags in the genome of Brachypodium distachyon community standard line Bd21. Plant Biotechnol J 8:734–747

22. Vera T, Sílvia CA, Barbara W, Michael WB, Philippe V (2009) A protocol for efficiently retrieving and characterizing flanking sequence tags (FSTs) in Brachypodium distachyon T-DNA insertional mutants. Nature Publishing Group

23. Johnson AA, Hibberd JM, Gay C, Essah PA, Haseloff J, Tester M, Guiderdoni E (2005) Spatial control of transgene expression in rice (*Oryza sativa* L.) using the GAL4 enhancer trapping system. Plant J 41:779–789

24. Hajdukiewicz P, Svab Z, Maliga P (1994) The small, versatile pPZP family of Agrobacterium binary vectors for plant transformation. Plant Mol Biol 25:989–994

25. Li XB, Fan XP, Wang XL, Cai L, Yang WC (2005) The cotton ACTIN1 gene is functionally expressed in fibers and participates in fiber elongation. Plant Cell 17:859–875

26. Paterson AH, Brubaker CL, Wendel JF (1993) A rapid method for extraction of Cotton Genomic DNA suitable for RFLP or PCR analysis. Plant Mol Biol Reptr 11:122–127

27. Thomas CM, Jones DA, English JJ, Carroll BJ, Bennetzen JL, Harrison K, Burbidge A, Bishop GJ, Jones JD (1994) Analysis of the chromosomal distribution of transposon-carrying T-DNAs in tomato using the inverse polymerase chain reaction. Mol Gen Genet 242:573–585

28. De Buck S, Jacobs A, Van Montagu M, Depicker A (1999) The DNA sequences of T-DNA junctions suggest that complex T-DNA loci are formed by a recombination process resembling T-DNA integration. Plant J 20:295–304

Estimating the Copy Number of Transgenes in Transformed Cotton by Real-Time Quantitative PCR

Chengxin Yi and Yan Hong

Abstract

Transgenic cotton has widely been employed both in commercial cultivation and basic research. It is essential to determine which plants contain the transgene and in how many copies after transgenic cotton plants are generated. A TaqMan quantitative real-time polymerase chain reaction (Tq RT-PCR) method is described here to examine transgene copy number in transgenic cotton plants. The estimation of two transgene elements, the target gene of green fluorescence protein (*GFP*) and the selective gene of neomycin phosphotransferase II (*NPTII*), is used as an example to detail each step in Tq RT-PCR procedure, including endogenous reference gene selection, reference plasmid construction, primer-probe design, DNA extraction, real-time PCR, and data analysis. Comparing with traditional approach-Southern hybridization analysis, this method can be used efficiently in screening large number of T0 transgenic cotton plants at early stage of transformation process as well as identifying transgene homozygotes in a segregation population.

1. Introduction

Upland cotton (*Gossypium hirsutum* L.) is an important fiber crop in the world. Transgenic technology has been applied to cotton for improving the agronomic traits, tolerance to insects, resistance to herbicides, and fiber qualities. Upland cotton varieties with genetically modified traits have been adopted by millions of farmers from 0.8 million hectares in 1996 up to 16 million hectares, nearly 50% of global cotton area in 2009 (1). Multiple copies of a transgene in plant genome often cause lower and/or unstable transgene expression and even transgene silencing, only plants with one or two copies of transgene are generally preferred for stable and high-level expression of an exogenous gene (2–5). For regulatory approval, clean insertion with a single copy transgene is generally required. Commonly used *Agrobacterium* mediated and

Baohong Zhang (ed.), *Transgenic Cotton: Methods and Protocols*, Methods in Molecular Biology, vol. 958,
DOI 10.1007/978-1-62703-212-4_9, © Springer Science+Business Media New York 2013

ballistic bombardment transformation methods frequently result in multiple transgene copies at the same or different integration sites (6, 7). During transgenic cotton plant development, it is essential to examine which plants integrate the target genes and in how many copies of the transgenes. And early screening of T0 transgenic plants and focusing on the transformation events with one to two copies of transgene would be highly desirable and beneficial, especially for upland cotton which requires big space to plant and long time to grow.

Traditionally, transgene copy number in cotton plants is estimated by means of Southern blot analysis. It is generally reliable but has a limit in resolution and can't differentiate fragments of similar sizes. It won't be able to tell exact number of tandem repeated T-DNA copies located in a complex locus; neither could it tell homozygote from heterozygote in a subsequent T1 and/or T2 segregation population. Moreover, the method is time consuming, laborious, and requires a relatively large amount of plant material and using hazardous radioisotopes in some cases (8). It is therefore impractical to use Southern blot analysis to test transgene copy number in the early transformation stage on a large scale.

Real-time quantitative PCR methods have been developed to monitor commercial transgenic cotton in an event-specific manner (9, 10). There are currently about 20 different real-time PCR chemistries on the market (11), only a few are applied in GMO detection methods. Five real-time PCR assays based on five different chemistries, AmpliFluor, Molecular Beacon, Minor Groove Binder (MGB), TaqMan, and SYBR-Green I, were comparatively evaluated in the same event-specific transgenic maize of Mon810 (12). Nine different real-time PCR chemistries for qualitative and quantitative applications in GMO detection were also compared and reviewed (13). SYBR-Green I, which is a kind of intercalating dye, binds to any double-stranded DNA including nonspecific double-stranded DNA sequences; it may generate false-positive signals (14). Primer-based chemistries—Lux fluorogenic primers (Invitrogen, USA) (15), Plexor technology (Promega, USA) (16), and AmpliFluor universal detection system (Chemicon International, USA) (17), Q-priming PCR (18)—enable dissociation curves to be analyzed and thus be possible to distinguish between specific and nonspecific amplicons. They are more reliable than those based on intercalating dyes (13). Probe-based chemistries-TaqMan and MGB probes (19), LNA probes (Sigma Proligo, France) (20), Molecular Beacons (21), and AllGlo probes (22)—improve detection specificity by using a fluorescence probe which is complementary to a target sequence lying between the PCR primers. Based on their comparison (12, 13), none of the chemistries seemed to be significantly better than any other except SYBR-Green. TaqMan quantitative real-time PCR, however, is currently the most broadly applied method. It had been used to check the amount of transgene component in GMOs or GMOs' products (9, 10, 23–25) and to determine transgene copy numbers in transgenic plants like maize, rice, wheat, rapeseed, and tobacco (26–29).

In our research on transgenic cotton (30), we develop a TaqMan quantitative real-time PCR detection system to examine transgene copy number during transgenic cotton development. This system can be used efficiently in screening large number of transgenic plant at early stage of transformation process as well as identifying transgene homozygote in a segregation population. In this chapter, this method is used to estimate the copy number of two transgenic elements, the target gene of green fluorescence protein (*GFP*) and the selective gene of neomycin phosphotransferase II (*NPTII*). *GhUBC1*, a gene present only in A subgenome of *G. hirsutum* (31), is validated as an internal copy number control.

2. Materials

2.1. Transgenic Cotton Plants

1. A binary vector pPZP-GFP (Fig. 1) containing the *GFP* and *NPTII* genes under control of CaMV 35S promoter was introduced into upland cotton (*Gossypium hirsutum* L.) variety Coker312 by means of *Agrobacterium tumefaciens*-mediated transformation (30).

2. Selected T0 and T1 plants were allowed to self in greenhouse. The transgenic T0 cotton plants and some T1 and T2 plants are used for molecular detection.

2.2. Endogenous Reference Gene

1. Genomic DNAs from several *Gossypium* species: species with A sub-genome [*G. hirsutum* (AD)$_1$, *G. arboretum* (A$_1$) and *G. herbaceum* (A$_2$)] and species with D sub-genome [*G. thurberi* (D$_1$) and *G. raimondii* (D$_5$)].

2. Primers for amplifying GhUBC1 gene fragment for construction of reference plasmid from *G. hirsutum* (AD)$_1$ (Table 1, *UBC1-f1/r1*).

2.3. Cotton DNA Isolation

1. Liquid nitrogen.
2. Pestle and Mortar.
3. 1.5 ml Eppendorf tubes.
4. Column-based QIAQEN DNeasy Plant Mini Kit and reagents (see Note 1).
5. Water bath or heating block (see Note 2).
6. 100% Ethanol.

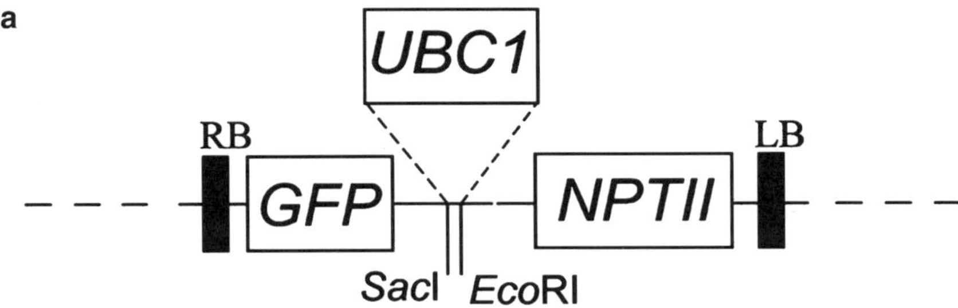

b

GAGCTCacaatcactgccgttactccccgagtttaattttctttctgggaattgaa

taattaatatcagaagagaatttctttctgggaattgaataattaatatcagaaga

gaaagaccaaccttgatcaaatatcat<u>tggcattatattgtcattgttactatcct</u>
UBC1-S

tctttgcatt|ggcttctgctgctgcttggaaatttaag|taatttttggtcttaaat
UBC1-Pro

atcatta<u>gagcttgtcttagaataagataacatggt</u>aaaaaaaaaaaaagtttaac
UBC1-R

tccaatgcaaaaccaggttctgctctcgatctgctcgttgttgactgatccaaacc

ctgatgacccacttgttccggagattgcacacatgtataagactgatcgggcaaag

tacgaagcgacagcatgtggctggacccagaagtatgccatgggatGAATTC

Fig. 1. Schematic diagram of pPZP-GFP-UBC1 reference plasmid and nucleotide sequences of the cloned UBC1 gene fragment. (**a**) Schematic diagram of pPZP-GFP-UBC1. (**b**) Sequence of the cloned fragment of *UBC1*. The *underlined* indicates primers for PCR and the *squared box* shows the TaqMan probe binding region.

7. Spectrophotometer, NanoDrop 1000 (Thermo Fisher Scientific, USA), is used to check DNA quality and measure DNA concentration in a volume of 1 µl.

8. Qubit™ fluorometer (Invitrogen, US) and Quant-iT™ PicoGreen dsDNA reagent to double check quantity of DNA (optional).

9. pH meter and calibration standards.

10. Stir plates with stir bars.

11. 50× TAE Buffer: to make 1 L of 50× TAE buffer, add 242 g Tris-base, 57.1 ml of glacial acetic acid, 100 ml of 0.5 M EDTA (pH 8.0) into 800 ml distilled deionized water, dissolve solids and adjust pH with HCl to 7.6–7.8, then bring up to 1 L with distilled deionized water (ddH$_2$O).

12. DNA molecular size marker (e.g., 100 bp DNA ladder, 0.5–10 kb DNA ladder).

Table 1
Primers and TaqMan probes for real-time PCR assay

Name	Orientation	Sequence 5′–3′	Size (nt)	Specificity	Amplicon size (bp)
UBC1-f1	Forward	ACATGCGAGCTCACAATCACTGCCGTTACTC	31	cotton GhUBC1	433
UBC1-r1	Reverse	ATCGCGGAATTCATCCCATGGCATACTTCTG	31		
UBC1-f	Forward	TGGCATTATATTGTCATTGTTACTATCC	28	cotton GhUBC1	121
UBC1-r	Reverse	ACCATGTTATCTTATTCTAAGACAAGCTC	29		
UBC1-Pro	Probe (reverse)	6JOE-CTT*AA*ATTTCCAAGCAGC*A*GCAGAAGCC-TMR	28		
GFP-f	Forward	GTGCAGGAAAGGACCATCTTCT	22	vector pPZP-GFP	108
GFP-r	Reverse	GATTCCCTTAAGCTCGATCCTGTT	24		
GFP-Pro	Probe (reverse)	6FAM-ACGTGTCTTGTAGTTCCCGTCGTCCTTG-TMR	28		
NPTII-f	Forward	GATAGCGGTCCGCCACAC	18	vector pPZP-GFP	113
NPTII-r	Reverse	CGAGGATCTCGTCGTGACACAT	22		
NPTII-Pro	Probe (forward)	6FAM-TTTCCACCATGATATTCGGCAAGCAGG-TMR	27		

13. 6× gel loading Buffer: 0.25% bromophenol blue (w/v), 0.25% xylene cyanol FF (w/v), 30% glycerol (v/v), 10 mM Tris–HCl (pH 8.0), and 60 mM EDTA (pH 8.0). The loading buffer stock is made up in a final volume of 10 ml.

14. 0.8% Agarose gel in 1× TAE.

15. 10 mg/ml of Ethidium bromide (EB).

16. A set of agarose gel electrophoresis equipment for DNA separation.

17. UV transilluminator.

18. General laboratory equipment, including vortex mixer, micro-centrifuge, desktop centrifuge.

2.4. Strand-Specific Primers and Probes

1. The oligonucleotide primers and TaqMan probes (Table 1) were designed by the PRIMER EXPRESS 2.0 software (Applied Biosystems, Foster City, USA) (see Note 3).

2.5. Reference Plasmid

1. A binary vector pPZP-GFP plasmid.

2. QIAquick plasmid DNA Extraction Kit.

3. QIAquick gel purification Kit.

4. Restriction enzymes: *Sac*I and *Eco*RI.

5. 10 mg/ml Bovine Serum Albumin (BSA).

6. T4 DNA ligase.

7. Heat shock or electroporation XL1 Blue competent cells.

8. Blade.

9. 100 mg/ml of Chloramphenical.

10. 15 g/L of LB agar plates containing 50 μg/ml of chloramphenical, freshly made or stored at 4°C no more than 1 month.

11. LB medium.

12. 37°C incubator.

13. 37°C shaking incubator.

14. Biosafety cabinet.

2.6. Tq RT-PCR

1. A dedicated set of pipettes and corresponding filtered tips.

2. ABI Prism 7900HT sequence detection system and SDS 2.3 software (Applied Biosysems, Foster City, USA) (see Note 4).

3. MicroAmp® Optical 384-Well Reaction Plate with Barcode, MicroAmp® Optical Adhesive Film Kit, TaqMan® Universal PCR Master Mix.

4. 1.5 ml Eppendorf tubes.

5. Centrifuge.

3. Methods

3.1. Cotton DNA Extraction

1. Grind young fresh cotton leaves (about 100 mg of wet weight) with mortar and pestle with liquid nitrogen to fine powder.

2. Add 400 μl buffer AP1 and 4 μl RNase A. Vortex and incubate in 65°C water bath for about 10 min. Invert tube two to three times during incubation.

3. Add 130 μl buffer AP2. Mix and incubate on ice for 5 min.

4. Centrifuge the lysate for 5 min at $20,000 \times g$ at room temperature.

5. Transfer the upper aqueous phase into a QIAshredder Mini spin column in a 2 ml collection tube. Centrifuge for 2 min at $20,000 \times g$ at room temperature.

6. Transfer the flow-through fraction into a new 1.5 ml Eppendorf tube without disturbing the pellet. Add 1.5 volumes of buffer AP3/E, and mix thoroughly.

7. Transfer 650 μl of the mixture into a DNeasy Mini spin column in a 2 ml collection tube. Centrifuge for 1 min at $8,000 \times g$ at room temperature. Discard flow-through. Repeat this step with the remaining sample.

8. Put the spin column into a new 2 ml collection tube. Add 500 μl of buffer AW. Centrifuge for 1 min at $8,000 \times g$ at room temperature. Discard the flow-through.

9. Add another 500 μl of buffer AW. Centrifuge for 2 min at $20,000 \times g$ at room temperature.

10. Transfer the spin column to a new 1.5 ml Eppendorf tube. Add 100 μl of buffer AE for elution. Incubate for 5 min at room temperature. Centrifuge for 1 min at $8,000 \times g$ at room temperature.

11. Take out 2 μl of stock DNA solution, and mix with 1 μl of 6× loading buffer and 3 μl of water. Check the DNA quality by running 0.8% agarose gel on electrophoresis equipment and monitor on UV transilluminator.

12. Measure the DNA concentration on the spectrophotometer with NanoDrop 1000.

3.2. Validation of an Endogenous Reference Gene (see Note 5)

1. Extraction genomic DNA from 5 *Gossypium* species: *G. hirsutum* $(AD)_1$, *G. arboretum* (A_1), *G. herbaceum* (A_2), *G. thurberi* (D_1), and *G. raimondii* (D_5) (see Note 6).

2. Design a universal *GhUBC1* primer-probe system (Table 1) to amplify and hybridize to *GhUBC1* gene sequence only based on the sequences comparison between *GhUBC1* and *GhUBC2*.

3. Normal PCR amplification to confirm that the *GhUBC1*-specific band can only be amplified in the species with

A sub-genome-*G. hirsutum* (AD)$_1$, *G. arboretum* (A$_1$), and *G. herbaceum* (A$_2$) but not in the species with D sub-genome-*G. thurberi* (D$_1$) and *G. raimondii* (D$_5$).

3.3. Construction of a Reference Plasmid (see Note 7)

1. Amplify the target fragment of *GhUBC1* gene from upland cotton using modified primer pair of *UBC1*-f1/r1 (Table 1) (see Note 8).

2. Purify the amplified fragment of *GhUBC1* gene with QIA quick PCR Purification Kit.

3. Digest purified UBC1 PCR product with restriction enzymes *Sac*I and *Eco*RI (see Note 9). At the same time digest pPZP-GFP plasmid with the same two enzymes separately (Table 2).

4. After digestion, load digestion products onto 1.5% agarose gel. Run the gel at 95 V for about 1 h.

5. Put the gel on a UV transilluminator, cut the correct DNA band with a clean blade.

6. Purify DNA fragments by using QIAGEN Gel Extraction Kit according to the manufacturer's instruction.

7. Take out 1 μl of purified DNA to check the concentration and purity with NanoDrop 1000.

8. pPZP-GFP DNA concentration is adjusted to about 25 ng/μl.

9. Set up ligation reaction (Table 3). Adjust to appropriate molecular ratio between vectors and inserts (see Note 10).

10. Incubate at 16°C overnight (see Note 11).

11. Heat shock transformation to XL1Blue competent cells.

    (a) Thaw 100 μl heatshock competent cells in a 1.5 ml Eppendorf tube on ice for 15 min.

Table 2
Restriction enzyme digestion reaction

Purified UBC1 PCR product or pPZP-GFP plasmid DNA	15 μl
*Sac*I	0.5 μl (4 units)
*Eco*RI	0.5 μl (400 units)
10× NE Buffer 4	2.5 μl (provided with enzymes)
BSA (10 mg/ml)	0.2 μl
ddH$_2$O	6.3 μl
Total	25 μl

Table 3
Ligation reaction

Digested pPZP-GFP vector	1.0 µl (about 25 ng)
Digested UBC1 PCR product	x µl (about 3 ng)
10× Ligase buffer	1.0 µl
T4 DNA ligase	1.0 µl (1–3 units)
ddH$_2$O	y µl ($x+y=7.0$ µl)
Total	10 µl

(b) Add 5 µl of ligation products into the tube and mix with competent cells by gentle pipetting.

(c) Keep on ice for 30 min.

(d) Heat shock in 42°C water bath around 45 s (see Note 12).

(e) Put on ice immediately and stay for 2–5 min.

(f) Add 1 ml LB liquid medium and invert the tube five times gently.

(g) Place in a shaking incubator at 250 rpm, at 37°C for 1 h.

(h) Spread 100 µl cell suspensions on LB agar plate containing 50 µg/ml chloramphenical, 20 µl of 20 mg/ml X-gal, and 5 µl of 1 M IPTG.

(i) Incubate the LB plate in 37°C oven overnight.

(j) Pick some white clones and do cloning PCR using three pairs of primer in Table 1 to verify whether the amplicons are of expected sizes.

(k) Conduct directly DNA sequencing using the PCR products to verify their sequences further.

3.4. Calculation of Transgene Copy Number

1. Relative quantitation of *GFP* and *NPTII* against the endogenous reference gene *UBC1* was used to calculate copy number of transgene in a similar way as those reported (11, 12). First of all, absolute quantity for the three genes was determined in reference to respective standard curves. Three standard curves for the three genes were obtained by plotting threshold cycle or C_T values (see Note 13) against log-transformed concentrations of serial tenfold dilutions (10^1, 10^2, 10^3, 10^4, and 10^5 copies per 1 µL) from the same reference plasmid pPZP-GFP-UBC1 DNA. Absolute copy numbers of *GFP*, *NPTII*, and *UBC1* in each transgenic cotton sample were calculated with C_T values based on their standard curves. The relative copy

number for a target gene was obtained by dividing the absolute amount of target gene by that of the endogenous gene in the same sample.

The initial amount of the target gene was calculated by the Eq. 1 based on the standard curve:

$$\text{Amount}_{\text{GFP}n} = S_{\text{GFP}}C_{\text{TGFP}n} + I_{\text{GFP}} \tag{1}$$

where $\text{Amount}_{\text{GFP}n}$ is the initial GFP gene amount of the nth sample, and $C_{\text{TGFP}n}$ is the threshold cycle of target gene GFP in the nth sample, S_{GFP} is the standard curve's slope of GFP, while I_{GFP} is the standard curve's intercept of GFP.

Similar to Eq. 1, the equation for the initial amount of reference gene in the nth sample was

$$\text{Amount}_{\text{UBC1}n} = S_{\text{UBC1}}C_{\text{TUBC1}n} + I_{\text{UBC1}} \tag{2}$$

where $\text{Amount}_{\text{UBC1}n}$ is the initial amount of and $C_{\text{TUBC1}n}$ is the threshold cycle of reference gene UBC1 in the nth sample, S_{UBC1} is the standard curve's slope of GFP, I_{UBC1} is the standard curve's intercept of UBC1.

The relative copy number of GFP in the nth sample were calculated by Eq. 3

$$\text{Copy}_{\text{GFP}n} = 2 \times \text{Amount}_{\text{GFP}n} / \text{Amount}_{\text{UBC1}n} \tag{3}$$

3.5. Real-time PCR

1. Primer combination optimization and amplification efficiency validation for target genes and internal reference gene (see Note 14).

    (a) For each target genes and reference gene, a suitable probe sequence is selected and three or four primer pairs are designed to amplify fragments including the probe region (see Note 15). The amplicon sizes for every primer pairs normally range from 80 to 200 bp.

    (b) A sensitive method for assessing whether two amplicons have the similar amplification efficiency is to look at whether their slopes of standard curves fall within the range between −3.1 and −3.6, preferably with the same template in the same serial dilutions. This is carried out by amplification of the target genes and reference gene over the same range of template dilutions in separate PCR reactions.

    (c) Prepare a mastermix of the reaction mix below (Table 4) so that there is enough for all samples to be investigated (see Note 16). PCR Master Mix contained No AmpErase UNG, hot-start AmpliTaq Gold DNA polymerase, deoxynucleoside triphosphates with dUTP and optimized reaction buffers (Applied Biosystems, Foster city, USA).

Table 4
PCR mastermix for primer optimization and amplification efficiency validation

Reagent	Volume (for 1 sample)
2×TaqMan universal PCR mix	5 μl
Forward primer (10 μM)	0.25 μl
Reverse primer (10 μM)	0.25 μl
Probe (10 μM)	0.25 μl
ddH$_2$O	2.25 μl
Total	8 μl

(d) Prepare serial tenfold dilutions of DNA templates (initial concentration for the reference plasmid is 5 ng/μl. For any transgenic cotton DNA, initial concentration is about 50 ng/μl).

(e) Add 8 μl of reaction mix above mentioned to each well in a 384-well plate.

(f) Add 2 μl of each DNA sample to individual wells in the plate.

(g) Replicate three times of each reaction on the same plate.

(h) Subject plate to ABI Prism 7900HT sequence detection system and conduct PCR reactions by following the standard program recommended by the manufacturer: 95°C for 10 min, followed by 40 cycles of 95°C for 15 sec and 60°C for 1 min.

(i) After amplification, use SDS 3.2 software to calculate Ct values and use average Ct values to build standard curves for a target gene and a chosen internal reference gene. One optimal primer pair should be decided and used for each gene (see Note 17).

2. Optimizing probe concentration (see Note 18).

(a) Prepare a PCR mastermix as shown below (Table 5). To determine the optimal probe concentration, various final concentrations between 50 and 250 nM (typically 50, 100, 150, 200, and 250 nM) should be tested (three replicates for each). A volume of 0.125 μl of probe is added into each PCR reaction. A control reaction without template should be conducted together.

(b) Transfer 384-well plate to ABI Prism 7900HT sequence detection system and conduct PCR following the standard

Table 5
PCR mastermix for probe concentration optimization

Reagent	Volume (for 1 sample)
2× TaqMan universal PCR mix	5 µl
Forward primer (10 µM)	0.25 µl
Reverse primer (10 µM)	0.25 µl
ddH$_2$O	2.375 µl
Template DNA (50 ng/µl)	2 µl
Total	9.875 µl

Table 6
PCR mastermix for real-time quantitative PCR

Reagent	Volume (for 1 sample)
2× TaqMan universal PCR mix	5 µl
Forward primer (10 µM)	0.25 µl
Reverse primer (10 µM)	0.25 µl
Probe (optimal concentration)	0.125 µl
ddH$_2$O	1.75 µl
Total	8 µl

program recommended by the manufacturer: 95°C for 10 min, followed by 40 cycles of 95°C for 15 sec and 60°C for 1 min.

(c) At the end of PCR, tabulate the results for C_T. Make a choice for the lowest probe concentration that gives the maximum ΔR_n (R_{n+} is the R_n value of a reaction containing all components, R_{n-} is the R_n value of an unreacted sample. ΔR_n is the difference between R_{n+} and R_{n-}. It is an indicator of the magnitude of the signal generated by the PCR).

3. Real-time quantitative PCR

(a) Prepare a mastermix for the reactions as shown below (Table 6). At least three replicates for each sample should be conducted on the same plate (batch).

(b) Pipette 8 µl of mastermix into each well on 384-well plate.

(c) Add 2 µl of DNA sample to each well

(d) Transfer the 384-well plate to ABI Prism 7900HT sequence detection system and conduct PCR reactions by following the standard program recommended by the manufacturer: 95°C for 10 min, followed by 40 cycles of 95°C for 15 s and 60°C for 1 min.

(e) Following amplification, C_T values of target genes and the internal reference gene for each sample can be calculated based on SDS 2.3 software.

4. Validation of standard curves by using a reference molecule.

(a) Our method of estimating transgene copy number depends on the assumption that all the three genes have almost equal PCR amplification efficiency. In order to validate this assumption, a standard reference molecule is constructed to include all target sequences in a single plasmid and their amplification efficiencies are compared. The reference plasmid pPZP-GFP-UBC1 containing *GFP*, *NPTII*, and a 433 bp fragment of *UBC1* genes (shown in Fig. 1) is constructed with sequences confirmed.

(b) The plasmid pPZP-GFP-UBC1 is 10,904 bp long and hence 1 pg equals to 88,500 copies. The plasmid DNA is diluted to 10^1, 10^2, 10^3, 10^4, and 10^5 copies per microliter, 2 μl is used to amplify each of the three genes and C_T values are plotted against initial copy numbers to obtain standard curves of the three genes.

(c) Three replicate reactions are conducted to construct standard curves thrice for each target gene. Efficiencies of amplification were calculated based on slopes of standard curves with the formula: $E = 10^{(-1/\text{slpoc})} - 1$.

(d) According to Eqs. 1–3, slopes and intercepts for the target genes, the efficiencies of *GFP* and *NPTII*, and endogenous gene *UBC1* are calculated. The average slopes for *GFP*, *NPTII*, and *UBC1* are −3.38, −3.29, and −3.42, corresponding to efficiencies of 0.977, 0.985, and 0.959 for *GFP*, *NPTII*, and *UBC1*, respectively. PCR reaction efficiencies for the three genes are all above 0.95 and close enough to assume equal efficiency for the purpose of copy number calculation. One the other hand, the average correlation coefficients (R^2) of the standard curves for *GFP*, *NPTII*, and *UBC1* are 0.999, 0.998, and 0.999, respectively (Table 7), indicating high level of accuracy in estimating absolute amount of the three genes based on the standard curves. Detailed data is shown in Table 2 with standard deviation (SD) values. These results demonstrate that the standard curves are robust enough to estimate the transgene copy numbers.

Table 7
Reproducibility of the real-time PCR systems for the *GFP*, *NPTII*, and *UBC1*genes

	Slope	Intercept	R^2	Efficiency
Genes	Mean ± SD	Mean ± SD	Mean ± SD	Mean ± SD
GFP	−3.38 ± 0.100	35.02 ± 0.670	0.999 ± 0.002	0.977 ± 0.041
NPTII	−3.29 ± 0.113	34.23 ± 0.201	0.998 ± 0.002	0.985 ± 0.050
UBC1	−3.42 ± 0.072	35.24 ± 0.319	0.999 ± 0.001	0.959 ± 0.028

5. Copy number estimation for transgenes in T0 lines.

   (a) A T0 transgenic cotton population containing 28 plants/ events is used as a case of quantitative real-time PCR to estimate their copy number of transgenes. Three batches of real-time PCR with three replicates within each batch are conducted for every T0 plant DNA. For each batch of PCR, the reference plasmid was included and batch-specific standard curves were constructed. C_T values of three replicates are averaged to calculate initial absolute amount of each gene based on their corresponding standard curves for the same batch. Relative copy numbers of GFP and NPTII are calculated by dividing twice of absolute amount by that of the internal reference gene UBC1. Average copy numbers of three batches are shown in Table 8 with their standard deviations given.

6. Determine homozygosity in a T1 population (see Note 19).

   (a) An interesting phenotypic mutation related to fruiting branch is identified from some T1 plants derived from T0 transgenic plant ID138, possible caused by insertion mutagenesis. Genomic Southern blot hybridization revealed two copies of *GFP* in these plants. However, the real-time PCR system detected three or four copies of *GFP* in these plants. We suspected that there were two unlinked transgene loci and the plants with four copies of *GFP* could be transgene homozygotes.

   (b) Several T1 plants were allowed to self and their T2 seeds were germinated and plants analyzed for *GFP* copies.

   (c) Real-time PCR analysis of 16T2 plants derived from one T1 plant having four copies of *GFP* found that they also had four copies of *GFP*, indicating this T1 plant a transgene homozygote.

Table 8
***GFP* and *NPTII* copy numbers in T0 transgenic plants deduced by real-time PCR (RT-PCR) and genomic Sothern blot analysis**

Lines	GFP			NPTII		
	$AvCopy_{GFPn} \pm$ SD by RT-PCR	Estimated copy number	Southern blot (*Eco*RI)	$AvCopy_{NpTIIn} \pm$ SD by RT-PCR	Estimated copy number	Southern blot (*Eco*RI)
ID12	1.10 ± 0.01	1	1	0.90 ± 0.15	1	1
ID33	0.96 ± 0.05	1	1	0.84 ± 0.11	1	1
ID38	0.86 ± 0.03	1	1	0.84 ± 0.09	1	1
ID44	0.84 ± 0.10	1	1	0.86 ± 0.10	1	1
ID51	0.96 ± 0.03	1	1	0.98 ± 0.00	1	1
ID72	1.06 ± 0.10	1	1	0.28 ± 0.06	1	1
ID83	1.08 ± 0.28	1	1	0.92 ± 0.07	1	1
ID129	0.86 ± 0.10	1	1	0.74 ± 0.11	1	1
ID131	0.84 ± 0.12	1	1	0.70 ± 0.09	1	1
ID133	0.64 ± 0.08	1	1	1.00 ± 0.12	1	1
ID139	1.16 ± 0.13	1	1	1.02 ± 0.13	1	1
ID145	0.96 ± 0.05	1	1	0.88 ± 0.09	1	1
ID135	0.66 ± 0.08	1	1	1.00 ± 0.11	1	2
ID56	0.00 ± 0.00	0	0	1.04 ± 0.19	1	1
ID137	0.00 ± 0.00	0	0	2.32 ± 0.15	2	1
ID104	1.68 ± 0.11	2	2	1.56 ± 0.10	2	2
ID141	1.64 ± 0.01	2	2	0.70 ± 0.03	1	2
ID144	2.20 ± 0.18	2	2	0.88 ± 0.09	1	1
ID27	3.04 ± 0.24	3	3	2.04 ± 0.08	2	2
ID122	3.02 ± 0.19	3	3	0.68 ± 0.05	1	2
ID134	2.98 ± 0.14	3	3	1.86 ± 0.08	2	1
ID85	2.50 ± 0.09	3	1	2.66 ± 0.24	3	3
ID98	4.02 ± 0.31	4	3	4.46 ± 0.71	4	3
ID130	4.40 ± 0.51	4	3	4.58 ± 0.18	5	5
ID95	4.80 ± 0.73	5	5	4.02 ± 0.23	4	4
ID118	4.92 ± 0.21	5	3	4.14 ± 0.22	4	4
ID110	5.58 ± 0.23	6	5	2.26 ± 0.15	2	2
ID138	8.06 ± 0.39	8	4	5.62 ± 0.29	6	6

(d) We further subjected another 21 T2 plants derived from the other T1 individual with 3 *GFP* copies to the same real-time PCR detection. Clear segregation of copy number is observed. The segregation ratio (Table 9) of T2 plants with two copies, three copies, and four copies of *GFP* was close to 1:2:1 (6:10:5), consistent with the Mendel segregation of two independent loci ($\chi_c^2 = 0.1428 < \chi_{0.01,2}^2$), one homozygous and one heterozygous in the T1 plant.

Table 9
Segregation of *GFP* loci in T2 plants as detected by real-time PCR

Plant ID	Copy$_{GFPn}$	Estimated copy number	Possible segregation pattern[a]
T1-13	2.88	3	AABb
T2-64	3.87	4	AABB
T2-65	3.63	4	AABB
T2-66	3.78	4	AABB
T2-67	4.00	4	AABB
T2-68	3.79	4	AABB
T2-69	1.97	2	AAbb
T2-70	1.99	2	AAbb
T2-71	2.88	3	AABb
T2-72	2.78	3	AABb
T2-73	2.76	3	AABb
T2-74	3.24	3	AABb
T2-75	2.92	3	AABb
T2-77	3.12	3	AABb
T2-78	2.97	3	AABb
T2-80	2.95	3	AABb
T2-81	3.85	4	AABB
T2-82	1.89	2	AAbb
T2-83	2.03	2	AAbb
T2-84	2.86	3	AABb
T2-86	2.18	2	AAbb
T2-87	2.84	3	AABb

[a]AA, homozygosity of the transgene at the first locus; BB, Bb, and bb represent homozygosity, heterozygosity, and absence of the transgene at the second locus

(e) In comparison, genomic Southern blot hybridization could only detect the same two bands for the above T1 and T2 plants. This example illustrates that, using real-time PCR method, we can reliably distinguish between homozygote and heterozygote, and hence this technique can be used in screening for homozygous in T1 plants.

4. Notes

1. If necessary, redissolve any precipitates in Buffers AP1 and AP3/E concentrate from QIAGEN DNeasy Plant Mini Kit. Ensure that ethanol (96–100%) has been added to Buffers AW and AP3/E. Do not heat Buffer AP3/E after ethanol has been added.

2. Preheat a water bath or heating block to 65°C before DNA isolation.

3. Primer and probe selection is based in similar T_m values, small amplicon size and a target region for specific detection. When ordering probe for target gene and probe for internal reference gene to be used in multiplex reactions, ensure they are labeled with different reporter dyes, such as 6-FAM and 6-JOE.

4. Other real-time PCR detection system can also be used but may not be able to accommodate 384 samples at the same time.

5. An appropriate endogenous reference gene is critical for quantitative real-time PCR to normalize errors and sample-to-sample variations. An ideal endogenous reference gene should be species-specific and have single or low copy numbers per haploid genome, as well as low heterogeneity across genotypes within the species (23, 32). Some other specific genes were also used as endogenous reference genes in quantitative real-time PCR to detect transgenic cotton (33).

6. Upland cotton (*G. hirsutum*) is an allotetraploid species (AADD) consisting of one A-subgenome and one D-subgenome per haploid genome. It is important to make sure that the primers and the probe used to detect the cotton endogenous gene should either discriminate A- and D-subgenome or match perfectly to copies in both subgenomes (23). In this case, *GhUBC1* (encoding an ubiquitin-conjugating enzyme, E2, GenBank No. AY082006) was selected as the endogenous reference gene. It is a member of *GhUBC1/2* gene family. Genomic origin analysis indicated that *GhUBC1* and *GhUBC2* (GenBank No. AY082007) are individually present in the A- and D-subgenome of *G. hirsutum* (31).

7. Our method of estimating transgene copy number depends on the assumption that all the three genes have almost equal PCR amplification efficiency. In order to validate this assumption, a standard reference molecule is constructed to include all target sequences in a single plasmid and their amplification efficiencies are compared. The reference plasmid pPZP-GFP-UBC1 containing *GFP*, *NPTII*, and 433 bp fragment of *UBC1* genes (shown in Fig. 1) is constructed with sequences confirmed.

8. A *Sac*I restriction site and six protective bases (ACATGCGA GCTC) is added to the 5′ end of sequence which hybrid to one end of target fragment of *GhUBC1* gene (ACAATCACTG CCGTTACTC), while a *Eco*RI restriction site and six protective bases (ATCGCGGAATTC) is added to the 5′ end of sequence of other end (ATCCCATGGCATACTTCTG) to form the primer pair of *UBC1*-f1/r1.

9. We have difficulty in cloning *Sac*I/*Eco*RI digested PCR fragment into pPZP-GFP, probably due to low efficiency double digestion with two enzymes or low quantity of PCR fragment after digestion. An alternative way is used to clone PCR fragment directly (without digestion) into pGEM-T Easy Vector System as an intermediate construct. Desired fragments are cut with *Sac*I and *Eco*RI from the intermediate plasmid DNA before further cloning.

10. The formula below is used to calculate molecular amount of UBC1 PCR product before cloning. The molecular ratio of vector/insert is set at around 3:1 for ligation.

    UBC1 PCR product (ng)=25 ng (pPZP-GFP DNA amount)×0.4 kb (insert length)×3/10 kb (pPZP-GFP length)=3 ng.

11. Store the ligation product at −20°C if not used to transform immediately.

12. Prepare water bath before reaction and keep the temperature at 42°C. The time of heat shock should be neither shorter than 30 sec nor longer than 1 min.

13. Threshold cycle or C_T value is the cycle at which a statistically significant increase in ΔR_n is first detected (Fig. 2). Threshold is defined as the average standard deviation of R_n for the early cycles, multiplied by an adjustable factor. On the graph of R_n versus cycle number shown below, the threshold cycle occurs when the Sequence Detection System begins to detect the increase in signal associated with an exponential growth of PCR product.

14. Before real-time PCR performance, a validation experiment must be performance to ensure that amplification efficiencies of the target genes and internal reference gene are approximately equal.

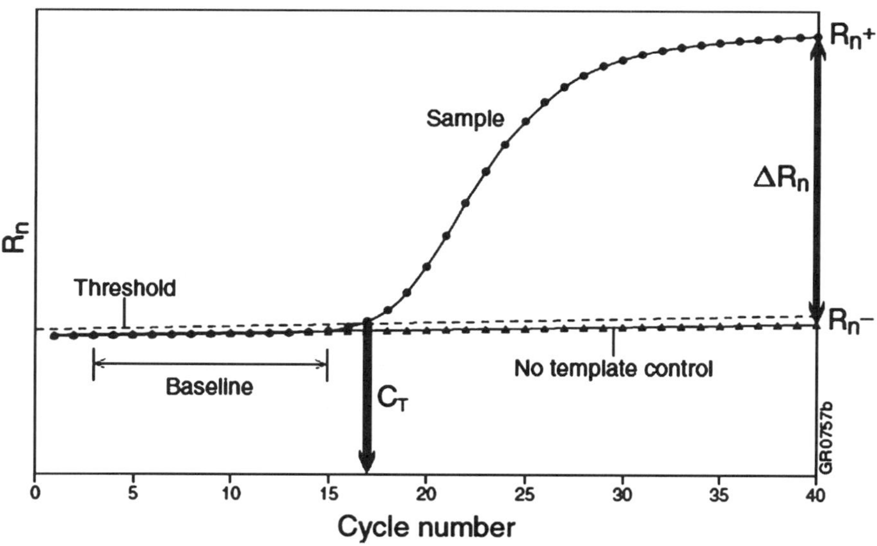

Fig. 2. Illustration of real-time PCR cycle number, threshold cycle, and ΔR_n.

15. For each probe (with a reporter fluorophore and a quencher), there is the need to test multiple primer pairs. One optimal primer pair should be identified and used for subsequent reactions. If no primer pair gives good efficiency of amplification, there is the need to change to another probe followed by testing optimal primer pair.

16. This method does not require optimizing primer concentration. A primer concentration similar to normal PCR reaction system will be fine (2.5–10 pmol per reaction).

17. If the target genes and internal reference gene have the similar amplification efficiency, their slopes of standard curves should remain nearly the same (ranging from −3.1 to −3.6) (Figs. 3 and 4). If value for slope of standard curve of one gene falls outside the range, a new probe or primer pair should be chosen.

18. The purpose to optimize probe concentration is to determine the minimum probe concentration that gives the maximum ΔR_n. A low but sufficient probe concentration will reduce nonspecific amplification and save operation cost.

19. In theory, a transgene homozygote can be obtained at T1 generation. However, genomic Southern blot hybridization is not suitable to identify a transgene homozygote plant. Neither can qualitative PCR method. Usually it takes one more generation to confirm homozygosity through analysis of transgene segregation in T2 population. It would be cost effective and time efficient if homozygosity can be checked at T1 generation.

Amplification Plot

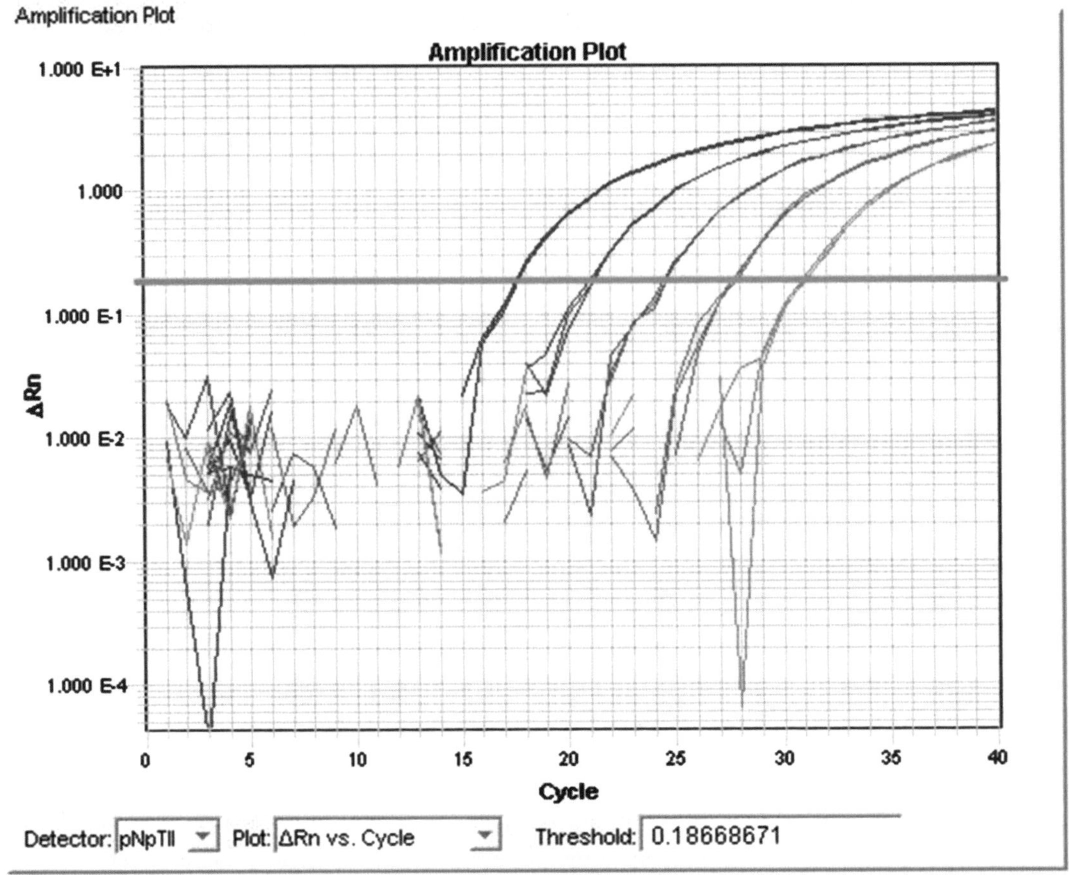

Fig. 3. Amplification plot for the standard curve of *NPTII* gene.

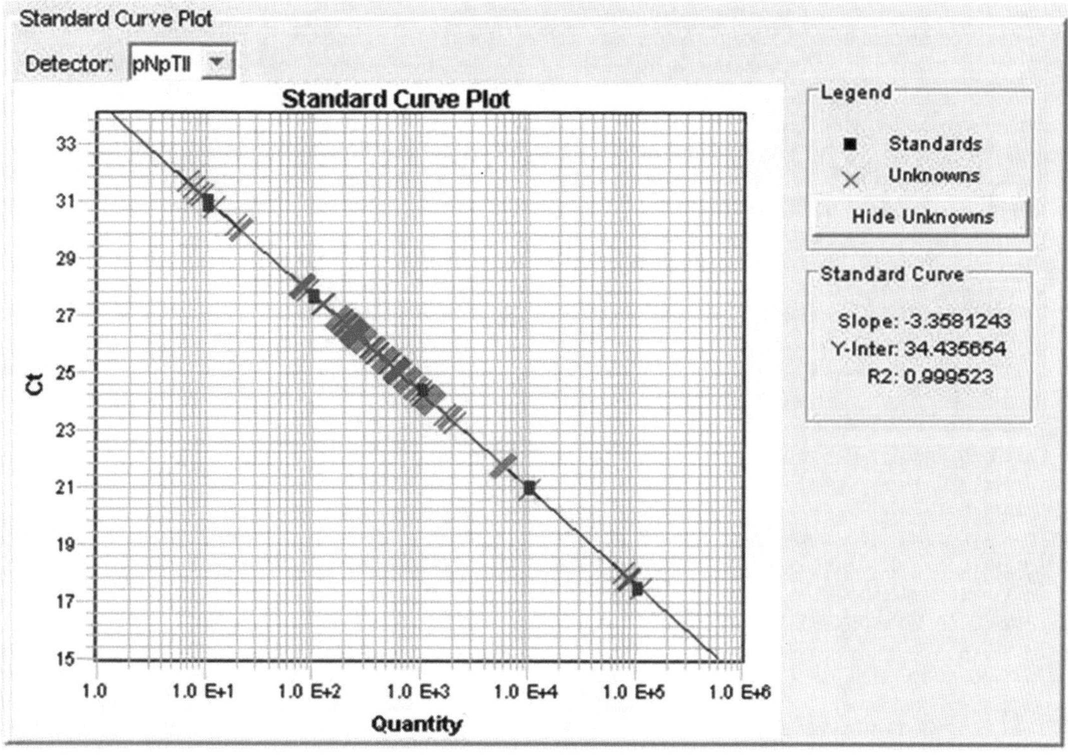

Fig. 4. Standard curve plot for *NPTII* gene in a test sample.

References

1. Clive J (2009) Global status of commercialized biotech/GM crops, in ISAAA brief 41. ISAAA, Ithaca, NY

2. Flavell RB (1994) Inactivation of gene expression in plants as a consequence of specific sequence duplication. Proc Natl Acad Sci USA 91:3490–3496

3. Kooter JM, Matzke MA, Meyer P (1999) Listening to the silent genes: transgene silencing, gene regulation and pathogen control. Trends Plant Sci 4:340–347

4. Morino K, Olsen OA, Shimamoto K (1999) Silencing of an aleurone-specific gene in transgenic rice is caused by a rearranged transgene. Plant J 17:275–285

5. Vaucheret H, Beclin C, Elmayan T, Feuerbach F, Godon C, Morel JB, Mourrain P, Palauqui JC, Vernhettes S (1998) Transgene-induced gene silencing in plants. Plant J 16:651–659

6. Kohli A, Leech M, Vain P, Laurie DA, Christou P (1998) Transgene organization in rice engineered through direct DNA transfer supports a two-phase integration mechanism mediated by the establishment of integration hot spots. Proc Natl Acad Sci USA 95:7203–7208

7. Srivastava V, Anderson OD, Ow DW (1999) Single-copy transgenic wheat generated through the resolution of complex integration patterns. Proc Natl Acad Sci USA 96:11117–11121

8. Mason G, Provero P, Vaira AM, Accotto GP (2002) Estimating the number of integrations in transformed plants by quantitative real-time PCR. BMC Biotechnol 2:20

9. Yang L, Guo J, Pan A, Zhang H, Zhang K, Wang Z, Zhang D (2007) Event-specific quantitative detection of nine genetically modified maizes using one novel standard reference molecule. J Agric Food Chem 55:15–24

10. Yang L, Pan A, Zhang K, Yin C, Qian B, Chen J, Huang C, Zhang D (2005) Qualitative and quantitative PCR methods for event-specific detection of genetically modified cotton Mon1445 and Mon531. Transgenic Res 14:817–831

11. (2011) The Reference in qPCR – academic & industrial information platform, http://gene-quantification.info/. Accessed on 3 June 2011

12. La Paz JL, Esteve T, Pla M (2007) Comparison of real-time PCR detection chemistries and cycling modes using Mon810 event-specific assays as model. J Agric Food Chem 55:4312–4318

13. Buh Gašparič M, Tengs T, La Paz J, Holst-Jensen A, Pla M, Esteve T, Žel J, Gruden K (2010) Comparison of nine different real-time PCR chemistries for qualitative and quantitative applications in GMO detection. Anal Bioanal Chem 396:2023–2029

14. Schneeberger C, Speiser P, Kury F, Zeillinger R (1995) Quantitative detection of reverse transcriptase-PCR products by means of a novel and sensitive DNA stain. PCR Methods Appl 4:234–238

15. Nazarenko I, Pires R, Lowe B, Obaidy M, Rashtchian A (2002) Effect of primary and secondary structure of oligodeoxyribonucleotides on the fluorescent properties of conjugated dyes. Nucleic Acids Res 30:2089–2195

16. Sherrill CB, Marshall DJ, Moser MJ, Larsen CA, Daude-Snow L, Jurczyk S, Shapiro G, Prudent JR (2004) Nucleic acid analysis using an expanded genetic alphabet to quench fluorescence. J Am Chem Soc 126:4550–4556

17. Nazarenko IA, Bhatnagar SK, Hohman RJ (1997) A closed tube format for amplification and detection of DNA based on energy transfer. Nucleic Acids Res 25:2516–2521

18. Liu XK, Hong Y (2007) Q-priming PCR: a quantitative real-time PCR system using a self-quenched BODIPY FL-labeled primer. Anal Biochem 360:154–156

19. Kutyavin IV, Afonina IA, Mills A, Gorn VV, Lukhtanov EA, Belousov ES, Singer MJ, Walburger DK, Lokhov SG, Gall AA, Dempcy R, Reed MW, Meyer RB, Hedgpeth J (2000) 3′-minor groove binder-DNA probes increase sequence specificity at PCR extension temperatures. Nucleic Acids Res 28:655–661

20. Costa JM, Ernault P, Olivi M, Gaillon T, Arar K (2004) Chimeric LNA/DNA probes as a detection system for real-time PCR. Clin Biochem 37:930–932

21. Tyagi S, Kramer FR (1996) Molecular beacons: probes that fluoresce upon hybridization. Nat Biotechnol 14:303–308

22. AlleLogic (2003) A breakthrough in PCR technology: AllGlo fluorogenic Reagents for RT-PCR

23. Baeumler S, Wulff D, Tagliani L, Song P (2006) A real-time quantitative PCR detection method specific to widestrike transgenic cotton (event 281-24-236/3006-210-23). J Agric Food Chem 54:6527–6534

24. Hernandez M, Pla M, Esteve T, Prat S, Puigdomenech P, Ferrando A (2003) A specific real-time quantitative PCR detection system for event MON810 in maize YieldGard based on the 3′-transgene integration sequence. Transgenic Res 12:179–189

25. Song P, Cai CQ, Skokut M, Kosegi B, Petolino J (2002) Quantitative real-time PCR as a screening tool for estimating transgene number in WHISKERS™-derived transgenic maize. Plant Cell Rep 20:948–954

26. Clive J (2006) Global status of commercialized biotech/GM crops, in ISAAA brief 35. ISAAA, Ithaca, NY

27. Li Z, Hansen JL, Liu Y, Zemetra R, Berger P (2004) Using real-time PCR to determine transgene copy number in wheat. Plant Mol Biol Rep 22:179–188

28. Subr Z, Novakova S, Drahovska H (2006) Detection of transgene copy number by analysis of the T1 generation of tobacco plants with introduced P3 gene of potato virus A. Acta Virol 50:135–138

29. Yang L, Ding J, Zhang C, Jia J, Weng H, Liu W, Zhang D (2005) Estimating the copy number of transgenes in transformed rice by real-time quantitative PCR. Plant Cell Rep 23:759–763

30. Zhang J, Cai L, Cheng J, Mao H, Fan X, Meng Z, Chan KM, Zhang H, Qi J, Ji L, Hong Y (2008) Transgene integration and organization in cotton (*Gossypium hirsutum* L.) genome. Transgenic Res 17:293–306

31. Zhang XD, Jenkins JN, Callahan FE, Creech RG, Si Y, McCarty JC, Saha S, Ma DP (2003) Molecular cloning, differential expression, and functional characterization of a family of class I ubiquitin-conjugating enzyme (E2) genes in cotton (Gossypium). Biochim Biophys Acta 1625:269–279

32. Zhang Z, Hu J (2007) Development and validation of endogenous reference genes for expression profiling of medaka (Oryzias latipes) exposed to endocrine disrupting chemicals by quantitative real-time RT-PCR. Toxicol Sci 95:356–368

33. GMDD: a database of GMO detection methods

Chapter 10

Development of Enzyme-Linked Immunosorbent Assay for the Detection of Bt Protein in Transgenic Cotton

Suchitra Kamle, Abhishek Ojha, and Arvind Kumar

Abstract

Enzyme-linked immunosorbent assays (ELISA) are being used extensively for the identification of Bt-protein in *Bt* transgenic crops. A sandwich ELISA test is the most preferable immunoassay for the quantification of Bt-protein in transgenic cotton plants. Here, we describe development of sandwich ELISA, employing polyclonal rabbit antibody as a capture antibody and HRP-labeled mouse anti-Bt protein-antibody as a detector antibody.

1. Introduction

During the past 14 years, the global area of GM crops has remarkably increased, year by year and covers 148 million hectares land area in developed and developing countries (1). Continuing this trend, insecticidal proteins of *Bacillus thuringiensis* (*Bt*) have emerged as the proteins of choice to be expressed in transgenic crops towards an environment-friendly mode of insect pest management in agriculture. *Bt*, a crystalliferous spore-forming gram-positive bacteria of the family *Bacillaceae*, which produce parasporal protein crystals having insecticidal property to kill Lepidoptera, Coleoptera, Diptera, Hymenoptera, Homoptera, and Mallophaga. In addition, *Bt* genes are constantly employed for crop protection from moths and butterflies also (2). Cry proteins classified on the basis of identity of amino acids and so far, more than 350 *cry* genes have been sequenced and the proteins classified in 53 groups. Proteins encoded by the *cry* gene family have a molecular weight 65 kDa and form crystals.

Currently, transgenic crop cotton (*Gossypium hirsutum* L.) expressing Cry proteins are being cultivated in more than a dozen

Baohong Zhang (ed.), *Transgenic Cotton: Methods and Protocols*, Methods in Molecular Biology, vol. 958,
DOI 10.1007/978-1-62703-212-4_10, © Springer Science+Business Media New York 2013

131

countries. A number of *Bt* crops are experiencing field testing (e.g., *Bt* cotton, *Bt* rice, *Bt* okra, *Bt* corn, *Bt* brinjal, *Bt* potato, and *Bt* tomato).

To prevent incidental commingling of the GM and non-GM contents, there is a pressing need for an efficient detection system to screen GMOs and also to meet the regulatory compliance. The enforcement of threshold values has created a demand for the development of reliable GMO analysis methods of a rapid and inexpensive character (3).

The tests, using an enzyme mediated immunoassay, have been widely used to detect GM protein in transgenic plants. Because, of its sensitivity, ruggedness, and inexpensive characteristics, it is a choice of breeders for testing of unapproved events, and determining GM content ensuring compliance with non-GM labeling requirements (4). On the basis of concentrations of transgenic material in plant tissue (>10 µg per tissue), the detection limits of protein immunoassays can predict the presence of modified proteins in the range of 1% GMOs (5).

Various sandwich enzyme-linked immunosorbent assay (ELISA) formats for detecting Cry1Ab, Cry1Ac, Cry2Ab, and EPSPS are already developed by commercialized companies and research laboratories (6–8).

Immunoassay technique is completely based on the specific binding between an antigen and antibody for qualitative as well as quantitative detection of proteins when the target analyte is known. Monoclonal antibodies (MAbs) are highly specific and polyclonal antibodies (PAbs) are often more sensitive that can be used for the protein detection system. Mainly there are two ELISA formats: competitive (higher the protein concentration, lower will be the color intensity) and sandwich ELISA (higher the protein concentration, higher will be the color intensity) (9, 10). Sandwich ELISA is more preferable immunoassay for protein detection in GM crops due to its sensitivity, in which the analyte is sandwiched in between the two antibodies: capture antibody and the detector antibody.

2. Materials

Prepare all dilutions, solutions, and buffers in triple distilled autoclaved filtered water.

1. 96-well microtiter plate (Nunc flat bottom).

2. *Bt* clone harboring desired transgene (store at −80°C).

3. Purified Bt protein (store at 4°C).

4. Purified antibodies develop against desired Bt protein (store at 4°C).

5. HRP-labeled anti-Bt protein detecting antibody (store at –20°C).

6. Transgenic cotton samples either leaf or seed (store at –80°C).

7. Multichannel pipette (8 and 12 Channels).

8. ELISA plate incubator shaker.

9. ELISA plate reader.

2.1. Buffers and Solutions

1. Phosphate buffer saline (PBS): 137 mM NaCl, 2.7 mM KCl, 1.4 mM KH_2PO_4, 8.0 mM Na_2HPO_4, pH 7.4.

2. Phosphate buffer saline and Tween-20 (PBST): 0.05% Tween-20 in PBS.

3. Blocking buffer: 5% skimmed milk powder dissolved in PBS (freshly prepare, kept at 37°C).

4. Lysis buffer: 80 mM Tris-HCl, 50 mM Sucrose, pH 8.0, Lysozyme 2 mg/mL.

5. Protein extraction buffer: 50 mM Sodium Phosphate Buffer, pH 8.0, 10 mM Dithio-thritol (DTT), 2% Polyvinylpyrrolidone-40 (PVP-40), Protease inhibitor cocktail 10 µL/mL (freshly prepare).

6. Substrate solution: 0.01% 3, 3′, 5, 5′ tetramethyl benzedine (TMB), 0.1 M sodium acetate, pH 5.1, 2.5 µL of 30% (v/v) hydrogen peroxide (H_2O_2)/10 mL.

3. Methods

3.1. Expression, Isolation, and Purification of Bt Protein

1. Under aseptic condition, inoculate recombinant *E. coli* clone harboring desired *Bt* transgene.

2. Propagate a culture to express the recombinant protein in Luria Bertani media 2% (w/v) containing appropriate antibiotic and incubate at 150 rpm for 16 h, 37°C.

3. Pellet the bacterial culture at $16,000 \times g$, 10 min, 4°C (see refs. (7–9)).

4. Resuspend the recovered pellet in 50 mL Lysis buffer. Add protease inhibitor.

5. Sonicate the pellet at 50 Hz, 6 cycles, 2 min rest, 4°C.

6. Centrifuge at $16,000 \times g$ and collect the supernatant.

7. Use this supernatant for protein purification (see ref. (9)).

8. Quantify purified Bt protein by Bradford's method (see ref. (11)).

3.2. Antibody Preparation

1. For immunization, use purified Bt protein to raise PAbs in Balb-c mice/New Zealand-white rabbit employing Freund's complete and incomplete adjuvant (see ref. (9)).

2. Give booster dose to each immunized animal at an interval of 8, 15, 21, 30 days.

3. Collect blood, left at room temperature for 2 h and then kept for overnight at 4°C.

4. Centrifuge at 2,500×*g*, 5 min, 4°C and collect the yellowish serum.

5. Purify the antiserum by using protein-A column as per manufacturer's instruction (see ref. (9)) (see Note 1).

3.3. Labeling of Antibody

1. If require, label purified anti-Bt protein-antibody with horse-radish peroxidase (HRP) by periodate oxidation method (see ref. (9)).

3.4. Optimization of Antigen and Antibody for Sandwich ELISA

Principle: Color intensity of the product formed is directly proportional to the amount of Bt protein present in the sample.

1. Take 96-well microtiter plate.

2. Coat 96-well plate with anti-Bt protein rabbit antibody (different dilutions) in vertical fashion and incubate the plate for overnight at 4°C.

3. Washed off, unabsorbed antibody with PBS (flicking and flapping manner (see Note 2)).

4. Block the unbound sites with blocking buffer for 2 h at 37°C (see Note 3).

5. Now, incubate the plate with purified Bt protein (1 ng/well) for 1 h at 37°C.

6. Again incubate the plate with anti-Bt protein-mouse antibody HRP-labeled (see Notes 4 and 5).

7. Add enzyme substrate solution (freshly prepared) into each well and immediately cover the plate with aluminium foil to avoid light (see Note 6).

8. Observe a gradual increase in brilliant blue color intensity, with the increasing concentration of Bt protein.

9. Stop the reaction after 15 min by adding 0.5 N Sulfuric acid (see Note 7).

10. Observe the brilliant blue color suddenly changes to fine yellow endpoint color and measure the absorbance at 450 nm using ELISA plate reader.

11. Select the optimum optical density in the range of 1.3–1.6.

12. Select the optimum concentration of antibody to produce further standard curve for quantifying Bt protein in transgenic cotton samples.

3.5. Development of Standard Curve for Quantification of Bt Protein

1. Coat a 96-well microtiter plate with protein "A" purified, rabbit polyclonal antibody and incubate for overnight at 4°C.

2. Block the unbound sites with blocking solution for 2 h at 37°C (see Note 2).

3. Now, incubate the wells with standards/samples (different concentrations) for 1 h at 37°C. Subsequent to this, incubate with mouse-monoclonal antibody (develop against desired Bt protein) for 1 h at 37°C (see Notes 3 and 4).

4. Again, incubate the wells with optimized HRP-labeled anti-Bt protein mouse antibody for 1 h at 37°C (see Note 5).

5. Add enzyme substrate solution (freshly prepared) into each well and immediately cover the plate with aluminium foil to avoid light (see Note 6).

6. Add 0.5 N Sulfuric acid to stop the reaction (see Note 7).

7. Read the absorbance at 450 nm using ELISA plate reader (Figs. 1 and 2).

3.6. Extraction of Bt Protein from GM Cotton Seed/Leaf Samples

1. Prepare a 10% homogenate. Collect about 100 mg seed powder from GM cotton seed/leaf samples and grind in 1 mL of protein extraction buffer.

2. Incubate this crude preparation of protein for 45 min, at 37°C with gentle shaking and then incubate on ice for 15 min and then centrifuge at $16,000 \times g$, for 10 min at 4°C.

3. Now, gently transfer the supernatant into a fresh tube without disturbing cell debris.

4. Use the supernatant for detection of Bt protein expression in transgenic cotton.

3.7. Detection of Bt Protein in Transgenic Cotton Sample

3.7.1. Assay Validation/ Verification

1. Evaluate the specificity (cross reactivity), sensitivity [the limit of detection (LOD) and limit of quantitation (LOQ)], precision (intra and inter-assay), accuracy (recovery), and sturdiness (CV%) of the developed ELISA.

3.7.2. Fortification

1. Evaluate the Bt protein standard ELISA graph with different concentrations of Bt protein in negative control (non-Bt) cotton plant matrices (Figs. 1 and 2).

2. Fortify Bt protein solution with an aliquot of negative control, cotton plant matrices. Centrifuge the mixture at $16,000 \times g$ for 10 min at 4°C (see ref. (8)).

3. Separate the transparent yellowish green supernatant to a fresh tube and use this supernatant as sample.

4. Do ELISA with the above sample and plot a graph.

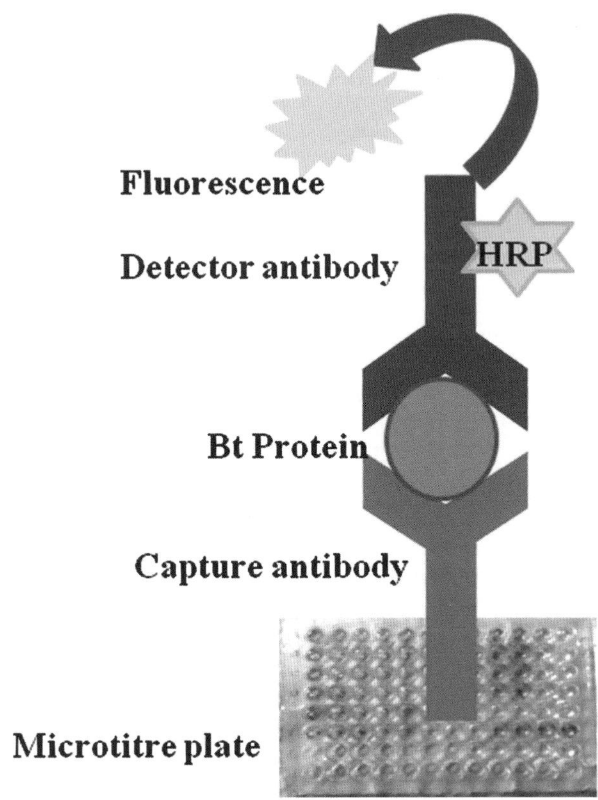

Fig. 1. A schematic view of sandwich ELISA.

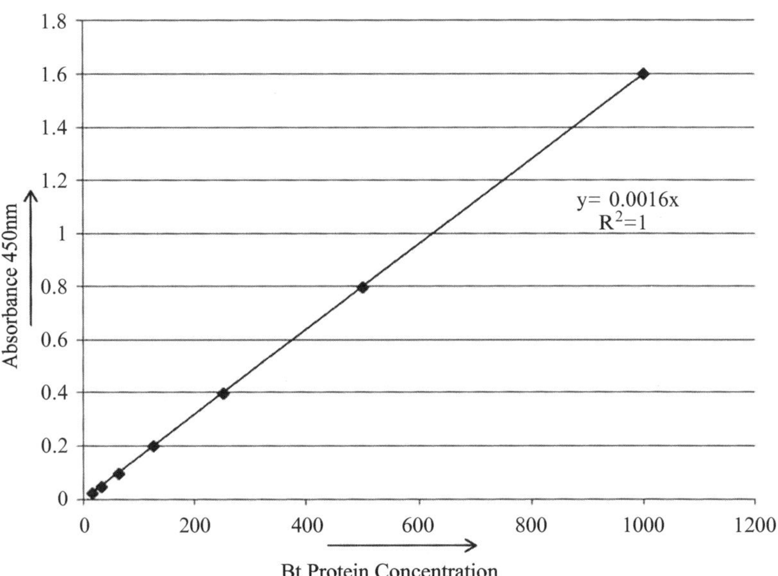

Fig. 2. Schematic representation of standard curve for the detection of Bt protein.

5. Examine and compare its O.D. with the standard ELISA graph.

6. Calculate the protein concentration by multiplying with the dilution factor.

7. Examine LOD and LOQ of the immunoassay.

3.7.3. Recovery

Recover the desire Bt protein with other nontarget Bt proteins to check and validate the standard assay. Steps are same as describe (see Subheading 3.7.2).

3.7.4. Statistical Analysis

1. Determine the false-positive and false-negative rates, analyze the specificity, sensitivity of the results and data, by statistical accuracy, and also deduce the inter-assay and intra-assay variation.

2. A false-positive result happens when residue at or above the established LOD is found in a sample known to be free of analyte. A false negative happens when no residue is detected in a sample fortified at the LOD.

3. Calculate LOD equal to 3 SD and LOQ to 10 SD.

4. Calculate % coefficient of variation (CVs) as $(SD/mean) \times 100$.

---

4. Notes

1. Availability of antibodies with the desired affinity and specificity is an important factor to develop an immunoassay.

2. After each step of incubation, wash the wells with PBS thrice.

3. Except blocking buffer (250 μL), volume level of wells do not exceed 100 μL, just to avoid cross-contamination.

4. Dilute primary and secondary antibodies in 0.05% blocking buffer.

5. After incubation with conjugated antibody, always wash the wells with 0.05% Tween-20 and PBS thrice. Lastly, wash with PBS only.

6. Keep the enzyme solution in dark bottle, if dispensing into wells immediately cover up the plate with aluminium foil.

7. Immediately read the plate, once reaction is stop.

8. Compare the optical density of transgenic cotton sample with standard curve for quantification of Bt protein.

9. Detailed types of ELISA in general (see ref. (10)).

References

1. James C (2010) Global status of commercialized biotech/GM crops. ISAAA, Ithaca, NY. http://www.isaaa.org/resources/publications/briefs/42/. Accessed 22 Feb 2011

2. Romeis J, Meissle M, Bigler F (2006) Transgenic crops expressing *Bacillus thuringiensis* toxins and biological control. Nat Biotechnol 24:63–71

3. European Committee for Standardization (2001) CEN/TC 275/WG 11N149

4. Stave JW (2002) Protein immunoassay methods for detection of biotech crops: applications, limitations, and practical considerations. J AOAC Int 85:780–786

5. Grothaus GD et al (2006) Immunoassay as an analytical tool in agricultural biotechnology. J AOAC Int 89:913–928

6. Chalam VC, Khetrapal RK. ELISA based detection of GMOs. http://www.envfor.nic.in/divisions/csurv/biosafety/Gef2/T5/12%20 Dr.%20Celia_ELISA%20based%20detection%20 of%20LMOs.pdf. Accessed 14 Jan 2012

7. Wang S et al (2007) Development of ELISA for the determination of transgenic Bt cottons using antibodies against Cry1Ac protein from Bacillus thuringiensis HD-73. Engg Life Sci 7:1–7

8. Kamle S, Ojha A, Kumar A (2011) Development of an enzyme linked immunosorbant assay for the detection of Cry2Ab protein in transgenic plants. GM Crops 2:118–125

9. Harlow E, Lane D (2010) Antibodies: a laboratory manual. Cold Spring Harbor Laboratory Press, New York

10. Crowther JR (2000) Methods in molecular biology: the ELISA guidebook. Humana Press, Totowa, New Jersey

11. Bradford MM (1976) A rapid sensitive method for the quantitation of microgram quantities of protein utilizing the principle of protein dye binding. Anal Biochem 72:248–254

Chapter 11

DNA-Based Diagnostics for Genetically Modified Cotton: Decaplex PCR Assay to Differentiate MON531 and MON15985 *Bt* Cotton Events

Gurinder Jit Randhawa, Monika Singh, and Rashmi Chhabra

Abstract

The adoption rate and global area under cultivation of genetically modified (GM) crops is dramatically increasing in recent past. GM cotton has occupied 25.0 million hectares (mha) comprising 15.6% of the global area under GM cultivation. *Bt* cotton, expressing delta-endotoxins from *Bacillus thuringiensis* (*Bt*), is the only commercialized crop in India that is planted on an area of 10.6 mha. With the increase in development and commercialization of GM crops, it is necessary to develop appropriate qualitative and quantitative methods for detection of different GM events. Robust diagnostics for GM detection need to be developed and implemented to monitor and detect different events of GM cotton in India. This chapter summarizes the methods based on polymerase chain reaction (PCR) being employed for detection of different GM events of cotton. We describe a decaplex PCR method for identification and differentiation of two major commercialized events of *Bt* cotton, i.e., MON531 and MON15985, in India.

1. Introduction

The innovation of biotechnology and genetic engineering techniques has led to the development of many genetically modified (GM) crops with improved insect resistance, herbicide tolerance, quality parameters, and production value. Cotton (*Gossypium hirsutum* L.) is an important fiber crop, which is being cultivated in the area of 12.1 million hectares (mha) in India, largest cotton growing country in the world. *Bt* cotton, which confers resistance to bollworm, a lepidopteron insect pest of cotton, was first adopted in India as hybrids in 2002. The number of events, as well as the number of *Bt* cotton hybrids and companies marketing approved hybrids have all increased significantly from 2002, the first year of commercialization of *Bt* cotton in India. In 2010, the adoption of

Baohong Zhang (ed.), *Transgenic Cotton: Methods and Protocols*, Methods in Molecular Biology, vol. 958,
DOI 10.1007/978-1-62703-212-4_11, © Springer Science+Business Media New York 2013

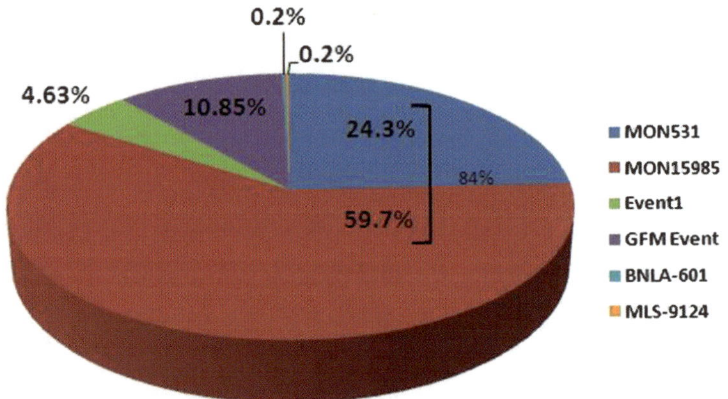

Fig. 1. Number of hybrids of commercially released six events of GM cotton (2010) *Source*: references (1, 2).

Bt cotton in India soared to a record 10.6 mha, equivalent to 86% of the record 12.1 mha cotton crop planted in the country. A total of 884 introductions of *Bt* cotton including 883 hybrids of MON531, MON15985, Event1, GFM-*cry1A*, and MLS-9124 events of *Bt* cotton and one *Bt* cotton variety, i.e., BNLA-601 were approved for planting in 2010 (1, 2). The details of commercialized events of *Bt* cotton in India are highlighted in Fig. 1.

1.1. Need for GM Detection

In order to facilitate effective regulatory compliance for identification of genetic traits, to assess possible effects of GM crops, for risk management and post-release monitoring, as well as to address the consumers' concerns, the safety of GM crops needs to be evaluated. The labeling threshold differs from country to country; in European Union, Korea, and Taiwan/Japan require mandatory labeling of GM food or products when GM content exceeds 0.9, 3.0, and 5.0%, respectively, in tested sample(s) (3, 4). So far, no labeling threshold has been implemented in India. However, to meet the regulatory obligations, as per the Supreme Court of India's instructions for conducting field trials of GM crops, a protocol for testing contamination up to 0.01% has to be established. (http://www.envfor.nic.in/divisions/csurv/geac/decision-jul-95.pdf). The safety of new GM crops is reviewed by an independent competent authority, Genetic Engineering Appraisal Committee (GEAC) in India. The next step is to monitor the genetically modified organisms (GMO) content in seeds and planting materials and to detect the specific GM event. GM testing is based on the detection of transgenic DNA elements or their protein products. As DNA is more stable molecule, hence DNA-based methods are widely accepted. The most preferred and robust DNA-based detection method is polymerase chain reaction (PCR)

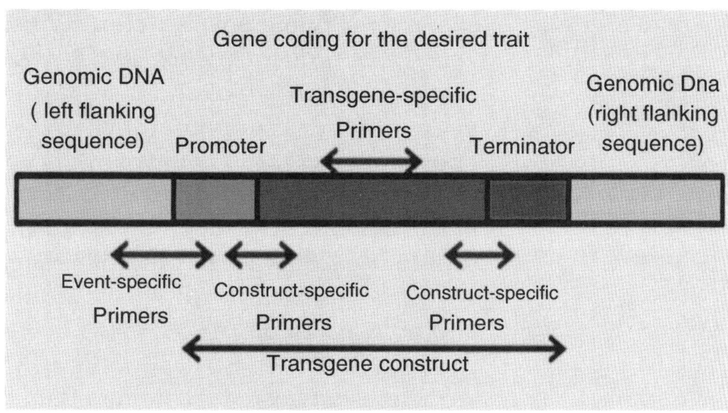

Fig. 2. The transgenic construct showing genetic elements addressed in GMO testing and primers' target positions (event-specific, construct-specific, and transgene-specific are shown by *double-headed arrows*).

because of its simplicity, specificity, and sensitivity (5). To date, TaqMan-based real-time PCR has been the method of choice for GMO quantification (5–7).

1.2. An Overview of GM Detection Using DNA-Based PCR Approach

Different types of genetic elements, viz., screening elements, transgenes for the desired trait, etc., are addressed in GM detection (Fig. 2). Screening elements are transgenic promoters, markers, and/or terminator genes, e.g., constitutively expressed *Cauliflower Mosaic Virus* (*CaMV*) 35S promoter sequence, selectable or scorable marker genes and *Agrobacterium tumefaciens nos* terminator gene and transgenes for the desired trait include genes that encode a certain trait, such as herbicide tolerance, e.g., *phosphinothricin-N-acetyltransferase* (*bar* and *pat*) and *5-enolpyruvylshikimate-3-phosphate synthase* (*EPSPS*), an insect resistance, e.g., δ-endotoxins (*cry* genes), or others. Plant-specific elements detect certain plant-specific endogenous genes, such as *invertase* for maize and *lectin* for soybean. A parallel examination of screening elements and plant-specific elements, increases the probability of detecting poorly characterized and/or non-certified GMOs. With commercialization of a range of GM crops carrying the same trait produced by different developers, event-specific quantitative PCR methods are required to differentiate between different events of the same trait.

PCR-based method for complete characterization of GM crops consists of three distinct steps.

Detection: It involves preliminary screening to determine whether a planting material is GM or not. The screening methods are usually based on the PCR detecting the screening elements such as promoters, terminators, or commonly present selectable or scorable marker genes in cloning vectors. A multiplex PCR approach simultaneously detecting the commonly present marker genes, i.e., *aadA*, *bar*, *hpt*, *nptII*, *pat*, and *uidA* encoding, respectively, for aminoglycoside-3′-adenyltransferase, *Streptococcus viridochromogenes* phosphinothricin-*N*-acetyltransferase, hygromycin phosphotransferase, neomycin phosphotransferase, *Streptococcus hygroscopicus* phosphinothricin-*N*-acetyltransferase, and β-D-glucuronidase, has been developed in our laboratory as a reliable tool for qualitative screening of a range of GM crops (8).

Identification: It involves confirming the presence of a specific GM trait, inserted gene construct or GM event. It may be transgene-specific PCR; construct-specific PCR targeting the junction sequences between the adjoining DNA segments for specific detection of inserted genetic construct; or event-specific PCR targeting the junction sequences in the integration site (plant-construct junction) to detect a specific transformation event.

Quantification: If a crop or its product has been shown to contain GM trait, then for determination of the amount of each of the transgene present, real-time PCR assays have been effectively used.

1.3. DNA-Based Analytical Methods for GM Cotton

Among conventional PCR technologies, multiplex PCR is less time-consuming and cost-effective, which can detect multiple target sequences of inserted gene construct in a single reaction. Several multiplex PCR methods (Table 1) have been developed and validated for precise and accurate monitoring, tracing, and regulation of GM cotton (9–11).

To enhance the accuracy, sensitivity, and reproducibility of GM detection and for automatic and high throughput, multiplex PCRs have been coupled with other methods. In 2009, Nadal et al. developed a multiplex PCR assay coupled with capillary gel electrophoresis for amplicon identification by size and color (multiplex PCR-CGE-SC) for simultaneous detection of cotton species and five events of GM cotton, viz., Bollgard®I, Bollgard®II, Roundup Ready, 3006-210-23, and 281-24-236 (12). Real-time quantitative PCR method specific to Widestrike GM cotton (event 281-24-236/3006-210-23) was established on the basis of the DNA sequences in the junction between the transgene insert and cotton genome (7). The reported method involves a DNA extraction method from cotton seeds and three PCR systems corresponding to a cotton-specific endogenous reference DNA sequence *SAH7* (Sinapis *Arabidopsis* Homolog 7) and specific detection of events 281-24-236 and 3006-210-23 (http://gmo-crl.jrc.it/statusof-doss.htm). Lee et al. (2007) reported the qualitative and quantitative detection methods for GM cotton events MON15985 and MON88913 using two kinds of specific primer pairs, probes, and one standard plasmid, and confirmed the applicability for practical use by in-house validation experiment (13). Real-time PCR assays have also been developed in our laboratory for quantification of

Table 1
Summary of DNA-based detection systems being employed for GM cotton

GM event/crop	PCR system	Target gene	References
Widestrike cotton (Event 281-24-236/3006-210-23)	Quantitative real-time PCR	A cotton-specific endogenous reference gene *SAH7* and event s281-24-236 and 3006-210-23	(7)
MON531 and MON15985	Decaplex PCR	*cry1Ac* and *cry2Ab* transgenes; *nptII, aadA,* and *uidA* marker genes; *CaMV* 35S promoter and *nos* terminator; two construct-specific sequences, i.e., *cry1Ac* transgene construct and *cry2Ab* transgene construct; and endogenous *Sad1* gene	(9)
GM cotton (VipCot14, VipCot29)	Multiplex; construct-specific	*vip-s* gene and *vip3A*-like genes, *CaMV* 35S promoter, *nos* terminator, and *npt*-II marker gene	(10)
Mon531, GK19, SGK321	Conventional as well as quantitative	Cowpea trypsin inhibitor (*CpTI*) gene of SGK321 cotton and the specific junction DNA sequences containing partial *Cry1A(c)* gene and *NOS* terminator of Mon531, GK19, and SGK321 cotton varieties	(11)
Bollgard I, Bollgard II, Roundup Ready, 3006-210-23, and 281-24-236	Multiplex PCR-CGE-SC	Bollgard, Bollgard II, Roundup Ready, 3006-210-23, and 281-24-236	(12)
MON15985, MON88913	Event-specific qualitative PCR and quantitative real-time PCR	MON15985, MON88913	(13)

cry1Ac and *cry2Ab* transgenes in MON531 and MON15985 events of *Bt* cotton (9).

In India, out of commercialized 780 hybrids of commercially released six events of *Bt* cotton, 743 hybrids (84.04%) belong to two major events, i.e., 215 hybrids of MON531 (Bollgard®I) and 528 hybrids of MON15985 (Bollgard®II), which are being grown in farmer's fields in the North, Central, and South zones of India. The MON531 event of *Bt* cotton, Bollgard®I, expressing single gene *cry1Ac* and MON15985 event, Bollgard®II expressing multiple genes *cry1Ac* and *cry2Ab*, were commercialized by Maharashtra Hybrid Seeds Company Ltd. (Mahyco). Based on the performance, farmers prefer multiple genes over a single gene *Bt* cotton hybrids as multiple gene *Bt* cotton hybrids provide additional protection to Spodoptera (a leaf eating tobacco caterpillar), while it also increases efficacy of protection to both American bollworm, Pink bollworm, and Spotted bollworm. Farmers adopting multiple gene *Bt* cotton earn higher profit through cost savings associated with fewer sprays for Spodoptera control as well as increasing yield by 8–10% over single gene *Bt* cotton hybrids (2). The market price of Bollgard®II is comparatively higher as compared to Bollgard®I. It is difficult to differentiate the seeds of two events so there are chances of adulteration or mixing of the events by traders to earn profits. To overcome such issue, a decaplex PCR method simultaneously detecting the transgenes, i.e., *cry1Ac* and *cry2Ab*; marker genes, i.e., *nptII*, *aadA*, and *uidA*; control elements, i.e., *CaMV* 35S promoter and *nos* terminator; two construct-specific sequences, i.e., *cry1Ac* transgene construct and *cry2Ab* transgene construct; and endogenous *Sad1* gene, has been reported for identification and differentiation of MON531 and MON15985 *Bt* cotton events (9). In this chapter, we describe this method in detail.

2. Materials

Prepare all solutions using MilliQ water and use analytical grade reagents. Prepare and store all reagents at room temperature or as per the instructions. Diligently follow all waste disposal regulations when disposing waste materials.

2.1. Planting Materials

Use the seeds of commercialized events of *Bt* cotton, i.e., MON531 (Bollgard®I) and MON15985 (Bollgard®II).

2.2. Genomic DNA Extraction Buffers and Reagents

1. CTAB extraction buffer: 2.0 g CTAB, 10.0 ml 1 M Tris–HCl pH 8.0, 4.0 ml 0.5 M EDTA pH 8.0, 28.0 ml of 5 M NaCl, 40.0 ml of H_2O. Adjust all to pH 5.0 with HCl and make up final volume to 100 ml with deionized water.
2. Chloroform: Iso Amyl Alcohol (24:1).
3. 3 M Sodium acetate.
4. Absolute ethanol and 70% ethanol.
5. RNaseA (10 µg/ml).

2.3. Primers

1. Use the sequences of primer pairs as listed in Table 2.
2. Dilute the oligonucleotides (primers) to have a final concentration of 10 µM with MilliQ water, which can be used for simplex PCR to check the amplification.

Table 2
Details of primer pairs to be used in decaplex PCR

Gene	Primer	Primer sequence (5′–3′)	Expected amplicon size (bp)	References
cry1Ac	Cry1Ac-F/R	F-GACCGCTTACAAGGAGGGATACG R-ACGGAGGCATAGTCAGCAGGACC	228	(9)
cry2Ab	Cry2Ab-F/R	F-CAGCGGCGCCAACCTCTACG R-TGAACGGCGATGCACCAATGTC	260	(9)
aadA	AadA-1-F/R	F-TCCGCGCTGTAGAAGTCACCATTG R-CCGGCAGGCGCTCCATTG	406	(8)
nptII	APH2 short/ APH2 reverse	F-CTCACCTTGCTCCTGCCCGAGA R-CGCCTTGAGCCTGGCGAACAG	215	(14)
uidA	Gus F/R	F-TTTCTTTAACTATGCCGGAATCCATC R-CACCACGGTGATATCGTCCAC	82	(15)
CaMV 35S promoter	SP1 F/R	F-TTGCTTTGAAGACGTGGTTG R-ATTCCATTGCCCAGCTATCT	196	(16)
nos terminator	NOS1/NOS3	F-GAATCCTGTTGCCGGTCTTG R-TTATCCTAGTTTGCGCGCTA	180	(17)
Sad1	S3F/S4 R	F-CCAAAGGAGGTGCCTGTTCA R-TTGAGGTGAGTCAGAATGTTGTTC	107	(18)
Cry1Ac transgene construct	Cry1Ac-35SF/R	F-CTTCGCAAGACCCTTCCTCTAT R-GAACTCTTCGATCCTCTGGTTG	326 bp region between *CaMV* 35S promoter and *cry1Ac* gene	(9)
Cry2Ab transgene construct	CTCR-F/ CTCR-2R	F-ATT GAA GAA GAG TGG GAT GAC GTT A R-GAC CAG AGT TCA GGA CGG AGT T	116 bp junction region between *CTP2* and *cry2Ab* genes	(13)

3. Then make 4× concentration of primer mix by mixing 1.6 µM of each primer pair to perform decaplex PCR. Primer mix must be freshly prepared (see Note 4).

2.4. PCR Reagents

Taq PCR buffer, $MgCl_2$, dNTP mix, primers, *Taq* DNA Polymerase (MBI Fermentas, Inc., USA), Hot Start *Taq* DNA Polymerase Polymerase (MBI Fermentas, Inc., USA).

2.5. Agarose Gel Electrophoresis: Buffer, Agarose Gel, and Metaphor Agarose Gel

1. Running buffer: 50× TAE (pH 8.0) 1,000 ml: 2 M Tris base, 1 M Glacial acetic acid (57.1 ml), 100 ml 0.5 M EDTA solution. Weigh 242 g of Tris base and transfer to the beaker. Add deionized water to a volume of 500 ml. Mix well and add 100 ml of 0.5 M EDTA solution and 57.1 ml of glacial acetic acid. Adjust the pH to 8.0. Add deionized water to a volume of 1,000 ml.

2. Preparation of 2% Agarose gel: Take 100 ml of 1× TAE buffer in a conical flask. Weigh 2 g of agarose and add it to 1× TAE buffer. Mix well and boil the mixture till agarose dissolves. Cool the gel and pour it in the gel casting tray.

3. Preparation of 4% Metaphor agarose gel (Cambrex Bioscience Rockland, Inc., Rockland, ME, USA): Take 100 ml of ice chilled 1× TAE buffer in a conical flask. Weigh 4 g of metaphor agarose and add it to 1× TAE buffer. Mix well and boil the mixture till metaphor agarose dissolves. Metaphor agarose boils vigorously, therefore do not shake the flask in between boiling. Cool the gel and pour it in the gel casting tray.

3. Methods

3.1. CTAB Extraction Method and DNA Quantification

1. Grind 200 mg of seeds to a fine powder in approximately 500 µl of CTAB buffer.

2. Transfer CTAB/plant extract mixture to a microfuge tube.

3. Incubate the CTAB/plant extract mixture for about 1 h at 60°C in a water bath (see Note 1).

4. After incubation, spin the CTAB/plant extract mixture at 9,000 ×*g* force for 5 min to spin down cell debris. Transfer the supernatant to clean microfuge tubes.

5. To each tube add 250 µl of chloroform: Iso amyl alcohol and mix the solution by inversion. After mixing, spin the tubes at 9,000 ×*g* force for 1 min.

6. Transfer the upper aqueous phase only (contains the DNA) to a clean microfuge tube.

7. To each tube add 50 µl of 3 M sodium acetate followed by 500 µl of ice cold absolute ethanol.

8. Invert the tubes slowly several times to precipitate the DNA. Generally the DNA can be seen to precipitate out of solution. Alternatively the tubes can be placed for 1 h at –20°C after the addition of ethanol to precipitate the DNA.

9. Following precipitation, the DNA can be pipetted off by slowly rotating/spinning a tip in the cold solution. The precipitated DNA sticks to the pipette and is visible as a clear thick precipitate. To wash the DNA, transfer the precipitate into a microfuge tube containing 500 µl of ice cold 70% ethanol and slowly invert the tube.

10. After the wash, spin the DNA into a pellet by centrifuging at $15,000 \times g$ force for 1 min. Remove all the supernatant and allow the DNA pellet to dry (approximately 15 min). Do not allow the DNA to over dry or it will be hard to redissolve.

11. Resuspend the DNA in sterile DNase-free water (approximately 50–400 µl H_2O; the amount of water needed to dissolve the DNA can vary, depending on the quantity of isolated DNA). RNaseA (10 µg/ml) can be added to the water prior to dissolving the DNA to remove any RNA in the preparation (10 µl RNaseA in 10 ml H_2O).

12. After resuspension, the DNA is incubated at 65°C for 20 min to destroy any DNases that may be present and store at 4°C.

13. Check the quality of extracted DNA using 0.8% agarose gels in 1× TAE buffer. Quantify the DNA by measuring UV absorbance at 260 nm and check further DNA purity based on UV absorption ratio at 260/280 nm (ranging from 1.6 to 1.9) using a UV spectrophotometer.

3.2. Simplex PCR to Check the Amplification Efficiency of Synthesized Primer Pairs

Before setting up any PCR with transgene/construct/event-specific primer pairs, it is necessary to check for the absence of any PCR inhibitors in isolated genomic DNA. Therefore, a simplex PCR with an endogenous reference gene/chloroplast t-RNA-specific primer pair is carried out. If the desired amplicon is obtained, it means that the isolated genomic DNA is free from any PCR inhibitors and further PCRs can be carried out with the same genomic DNA (see Note 3).

Make a final volume of 25 µl PCR mixture with the following reagent concentrations:

100 ng of template DNA (see Note 2)
1× Taq PCR buffer
1.5 mM of $MgCl_2$
200 µM of dNTP mix
0.4 µM of each primer
0.5 U of Taq DNA Polymerase

1. Perform the PCRs in a programmable thermal cycler using the following program: one cycle of initial denaturation at 95°C for 5 min; 35 cycles of denaturation at 94°C for 30 s, annealing at 59°C for 1 min, and extension at 72°C for 1 min; followed by a final extension at 72°C for 8 min.

2. Analyze the PCR amplicons on horizontal gel electrophoresis using 2.0% agarose, and then visualize the results under UV light using a Gel Documentation System (see Note 6).

3.3. Decaplex PCR for Detection of MON531 and MON15985 Events

Decaplex (10-plex) PCR is being designed to amplify nine different elements of the inserted gene constructs, i.e., *cry1Ac*, *cry2Ab* gene, *CaMV* 35 promoter, *nos* terminator, *nptII*, *aadA*, *uidA*, *cry1Ac* transgene construct, MON15985 transgene construct and an endogenous gene *Sad1*, simultaneously.

1. Prepare the reaction mixture (25 μl) containing:

    175 ng of template DNA

    1× Hot Start Taq PCR buffer

    3.2 mM of MgCl$_2$

    600 μM of dNTP mix

    1× primer mix

    0.2 U/μl of Hot Start Taq DNA Polymerase (see Note 5)

2. Perform the PCR in a programmable thermal cycler using the amplification conditions: initial denaturation at 95°C for 10 min, 40 cycles consisting of denaturation at 95°C for 50 s, primer annealing at 59°C for 50 s, primer extension at 72°C for 50 s, and final extension at 72°C for 5 min.

3. Resolve the PCR amplicons on horizontal gel electrophoresis using 4.0% Metaphor® agarose and then visualize the resolved fragments under UV light using a Gel Documentation System (see Notes 6 and 7)

4. Notes

1. Mix the contents gently while incubation step of DNA extraction by inverting the tubes. This may minimize the shearing of DNA.

2. During PCR, always have a positive amplification control or/ and a negative amplification control with the test samples so that a problem with the primers, enzyme or a machine setting can be ruled out.

3. In order to make PCR assays more reliable, the target sequences and plant species specific endogenous reference genes may be

detected. The endogenous reference genes are required to be included as internal control targets to assess the efficiency of PCR reactions by eliminating any false negatives. So it is always better to run PCR along with amplification of endogenous reference gene, which would also be used as amplification control for the PCR.

4. As the number of primers increases in a decaplex PCR, the possible sequence-dependent interactions between primers of different primer pairs also increase, which results in the formation of primer-dimers. Small differences in amplification efficiencies for the different primer pairs might result in the preferential amplification of some of the PCR products, leaving other PCR products at sub-detectable levels. Hence, primer design, PCR cycling conditions, and the concentration of each reaction component need to be cautiously optimized in order to avoid the formation of primer-dimers and to detect all DNA targets simultaneously without any primer interference.

5. The choice of DNA polymerase is very important for the optimum performance of the PCR. In a multiplex PCR, the Hot Start DNA polymerase, coupled with a preoptimized primer mix for different multiplex reactions, gave the best results both in terms of reproducibility and robustness. Use of Hot Start DNA polymerase prevents the formation of misprimed products and reduces primer-dimer formation.

6. In simplex PCR, the specific amplicons of desired size for all the ten target sequences, i.e., (a) 228 bp for *cry1Ac* gene, (b) 260 bp for *cry2Ab* gene, (c) 196 bp for *CaMV* 35 promoter, (d) 180 bp for *nos* terminator, (e) 215 bp for *nptII*, (f) 406 bp for *aadA*, (g) 82 bp for *uidA*, (h) 326 bp for *cry1Ac* construct-specific (targeting the desired region between *CaMV* 35 S promoter and *cry1Ac* gene), (i) 116 bp for MON15985 construct-specific (targeting the junction region between chloroplast transit peptide *CTP2* and *cry2Ab* genes), (j) 107 bp for *Sad1* should be detected in the sample of MON15985 cotton. Whereas only seven of the target sequences, i.e., (a) 228 bp for *cry1Ac* gene, (b) 196 bp for *CaMV* 35 promoter, (c) 180 bp for *nos* terminator, (d) 215 bp for *nptII*, (e) 406 bp for *aadA*, (f) 326 bp for *cry1Ac* construct specific, (g) 107 bp for *Sad1* should be detected in the sample of MON531 event of cotton. Amplicon of 107 bp for endogenous reference gene *Sad1* should also be detected in non-*Bt* cotton sample.

In decaplex PCR (Fig. 3), for MON15985 event, all the seven inserted gene sequences, i.e., *cry1Ac*, *cry2Ab*, *CaMV* 35 S promoter, *nos* terminator, *uidA*, *nptII*, and *aadA*, two transgene constructs, i.e., *cry1Ac* gene construct and *cry2Ab* gene construct along with an endogenous reference gene should be amplified with desired band size of 228, 260, 196,

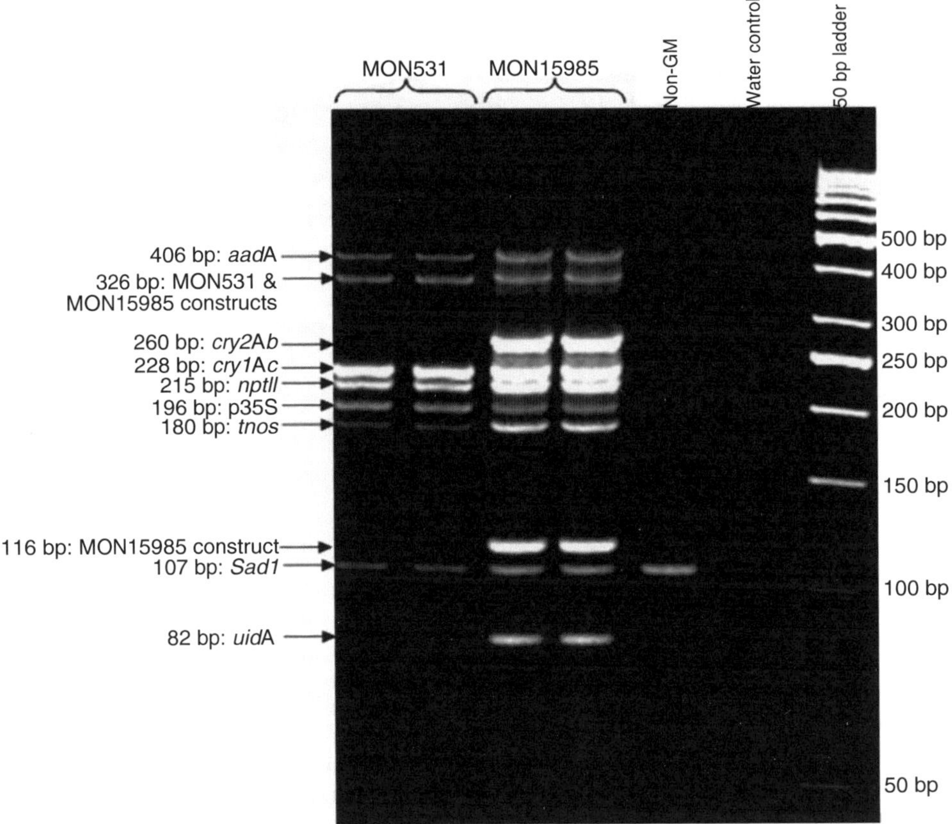

Fig. 3. Transgene- and construct-specific decaplex PCR amplification for discrimination of two *Bt* cotton events, i.e., MON531 and MON15985 using primer pairs for *cry1Ac* and *cry2Ab* transgenes, *nptII, aadA,* and *uidA* marker genes, *CaMV* 35S promoter, *nos* terminator, endogenous *Sad1* gene and specific gene constructs in MON531/MON15895 and MON15985 *Source*: reference (9).

180, 82, 215, 406, and 107 bp, respectively. In the samples of MON531 event, five inserted genes, i.e., *cry1Ac, CaMV* 35S promoter, *nos* terminator, *nptII*, and *aadA* and *cry1Ac* gene construct along with an endogenous reference gene should be amplified with the desired amplicons of 228, 196, 180, 215, 406, and 107 bp, respectively. In non-GM cotton sample, only endogenous reference gene should be amplified and in water sample taken as negative amplification control, no gene should be amplified showing specificity of the developed PCR protocol.

7. Since no amplification was detected for *cry2Ab* gene, *uidA* reporter gene, and *cry2Ab* transgene construct in MON531 event of *Bt* cotton, hence, the developed decaplex method can be used as an efficient tool for differentiating the MON531 and MON15985 events of *Bt* cotton.

Acknowledgments

The funds provided by the Department of Biotechnology are duly acknowledged. We also thank Director, NBPGR, New Delhi for providing necessary facilities.

References

1. Clives J (2011) Global status of commercialized biotech/GM crops: 2011. ISAAA brief no. 43. ISAAA, Ithaca, NY

2. Choudhary B, Gaur K (2010) Bt cotton in India: a country profile. ISAAA series of biotech crop profiles

3. Gruère GP, Rao SR (2007) A review of international labeling policies of genetically modified food to evaluate India's proposed rules. AgBioForum 10(1):51–64

4. Vilijoen CD (2005) Asian Biotechnol Dev Rev 7(3):55–69

5. Lipp M, Shillitto R, Giroux R, Spiegelhalter F, Charlton S, Pinero D, Song P (2005) Polymerase chain reaction technology as analytical tool in agricultural biotechnology. J AOAC Int 88:136–155

6. Miraglia M, Berdal KG, Brera C, Corbisier P, Holst-Jensen H, Kok EJ, Marvin HJP, Schimmel H, Rentsch J, Van Rie JPPF, Zagon J (2004) Detection and traceability of genetically modified organisms in the food production chain. Food Chem Toxicol 42:1157–1180

7. Baeumler S, Wulff D, Tagliani L, Song P (2006) A real-time quantitative PCR detection method specific to Widestrike transgenic cotton (event 281-24-236/3006-210-23). J Agric Food Chem 54:6527–6534

8. Randhawa GJ, Chhabra R, Singh M (2009) Multiplex PCR-based simultaneous amplification of selectable marker and reporter genes for screening of genetically modified crops. J Agric Food Chem 57:5167–5172

9. Randhawa GJ, Chhabra R, Singh M (2010) Decaplex and real-time PCR based detection of MON531 and MON15985 *Bt* cotton events. J Agric Food Chem 58(18):9875–9881

10. Singh KC, Ojha A, Kachru DN (2007) Detection and characterization of *cry1Ac* transgene construct in *Bt* cotton: multiplex polymerase chain reaction approach. J AOAC Int 90:1517–1525

11. Yang L, Pan A, Zhang K, Guo J, Yin C, Chen J, Huang C, Zhang D (2005) Identification and quantification of three genetically modified insect resistant cotton lines using conventional and TaqMan real-time polymerase chain reaction methods. J Agric Food Chem 53(16):6222–6229

12. Nadal A, Esteve T, Pla M (2009) Multiplex polymerase chain reaction-capillary gel electrophoresis: a promising tool for GMO screening-assay for simultaneous detection of five genetically modified cotton events and species. J AOAC Int 92(3):765–772

13. Lee SH, Kim JK, Yi BY (2007) Detection methods for biotech cotton MON15985 and MON88913 by PCR. J Agric Food Chem 55:3351–3357

14. Ding J, Jia J, Yang L, Wen H, Zhang C, Liu W, Zhang D (2004) Validation of a rice specific gene, *sucrose phosphate synthase*, used as the endogenous reference gene for qualitative and real-time quantitative PCR detection of transgenes. J Agric Food Chem 52:3372–3377

15. ISO 21569:2005 Foodstuffs – methods of analysis for the detection of genetically modified organisms and derived products – qualitative nucleic acid based methods

16. Randhawa GJ, Sharma R, Singh M (2009) Multiplex polymerase chain reaction for detection of genetically modified potato with *cry1Ab* gene. Indian J Agric Sci 79(5):368–371

17. Hardegger M, Brodmann P, Herrmann A (1999) Quantitative detection of the 35S promoter and the NOS terminator using quantitative competitive PCR. Eur Food Res Technol 209:83–87

18. Yang L, Chen J, Huang C, Liu Y, Jia S, Pan L, Zhang D (2005) Validation of a cotton-specific gene, *Sad1*, used as an endogenous reference gene in qualitative and real-time quantitative PCR detection of transgenic cottons. Plant Cell Rep 24:237–245

Chapter 12

A Simple and Rapid Method for Determining Transgenic Cotton Plants

Baohong Zhang, Hongmei Wang, Fang Liu, and Qinglian Wang

Abstract

Determining transgenic events is a critical step for obtaining transgenic plants as well as the later stage of application. Traditional methods, such as Northern blotting and qRT-PCR, for determining transgenic events either require radioactively labeled substrates, expensive instruments, or long-time commitments, which result in lab and time-consuming as well as expensive costs. These methods also require destroying the transgenic events. In this chapter, we present a simple and rapid method for determining transgenic cotton plants in both laboratory and field conditions. This method is based on the sensitivity of transgenic and non-transgenic plants to a specific chemical, such as antibiotics or herbicides. This method will facilitate the screening of transgenic events, save time, reduce cost, and speed up the application of transgenic technology on cotton breeding and production. More important, this is a nondestructive bioassay method; the transgenic plants can be transferred into greenhouse or field for the later study after the detection process.

1. Introduction

In the past several decades, genetic engineering has been making significant progress on the improvement of cotton on various aspects, particularly on cotton resistance to insects and herbicides (1–5). Cotton is among the first transgenic crops to be commercialized and has been widely adopted around the world, including the USA and China (1–5). Transgenic cotton offers cotton farmers many economic benefits, including increasing yield and decreased usage of toxic pesticide; reduced usage of pesticides also produces significant benefits for human and environmental health because of the less pesticide exposure and less residue in both field and air. Except the direct application of genetic engineering on agricultural practices, transgenic cotton is also widely used to some fundamental biological aspects, such as investigating the mechanism of

Baohong Zhang (ed.), *Transgenic Cotton: Methods and Protocols*, Methods in Molecular Biology, vol. 958,
DOI 10.1007/978-1-62703-212-4_12, © Springer Science+Business Media New York 2013

cellulose biosynthesis (6, 7) and tolerance to abiotic stress. Thus, transgenic technology is a powerful tool for improving cotton yield, quality, and tolerance to abiotic and biotic stress as well as basic biology study.

Agrobacterium-mediated, gene gun (particle bombardment), and pollen-tube pathway are three major strategies for obtaining transgenic cotton plants. After the transformation, a rigorous screening and detection process is needed to undertake for identifying transgenic plant events. Traditionally, Southern blotting, Northern blotting, PCR, qRT-PCR, and Western blotting are being used for determining the authentic transformants by assaying the existence and/or expression of the transformed target genes or reported genes. However, all these methods either require radioactively labeled substrates, expensive instruments, or long-time commitments, these methods also require to destroy the transgenic events; thus the current methods are expensive or lab and time-consuming.

In this chapter, we present a simple and rapid method for determining transgenic plants without destroying the transgenic events (8). This method is based on the sensitivity of transgenic and non-transgenic plants to a specific chemical, such as antibiotics or herbicides. This method will facilitate the screening of transgenic events, save time, reduce cost, and speed up the application of transgenic technology on cotton breeding and production. More important, this is a nondestructive bioassay method; the transgenic plants can be transferred into greenhouse or field for the later study after the detection process.

2. Materials

2.1. Laboratory Bioassay

1. Seeds of the transgenic cotton line 508. The transgenic line was generated by scientists in the Cotton Research Institute of the Chinese Academy of Agricultural Sciences through a cross-breeding process using a fine cotton cultivars and a 2,4-D-resistant transgenic cotton plants that were produced by *Agrobacterium*-mediated gene transformation using *Agrobacterium tumefaciens* LBA 4404 with the plasmid pBI 121 carrying npt II and tfd A genes. This transgenic line is resistant to both antibiotics kanamycin and herbicide 2,4-Dichlorophenoxyacetic acid (2,4-D).

2. Seeds of the non-transgenic cotton cultivar CRI 34, which was generated by scientists in the Cotton Research Institute of the Chinese Academy of Agricultural Sciences.

3. Agar.

4. Murashige and Skoog (MS) inorganic salts (9).

5. B$_5$ vitamins (10).
6. Sucrose.
7. Kanamycin. Store at –20°C.
8. 2,4-D. Store at 4°C.
9. Sterilized deionized water (ddH$_2$O).
10. Bleach.
11. Ethanol.
12. 1 N KOH.
13. 1 N HCl.
14. ProX 0.22 μm membrane filter.
15. Syringe.
16. Hood.
17. 250×30 mm test tubes.
18. 1.5 mL centrifuge tubes.
19. Filter paper.

2.2. Field Bioassay

1. Seeds of the transgenic cotton line 508.
2. Seeds of the non-transgenic cotton cultivar CRI 34.
3. 2,4-D.
4. Field suitable for growing cotton.
5. Pesticides and herbicides potentially used for pest, disease and weed control.

3. Methods

3.1. Laboratory Bioassay

3.1.1. Prepare the Medium

1. Prepare antibiotic kanamycin stock solution. 3 g/L kanamycin stock solution is prepared by adding 3 g kanamycin into 10 mL ddH$_2$O.
2. Before used, kanamycin stock solution is sterilized by filtering through a ProX 0.22 mm membrane filter into a 1.5 mL centrifuge tube in a clean hood (see Note 1).
3. Both kanamycin stock solutions are stored at –20°C (see Note 2).
4. Prepare 2,4-D stock solution. 1 g/L 2,4-D stock solution is prepared by adding 100 mg kanamycin into 100 mL ddH$_2$O.
5. The 2,4-D stock solution is stored at 4°C.
6. 100 mL screening medium (SM) is prepared, which contain MS inorganic salts and B$_5$ vitamins.
7. The pH of the media is adjusted to 5.8 using 1 N KOH or HCl.
8. Add 0.7 g agar to 100 mL medium.

9. 2,4-D is added into the SM medium to make the following final concentrations: 0, 2, 4, 6, 8, and 10 mg/L 2,4-D.

10. SM medium is autoclaved under 121°C for 15 min.

11. Add kanamycin into the SM medium to the final concentrations of 0, 75, 150, and 300 mg/L when the medium is cooled down about 60–70°C (see Note 3).

12. Aliquot the medium into the 250×30 mm test tubes with 20 mL per each tube.

13. Solidified the medium at room temperature.

3.1.2. Sterilize and Culture Cotton Seeds

All following operation should be in a clean hood.

1. Select the mature cotton seeds.

2. Manually remove the seed coats.

3. Soak the mature seeds without seed coats (kernels) in 70% ethanol for 2 min.

4. Soak the kernels in ten bleach for 15 min.

5. Rinse the seeds with sterilized ddH_2O for at least three times (see Note 4).

6. During rinsing with water, the immature seeds that floated and bad seeds are removed and discarded to ensure mature seeds for later study.

7. Dry seeds on sterilized filter paper.

8. Place seeds on the SM media with different concentrations of kanamycin or 2,4-D as well as the control without any kanamycin or 2,4-D. Two seeds are placed in each tube (see Note 5).

9. Allow the seed germinate under a 14 h day/10 h night cycle with a light intensity of 3,000 lux at 28 ± 2°C.

10. After 2 days of culture, the number of germinated seeds is recorded for each treatment for transgenic line and non-transgenic cultivar, respectively (see Note 6).

11. The germination rate is calculated.

12. After 7 days of culture, the growth of roots, hypocotyls and cotyledons, and the number of seeds converted into plants are recorded for each treatment for transgenic line and non-transgenic cultivar, respectively.

13. After 14 days of culture, the seedlings are measured for seedling height, fresh weights of roots, stems, cotyledons, and whole plants (see Notes 7 and 8).

14. Roots, stems, and cotyledons are then dried for 24 h at 121°C, and dry weights are measured.

15. Date are transformed to percentage inhibition compared to the control without kanamycin or 2,4-D.

16. Analysis of variance (ANOVA) is used to determine if there are interactions between kanamycin and 2,4-D treatment, so that a suitable concentration is identified for determining whether or not a plant is transgenic plant.

17. Based on the ANOVA, a concentration is determined for 2,4-D or kanamycin which causes 50% and 100% reduction in growth.

3.2. Field Bioassay

1. Transgenic and non-transgenic cotton are sown by conventional methods at a rate of approximately 15 seed/m of row on a 1.0 m row spacing.

2. Final plant density is about 9–10 plants/m^2.

3. Standard cultivation practices and insect control measures are used.

4. About 6 weeks after planting, 200 mg/L 2,4-D is sprayed on the transgenic line 508 and CRI 34.

5. After 2 weeks after spraying, the growth and development of cotton plants are observed. The damage of the cotton leaves is determined and recorded (see Note 9).

6. The plants whose leaves are twisted and damaged are non-resistant plants that are non-transgenic cotton (see Note 10).

7. The plants whose leaves are normal are resistant plants that are transgenic plants.

8. The numbers of resistant and non-resistant plants are recorded.

4. Notes

1. Kanamycin is unstable at high temperature. It cannot be autoclaved and only use filter to sterilize kanamycin.

2. Kanamycin must be stored at –20°C; otherwise, it is easy to be degraded.

3. As in Note 1, kanamycin is unstable at high temperature. It is easy to be degraded if immediately adding kanamycin into medium after autoclave. If letting it cooled down to room temperature, the medium will be solidified.

4. Bleach can damage the seed, cause low generation rate and later damaged leaves. Thus, bleach should be completely removed through rinsing using sterilized ddH$_2$O until no bleach odor

remained. Usually three times is enough, if not, more rinse should be performed.

5. Seed should be placed in a right direction, the bottom of seeds (roots) should face downside and it is better to insert into the medium about 2–3 mm. This way will help the seed germination and later development.

6. Kanamycin does not affect the seed germination of both transgenic and non-transgenic plants (Fig. 1); however, 2,4-D significantly inhibited seed germination of non-transgenic cotton but little effect on transgenic plants (Fig. 2).

7. Kanamycin inhibited the chlorophyll biosynthesis on non-transgenic cotton CRI 34 and resulted in albino cotyledons after cotton seed germinated. Although kanamycin also affected the transgenic line 508, the effect is significantly smaller than that on non-transgenic cotton.

8. 2,4-D significantly inhibited the growth and development of non-transgenic cotton seedlings, but no effect in low concentration and little effect under high concentration on transgenic cotton seedlings (Fig. 3).

9. 2,4-D damage cotton leaves and growth at a very low concentration. Normally, 2,4-D drafted from adjacent field spray can cause cotton leaves twisted and inhibited cotton growth and development.

10. If it is necessary, molecular detection, such as Northern blotting or qRT-PCR, is need for confirmation of the transgenic plants.

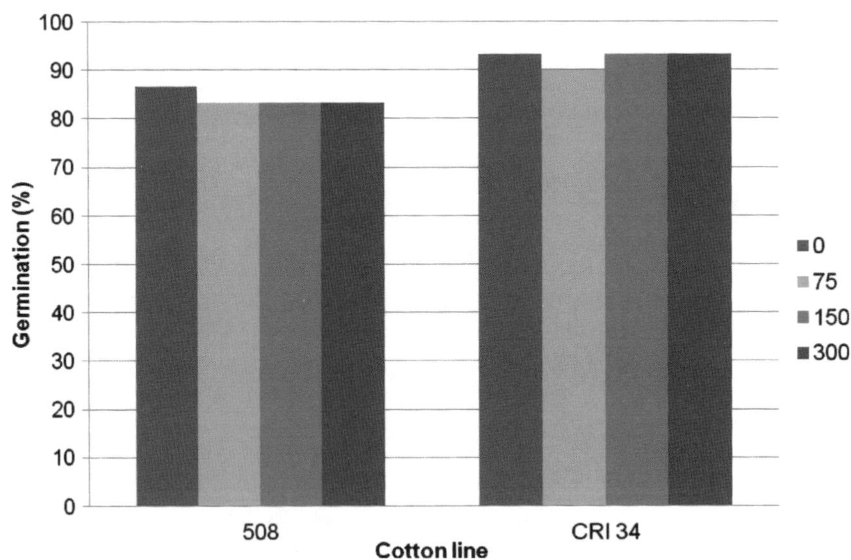

Fig. 1. Effect of kanamycin on the germination rate of transgenic cotton line 508 and non-transgenic cotton cultivar CRI 34.

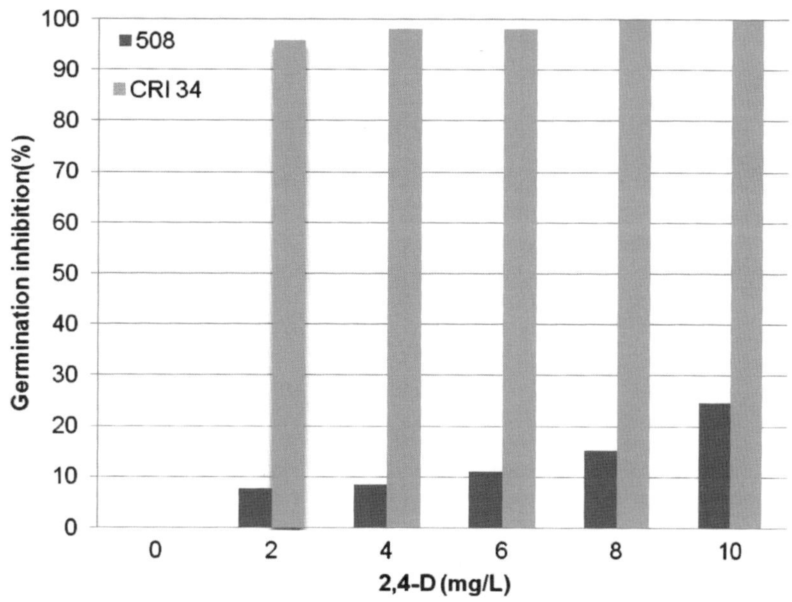

Fig. 2. Effect of 2,4-D on the seed germination of transgenic line 508 and non-transgenic cotton cultivar CRI 34.

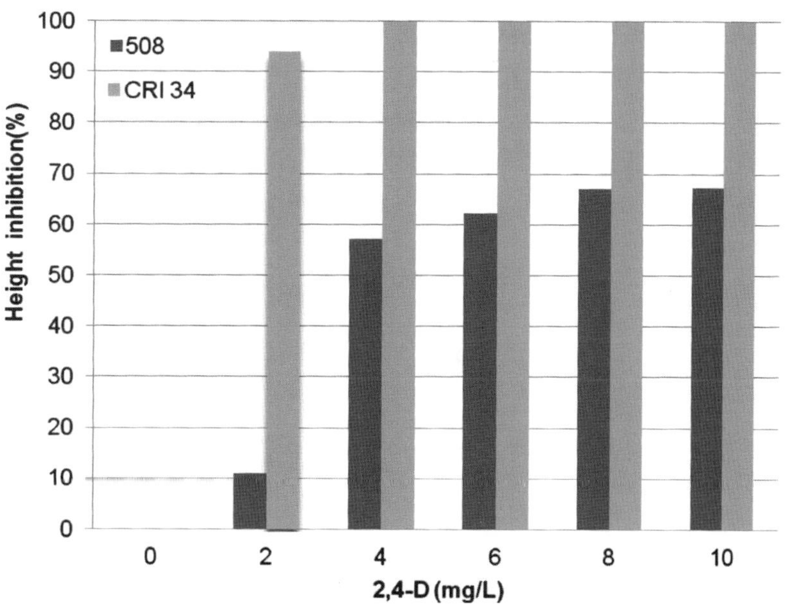

Fig. 3. Effect of 2,4-D on the seedling growth of transgenic line 508 and non-transgenic cotton cultivar CRI 34. This data shows that non-transgenic cotton is very sensitive to 2,4-D and 2 mg/L 2,4-D almost completely inhibited seedling growth and development.

References

1. Showalter AM, Heuberger S, Tabashnik BE, Carriere Y (2009) A primer for using transgenic insecticidal cotton in developing countries. J Insect Sci 9:22
2. Zhang BH, Feng R (2000) Cotton resistance to insect and transgenic cotton. China Agricultutal Science and Technology Press, Beijing
3. Zhang BH, Liu F, Yao CB, Wang KB (2000) Recent progress in cotton biotechnology and genetic engineering in China. Curr Sci 79: 37–44
4. John ME (1997) Cotton crop improvement through genetic engineering. Crit Rev Biotechnol 17:185–208
5. Wilkins TA, Rajasekaran K, Anderson DM (2000) Cotton biotechnology. Crit Rev Plant Sci 19:511–550
6. Kim HJ, Murai N, Fang DD, Triplett BA (2011) Functional analysis of *Gossypium hirsutum* cellulose synthase catalytic subunit 4 promoter in transgenic Arabidopsis and cotton tissues. Plant Sci 180:323–332
7. Wu AM, Hu JS, Liu JY (2009) Functional analysis of a cotton cellulose synthase A4 gene promoter in transgenic tobacco plants. Plant Cell Rep 28:1539–1548
8. Zhang BH, Wang HM, Liu F, Li YH, Liu ZD (2001) In vitro assay for 2,4-D resistance in transgenic cotton. In Vitro Cell Dev Biol Plant 37:300–304
9. Murashige T, Skoog F (1962) A revised medium for rapid growth and bioassays with tobacco tissue cultures. Physiol Plant 15: 473–497
10. Gamborg OL, Miller RA, Ojima K (1968) Nutrient requirements of suspension cultures of soybean root cells. Exp Cell Res 50: 151–158

Chapter 13

An Efficient Grafting Technique for Recovery of Transgenic Cotton Plants

Min Wang, Qinglian Wang, and Baohong Zhang

Abstract

Recovery of transgenic cotton plants from tissue culture condition to greenhouse condition is a critical step for improving cotton through genetic engineering. Traditional methods always cause low survival rate of transplanted plants. In 1998, we developed an efficient grafting technique for recovery of transgenic cotton plants, which significantly increased the survival rate of the transplanting regeneration plants. In this chapter, we present a detailed protocol for grafting transgenic cotton plants obtaining somatic embryogenesis.

1. Introduction

Transplanting transgenic plants from tissue culture to greenhouse and then to a field is a critical step for improving crop through genetic engineering technique. Traditional method is always involved in multiple complicated steps for transplanting regeneration plants to the greenhouse condition, which include rooting, hardening, sterilization, and transplanting. Both rooting and transplanting are difficult for cotton regeneration plants. Although it is time consuming and lab extensive, the survival rate of transplanted regenerated cotton plants is low and about more than 50% plants are died during that process. Thus, low recovery rate of transgenic cotton plants is a significant bottleneck in transgenic cotton application.

In 1998, we adopted grafting technique to regenerated transgenic cotton plants obtaining from somatic embryogenesis and obtained 70–95% of recovery rate of transgenic cotton plants (1). Since then, several other laboratories also adopted the grafting technique for transplanting regenerated plants to the greenhouse (2–8). In this chapter, we present in step by step the detailed protocol for transplanting transgenic cotton plants using grafting technique.

Baohong Zhang (ed.), *Transgenic Cotton: Methods and Protocols*, Methods in Molecular Biology, vol. 958,
DOI 10.1007/978-1-62703-212-4_13, © Springer Science+Business Media New York 2013

2. Materials

1. Transgenic cotton plants. They are obtained through somatic embryogenesis after *Agrobacterium*-mediated transformation.
2. Seeds of Beimian 3. Beimian 3 is bred by scientists at Henan Institute of Sciences and Technology.
3. Parafilm.
4. Scalpel.
5. Small pots with about 10 cm in diameter.
6. Large pots with about 30 cm in diameter.
7. Soil mixture.
8. Murashige and Skoog (MS) inorganic salts (9).
9. Plastic bags.

3. Methods

1. Prepare rootstock.
   (a) Sterilize the soil mixture under 121°C for 25 min.
   (b) Fill the small pots with the sterilized soil.
   (c) Beimian 3 is cultured in the small pots in the greenhouse.
   (d) After about 2 weeks, move the healthy Beimian 3 into a plant tissue culture room.
   (e) Remove the parts above the cotyledons.
   (f) Vertically split the stem to about 2–3 cm in depth (see Note 1).
2. Prepare scion.
   (a) Select the transgenic plants that are ready to be transplanted.
   (b) Remove the transgenic plants from the test tubes.
   (c) Keep only one opened leaf and remove the rest (see Note 2).
   (d) Cut the bottom of the transgenic plants to form a deep "V" shape (see Notes 3 and 4).
3. Grafting.
   (a) The scion is inserted into the base of the vertically slit rootstock until it fits securely.
   (b) The graft section is secured using Parafilm.

(c) Grafted plants are covered with a plastic bag or cap (see Note 5).

(d) Grafts are grown with 16 h light/8 h dark cycle with a 1,500 lux illumination under $28 \pm 2°C$.

(e) Shoots arising from the node of the rootstock are removed.

(f) Daily monitor the soil and add water and 1/2 MS medium if it is necessary (see Note 6).

(g) After 1 week, remove the plastic bag or cup from the grafts.

(h) After about 2 weeks, about two new leaves arise from the scion; grafts are translocated to the greenhouse.

(i) After another 3 weeks, plants are transferred to a large pot (see Note 7).

4. Notes

1. It is important to split the stem in the middle to make the both sides in balance. Cut the stem until the stem vascular tissue is met.

2. Multiple leaves will need more water and nutrients, which may cause the scion died because of water and nutrient limited.

3. It is very important to cut the scion in a smooth V-shaped base. Otherwise, it is hard to secure the scion and rootstock tightly.

4. The size of scions affected the survival and growth of the scions. Based on a study, small-sized (0.1–0.2 cm) scions had a lower survival rate and big-sized (0.8–1.0 cm) scions had up to 95% survival rate (3).

5. The major purpose of covering the grafted plants is to keep the humidity because the new vascular tissues have not formed between the scion and the rootstock. Water loss may cause the scion death.

6. It is very critical for the first 1 week after grafting. The humidity should be carefully monitored; low humidity usually results in low survival rate of grafts.

7. Grafting technique significantly increased the survival rate of transplanting transgenic cotton plants (Table 1). For healthy regenerated plants, the survival rate can be up to 94%.

Table 1
Application of grafting on transgenic cotton plants

Type of regenerated plants	Method for transplanting	Number of seedlings	Number of survival seedling	Survival rate (%)
Strong plants	Grafting	35	33	94.29
	Traditional method	50	24	48.00
	Direct transferring	21	0	0.00
Weak plants	Grafting	30	21	70.00
	Traditional method	27	2	7.41
	Direct transferring	15	0	0.00

References

1. Wang QL, Zhang BH, Liu F (1999) A new method for transplanting transgenic cotton plants. J Henan Sci Technol 27:27–29

2. Jin SX, Liang SG, Zhang XL, Nie YC, Guo XP (2006) An efficient grafting system for transgenic plant recovery in cotton (*Gossypium hirsutum* L.). Plant Cell Tiss Org Cult 85: 181–185

3. Luo JH, Gould JH (1999) In vitro shoot-tip grafting improves recovery of cotton plants from culture. Plant Cell Tiss Org Cult 57: 211–213

4. Wang W, Chen WX, Zhu Z, Xu HL, Gao YF, Wu P, Zhu Y, Guo ZS, Li XH (1999) Studies on highly efficient planting of transgenic cotton. Acta Bot Sin 41:1072–1075

5. Zhu SW, Sun JS (2000) Rapid plant regeneration from cotton (*Gossypium hirsutum* L.). Chinese Sci Bull 45:1771–1773

6. Zhang HS, Li JL, Zhao JL, Fu SP (2003) Study on graft technique of regenerated in cotton. China Cotton 7:29

7. Wang YX, Wang XF, Ma ZY, Zhang GY, Zhao JF (2007) A new high-efficient cotton graft technique and application. Scientia Agricultura Sinica 40:264–270

8. Wu SJ, Li FF, Zhang TZ (2006) Application of bark grafting method on grafting of regeneration cotton. Cotton Sci 18:347–351

9. Murashige T, Skoog F (1962) A revised medium for rapid growth and bioassays with tobacco tissue cultures. Physiol Plant 15:473–497

Chapter 14

Inheritance of Transgenes in Transgenic *Bt* Lines Resistance to *Helicoerpa armigera* in Upland Cotton

Baolong Zhang, Wangzhen Guo, and Tianzhen Zhang

Abstract

Six transgenic *Bt* cotton cultivars (lines) including GKsu12, GK19, MR1, GK5, 109B, and SGK1 are highly resistant to bollworm from the seedling to boll-setting stages in bioassays with detached cotton leaves, though there are differences in resistant level and Bt toxin content in these transgenic cottons. Genetics analysis reveals that the resistance to *Helicoverpa armigera* in these six transgenic *Bt* cotton cultivars (lines) are controlled by one pair of dominant genes. Allelic tests further demonstrate some populations are in Mendel segregation for two nonallelic genes, i.e., the inserted *Bt* gene in GKsu12 is nonallelic to that of SGK1, GK5, 109B, and GK19 and *Bt* genes in GK19 and SGK1 are likely inserted in the same or in close proximity (genetically closely linked), while some F_2 produce abnormal segregation patterns, with a segregation of resistance to *Helicoerpa armigera* vary between 15:1 and 3:1, though their *Bt* segregation fit into 15:1 by PCR analysis, suggesting *Bt* gene silence in these populations. Two genes silence may occur in these populations due to the homologous sequence by crossing since the silenced individuals accounted for 1/16 of the F_2 populations for allelic test. To those silenced populations, one of their parents all showed high resistance to bollworm.

1. Introduction

Cotton is an important and worldwide fiber crop that is mainly damaged by *Helicoverpa armigera* Hubner (cotton bollworm). The development of transgenic *Bt* cultivars is of paramount importance to cotton production. Since 1987, great progress has been made through transferring the CrylA(b) δ-endotoxin gene from *Bacillus thuringiensis* (Bt) to cultivars via *Agrobacterium tumefaciens* (1–3). In China, transgenic plants expressing GFM *CrylA* (*b*+*c*) gene presented excellent control of lepidopteran pests and reduces the use of broad spectrum insecticides (4, 5). However, *Bt* gene expression was influenced by many factors: gene construct, transformation methods, insertion site, gene copy number, gene structure, and environmental element (6, 7). In transgenic cereals, many published papers

Baohong Zhang (ed.), *Transgenic Cotton: Methods and Protocols*, Methods in Molecular Biology, vol. 958,
DOI 10.1007/978-1-62703-212-4_14, © Springer Science+Business Media New York 2013

165

reported that the plants generated by *Agrobacterium tumefaciens* mediated transformation usually contained a single insert and segregated in a Mendelian fashion, while those generated by direct transformation system such as particle bombardment usually resulted in integration of multiple transgene copies at single loci and segregated in a complex fashion (8, 9). The loss of the gene, poor transmission of the gene, gene rearrangement, and gene silencing was responsible for the distorted segregation ratio (8, 10–13). Transgenic CrylA cotton lines segregated as Mendelian fashion regardless of transformation methods in most cases (1, 14, 15). Only one expression deviated Mendelian segregation in backcross due to poor seed germination (16).

To understand the inheritance and interaction patterns, and to explore new germplasm resources, we analyzed five transgenic *Bt* gene lines generated via pollen-tube pathway and one transgenic *Bt* gene cultivar generated through *Agrobacterium* mediated transformation. Inheritance pattern of CrylA phenotype and genotype, the allelic relationship among six transgenic lines, the gene silencing phenomenon, and the suppress level in F_2 were presented. The results presented here will provide useful information in developing future insect management programs in China for transgenic cottons containing the CrylA (b + c) δ-endotoxin.

2. Materials

2.1. Transgenic Cottons

Transgenic cotton lines GKsu12, GK19, MR1, and GK5 which contain a single *Bt* gene and line SGK1 which containing *Bt* and *CpTI* genes were obtained from Agrobiological Genetics and Physiological Institute, Jiangsu Academy of Agricultural Sciences (see Note 1). Transgenic *Bt* lines 109B, which was developed and commercially released by Deltapline Inc., was purchased in China. Details about these six transgenic cotton cultivars (lines) are presented in Tables 1, 2, and 3.

2.2. Non-transgenic Cottons

Commercial non-transgenic cotton cultivars Tong9450 and Xu244 were obtained from Agrobiological Genetics and Physiological Institute, Jiangsu Academy of Agricultural Sciences.

2.3. Cotton Bollworm

1st instar cotton bollworm larvae are fed with detached cotton leaves in bioassay.

2.4. Other Materials

1. Incubator with temperature and humidity control.
2. 120 mm Petri dishes.
3. Commercial kanamycin sulfate available in market.
4. Cotton wool.
5. Distilled H_2O.

Table 1
Transgenic *Bt* cotton cultivars and lines used in the present research

Materials	Receptor	Inserted gene	Transformation method	Commercial status
GK19	Simian 3	*CrylA (b+c)*	Pollen-tube	Commercial
GKsu12	Sumian 12	*CrylA (b+c)*	Pollen-tube	Trial
GK5	Simian 3	*CrylA (b+c)*	Pollen-tube	Trial
MR1	Unknown	*CrylA (b+c)*	Pollen-tube	Trial
109B	Coker 312	*CrylA (b)*	*Agrobacterium*	Commercial
SGK1	Wanmian 916	*Bt+ CpTI*	Pollen-tube	Trial

Table 2
Resistance of six transgenic *Bt* cotton cultivars (lines) to bollworm

| Cotton materials | Mortality ± SE% | | | |
	Seedling stage	Squaring stage	Flowering stage	Boll-setting stage
GKsu12	89.20 ± 5.16	65.00 ± 3.16	60.20 ± 4.58	35.46 ± 3.59
GK19	85.86 ± 4.56	59.48 ± 2.97	56.25 ± 1.52	38.27 ± 3.21
SGK1	82.63 ± 2.58	70.32 ± 4.83	61.25 ± 2.74	39.84 ± 4.60
GK5	100 ± 0	100 ± 0	100 ± 0	56.32 ± 4.82
109B	95.37 ± 3.16	88.75 ± 2.80	80.76 ± 5.11	46.28 ± 3.95
MR1	100 ± 0	81.25 ± 3.10	77.25 ± 4.34	42.47 ± 5.78
Xu244(CK)	3.75 ± 1.90	2.51 ± 1.76	3.75 ± 1.91	2.46 ± 1.67

Table 3
Content of CrylA toxin protein in six transgenic lines in cotton

Cotton materials	Bt toxin ± SE at budding stage (ng/g.FW)	Dif[a]	Bt toxin ± SE at flowering stage (ng/g.FW)	Dif
GKsu12	96.19 ± 15.36	c	88.89 ± 12.33	c
GK19	83.71 ± 14.95	c	78.93 ± 18.39	c
SGK1	90.18 ± 11.09	c	87.81 ± 23.04	c
GK5	164.58 ± 26.22	a	132.18 ± 18.94	c
109B	152.24 ± 22.50	a	121.19 ± 25.04	ab
MR1	136.36 ± 14.52	b	107.13 ± 25.23	b

[a]The different lowercase represents significant difference between materials

3. Method

3.1. Insect-Resistant Test of Transgenic Cotton

1. Detach the top 4th leaf from cotton plant and put in 120 mm Petri dish, one leaf per dish.

2. Add five 1st instar larvae per leaf.

3. Keep the Petri dishes in an incubator with constant temperature at $27 \pm 1°C$ and relative humidity at 60–85%.

4. After 3 days, record the number of surviving larvae, larvae age and the leaf damage index.

5. Four degrees of leaf damage index: Degree 1: leaf damaged area below 10%, with only small needle-like damaged parts could be observed. Degree 2: damaged area about 11–50%, damaged parts distributed in small patches. Degree 3: damaged area 51–90%, but the mesophyll tissues still present in connected patches. Degree 4: damaged area over 90%, mesophyll tissues do not present in connected patches.

3.2. Kanamycin-Resistant Test of Transgenic Cotton

1. Prepare 500 mg/L kanamycin sulfate solution with commercial kanamycin sulfate and distilled H_2O.

2. Tear the cotton wool into small stripes, dip them into kanamycin sulfate solution.

3. Put the moistened cotton stripes onto the top 2nd leaf of cotton plant.

4. 7 day later, examine the syndrome of kanamycin-treated leaf (see Note 2).

3.3. Genetic Analysis of Transgenic Cotton

1. Carry out 12 crossings between six transgenic *Bt* strains and two commercial cultivars (GKsu12×Xu244, GK19×Xu244, SGK1×Xu244, 109B×Tong9450, MR1×Tong9450, GK5×Tong9450, Xu244×GKsu12, Xu244×GK19, Xu244×SGK1, Tong9450×109B, Tong9450×MR1, and Tong9450×GK5).

2. Self-cross and back-cross to acquire F_2 and BC_1 populations, respectively.

3. Insect-resistant test (see Note 3) and kanamycin-resistant test (see Note 4) with F_2 and BC_1 populations.

3.4. Allelic Test of Transgenic Cotton

1. Carry out a diallel cross among six transgenic cotton lines.

2. Self-cross F_1 (see Note 5) to obtain F_2.

3. Insect-resistant test and kanamycin-resistant test with F_2 populations (see Note 6).

3.5. Bt Silence Test of Transgenic Cotton

1. PCR (Polymerase Chain Reaction) test the presence of *Bt* gene in F_2 population segregated varied between 15:1 and 3:1 ratio (see Note 7).

2. Insect-resistant test in *Bt* gene silence individuals (see Note 8).

4. Notes

1. These transgenic *Bt* cotton lines had been bred-true for agronomic traits and resistant to bollworm by pedigree selection via individual plant bioassay and PCR analysis before using in the present research (Tables 1, 2, and 3).

2. Leaf of kanamycin-resistant plant shows no symptom, while the counterpart of kanamycin-sensitive plant shows yellow spot at the kanamycin-treated area.

3. In Insect-resistant test, the insect-resistant and susceptible plants segregate in F_2 and BC_1 populations (Table 4). The leaf damage index of susceptible plants is generally in degree 3 and 4, and larvae mortality is the same as the check cultivars Xu244 and Tong 9450. However the leaf damage index of resistant plants is in degree 1 and 2, and the larvae mortality account for 50–100% after 7 days.

4. In kanamycin-resistant test, all the treated leaves of the susceptible plants should show apparent yellow pots, while those of resistant plants remain green. Segregation of resistant and susceptible plants fits 3:1 ratio in 12 F_2 populations and 1:1 ratio in 12 BC_1 populations ($\chi_c^2 < 3.84$), suggesting that the resistance of GKsu12, GK19, MR1, GK5, 109B, and SGK1 to *Helicoverpa armigera* is controlled by one pair of dominant genes.

5. All of the F_1 hybrids from crossing among these six transgenic *Bt* lines should have good resistance level to cotton bollworm.

6. Four types of segregation present in both reciprocal cross F_2 populations by chi-square test (Table 5): Type I: Segregation value of resistant plants and susceptible plants fit into 15:1 ratio in both reciprocal cross F_2 populations, indicating independent insertion loci of *Bt* genes in the parents. *Bt* insertion locus in GKsu12 is nonallelic to those in SGK1, GK5, 109B, and GK19. *Bt* in GK19 is nonallelic to those in 109B and GK5. *Bt* in MR1 is nonallelic to those in SGK1 and 109B. Type II: No segregation in F_2, indicating *Bt* gene is possible to be inserted in the near site of a same chromosome, such as F_2 populations from reciprocal cross between GK19 and SGK1.

Table 4
Segregation of insect-resistance in F_2s crossed between transgenic *Bt* lines and commercial cultivars and BC populations

Cross combinations	No. resistant plants	No. susceptible plants	Chi-square value	*P*-value
(GKsu12 × Xu244)F_2	302	94	0.2727	0.601508
(GK19 × Xu244)F_2	250	72	1.060	0.303206
(SGK1 × Xu244)F_2	224	74	0	1
(109B × Tong9450)F_2	201	66	0.0012	0.971814
(MR1 × Tong9450)F_2	205	70	0.0109	0.916815
(GK5 × Tong9450)F_2	273	90	0.0009	0.975825
(Xu244 × GKsu12)F_2	270	95	0.1543	0.694424
(Xu244 × GK19)F_2	328	104	0.1512	0.697358
(Xu244 × SGK1)F_2	291	98	0.0008	0.976647
(Tong9450 × 109B)F_2	320	98	0.4593	0.497937
(Tong9450 × MR1)F_2	346	119	0.0581	0.809582
(Tong9450 × GK5)F_2	245	79	0.0370	0.84739
(GKsu12 × Xu244) BC_1	131	136	0.0599	0.806614
(GK19 × Xu244) BC_1	166	160	0.0767	0.781838
(SGK1 × Xu244) BC_1	115	124	0.2678	0.604823
(109B × Tong9450) BC_1	148	146	0.0034	0.953493
(MR1 × Tong9450) BC_1	123	134	0.3891	0.53277
(GK5 × Tong9450) BC_1	139	148	0.2230	0.636766
(Xu244 × GKsu12) BC_1	130	121	0.2550	0.61359
(Xu244 × GK19) BC_1	151	157	0.0812	0.77572
(Xu244 × SGK1) BC_1	149	143	0.0856	0.769826
(Tong9450 × 109B) BC_1	157	144	0.4784	0.489146
(Tong9450 × MR1) BC_1	131	134	0.0151	0.902219

Table 5

Segregation of insect-resistance in F_2 crossed among six transgenic *Bt* strains

Combinations	No. resistant plants	No. susceptible plants	Chi-square value	Probability
(GKsu12×SGK1)F_2	232	16	0	>0.900
(GKsu12×GK5)F_2	231	16	0.001	>0.900
(GKsu12×109B)F_2	286	20	0.008	>0.900
(GKsu12×GK19)F_2	294	11	3.2	0.050–0.100
(GK19×109B)F_2	264	17	0.068	0.750–0.900
(GK19×GK5)F_2	293	20	0	>0.900
(SGK1×MR1)F_2	217	13	0.057	0.750–0.900
(109B×MR1)F_2	371	24	0.001	>0.900
(SGK1×GKsu12)F_2	214	15	0.004	>0.900
(GK5×GKsu12)F_2	496	39	0.987	0.250–0.500
(109B×GKsu12)F_2	312	22	0.044	0.750–0.900
(GK19×GKsu12)F_2	236	16	0.001	>0.900
(109B×GK19)F_2	358	24	0.006	>0.900
(GK5×GK19)F_2	277	19	0	>0.900
(MR1×SGK1)F_2	278	22	0.43	0.500–0.750
(MR1×109B)F_2	331	21	0	>0.900
(SGK1×GK19)F_2	194	0	–	–
(GK19×SGK1)F_2	408	0	–	–
(GK19×MR1)F_2	344	39	9.45	<0.01
(SGK1×109B)F_2	307	45	24.34	<0.01
(SGK1×GK5)F_2	225	29	10.3	<0.01
(MR1×GK5)F_2	355	54	32.569	<0.01
(MR1×GK19)F_2	335	46	21.069	<0.01
(109B×SGK1)F_2	273	40	21.674	<0.01
(GK5×SGK1)F_2	137	24	25.00	<0.01
(GK5×MR1)F_2	407	60	33.58	<0.01
(MR1×GKsu12)F_2	313	24	0.301	0.950–0.975
(109B×GK5)F_2	292	19	0.021	0.750–0.900
(GKsu12×MR1)F_2	187	30	19.976	<0.01
(GK5×109B)F_2	232	32	14.55	<0.01

Type III: Segregation values vary between 15:1 and 3:1 ratio in both reciprocal cross F_2 and allelic relationship could not be deduced, such as (GK19×MR1) F_2, (SGK1×109B)F_2, (SGK1×GK5)F_2, and (MR1×GK5)F_2. And Type IV: One F_2 population fit into 15:1 segregation ratio, while its reciprocal cross F_2 population do not fit into 15:1 ratio, (MR1×GKsu12) F_2 fit into 15:1 ratio, while (GKsu12×MR1)F_2 do not fit into 15:1 ratio; (109B×GK5) F_2 fit into 15:1 ratio, in contrast that (GK5×109B)F_2 do not fit into 15:1 ratio.

7. In (GKsu12×MR1)F_2, (SGK1×109B)F_2, and (MR1×GK19) F_2, all the individuals are tested by PCR, while in (GK19×MR1) F_2, (SGK1×GK5)F_2, (109B×SGK1)F_2, (MR1×GK5)F_2, (GK5×MR1)F_2, (GK5×109B)F_2, and (GK5×SGK1)F_2, only the plants susceptible to bollworm and Kanamycin are checked by PCR. PCR results demonstrate that some bollworm-susceptible individuals contain the *Bt* gene, suggesting the silence of *Bt* gene in these individuals. However, the segregation of *Bt* gene fit into 15:1 ratio ($\chi_c^2 < 3.84$), indicating individual insertions of *Bt* gene among these transgenic cotton (Table 6).

8. Sixteen bollworm-susceptible individuals containing *Bt* gene from (GKsu12×MR1)F_2 are subjected to insect-resistant test. After 7-day culture, in contrast to high resistance to bollworm in resistant parent, damage index of leaves from bollworm-susceptible individuals is in degree 3 or degree 4, larvae mortality is 0 or very low, and 3rd larvae survive, the same as negative check control (Table 7), indicating *Bt* gene silence in these F_2 individuals.

Acknowledgments

We thank Dr. John Z Yu, USDA-ARS, Southern Plains Agriculture Research Center, Crop Germplasm Research Unit, USA, for the critical reviewing this manuscript. This project was financially supported in part by the National Transgenic R&D Project in China (2011ZX08005-001) and the Priority Academic Program Development of Jiangsu Higher Education Institutions.

Table 6
Segregation of bollworm resistance in bioassay and *Bt* gene by PCR amplification in F_2 populations

Combinations	No. resistant plants	No. susceptible plants	Chi-square value (15:1)	*P*-value	No. plants with *Bt*	No. plants without *Bt*	Chi-square value	*P*-value
$(GK19 \times MR1)F_2$	344	39	9.450	0.0021115	362(344 + 18[a])	21(39 − 18[a])	0.265	0.6067057
$(SGK1 \times 109B)F_2$	307	45	24.340	8.074E-07	330(307 + 23[a])	22(45 − 23[a])	0	1
$(SGK1 \times GK5)F_2$	225	29	10.300	0.0013303	238(225 + 13[a])	16(29 − 13[a])	0.009	0.9244194
$(MR1 \times GK5)F_2$	355	54	32.569	1.15E-08	381(355 + 26[a])	28(54 − 26[a])	0.157	0.6919337
$(MR1 \times GK19)F_2$	335	46	21.069	4.43E-06	359(335 + 24[a])	22(46 − 24[a])	0.077	0.7814049
$(109B \times SGK1)F_2$	273	40	21.674	3.231E-06	284(273 + 21[a])	19(40 − 21[a])	0	1
$(GK5 \times SGK1)F_2$	137	24	25.000	5.733E-07	148(137 + 11[a])	13(24 − 11[a])	0.630	0.4273553
$(GK5 \times MR1)F_2$	407	60	33.580	6.839E-09	435(407 + 28[a])	32(60 − 28[a])	0.195	0.6587873
$(MR1 \times GKsu12)F_2$	313	24	0.301	0.5832562	–	–	–	–
$(GKsu12 \times MR1)F_2$	187	30	19.976	7.842E-06	203(187 + 16[a])	14(30 − 16[a])	0	1
$(109B \times GK5)F_2$	292	19	0.021	0.884779	–	–	–	–
$(GK5 \times 109B)F_2$	232	32	14.550	0.0001365	247(232 + 15[a])	17(32 − 15[a])	0	1

[a]Individuals susceptible to bollworm and Km by bioassay but with *Bt* gene inserted by PCR analysis

Table 7

The insect-resistant performance of individuals with unexpressed *Bt* gene in (GKsu12 × MR1)F$_2$

| | 3-Days results fed with larvae | | | | | 7-Days results fed with larvae | | | | |
| | Mortality larvae | Survived larvae | | Damaged degree | Mortality instar % | Mortality larvae | Survived larvae | | Damaged degree | Mortality instar % |
		1st instars	2nd instar				3rd instar	4th instar		
B30-26			3	4	0		3		4	0
B30-40			4	4	0		4		4	0
B30-74			4	4	0	1	3		4	25
B30-79			3	4	0	1	1		4	50
B30-80			4	4	0		3		4	0
B30-81			2	3	0		2		4	0
B30-96			4	4	0		3		4	0
B30-107			4	4	0		3		4	0
B30-112		1	3	4	0		2		4	0
B30-117			3	3	0		1		4	0
B30-139			4	4	0		2		4	0
B30-143	1		3	3	25	0	3		4	0
B30-175		2	1	4	0	0	1	1	4	0

B30-187			3	3	0	0	2	1	4	0
B30-190			4	3	0	0	3		4	0
B30-217			4	4	0	0	3		4	0
Xu244		0	3	4	0		2		4	0
Xu244			4	4	0		2		4	0
Xu244			3	4	0		1		4	0
Xu244			3	3	0	1	2		3	33.33
Xu244			3	4	0	1	2		3	33.33
MR1	5				1	100				
GKsu12	4	1			1	80	1		1	100

Xu244 is a commercial cultivar as CK

References

1. Hofte H, Whiteley HR (1989) Insecticidal crystal proteins of *Bacillus thuringiensis*. Microbiol Rev 53(242):255

2. Benedict JH, Sachs ES, Altman DW, Ring DR, Stone TB, Sims SR (1993) Impact of endoxin-producing transgenic cotton on insect-plant interaction with *Heliothis virescens* and *Helicoverpa zea* (Lepidoptera: Noctuidae). Environ Entomol 22(1):9

3. James C (2010) Global status of commercialized biotech/GM crops: 2010. ISAAA brief no. 42. ISAAA, Ithaca, NY

4. Sun J, Tang CM, Zhu XF, Guo WZ, Zhang TZ, Zhou WJ, Meng FX, Sheng JL (2002) Characterization of resistance to *Helicoverpa armigera* in three lines of transgenic *Bt* upland cotton. Euphytica 123(343):351

5. Wu K, Lu Y, Feng H, Jiang Y, Zhao J (2008) Suppression of cotton bollworm in multiple crops in China in areas with Bt toxin-containing cotton. Science 321:1676–1678

6. Dong H, Li W (2007) Variability of endotoxin expression in Bt transgenic cotton. J Agron Crop Sci 193:21–29

7. Tabashnik BE, Sisterson MS, Ellsworth PC, Dennehy TJ, Antilla L, Liesner L, Whitlow M, Staten RT, Fabrick JA, Unnithan GC (2010) Suppressing resistance to Bt cotton with sterile insect releases. Nat Biotechnol 28:1304–1307

8. Register JC, Peterson DJ, Bell PJ, Bullock WP, Evans IJ, Frame B, Greenland AJ, Higgs NS, Jepson I, Jiao S, Lewnau CJ, Sillick JM, Wilson HM (1994) Structure and function of selectable and non-selectable transgenes in maize after introduction by particle bombardment. Plant Mol Biol 25(951):961

9. Cheng M, Fry JE, Pang S, Zhou H, Hironaka CM, Duncan DR, Conner TW, Wan Y (1997) Genetic transformation of wheat mediated by *Agrobacterium tumfaciens*. Plant Physiol 115(971):980

10. Spencer TM, O'Brien JV, Start WG, Adams TR, Gordon-Kamm WJ, Lemaux PG (1992) Segregation of transgenes in maize. Plant Mol Biol 18(201):210

11. Pawlowski WP, Somers DA (1996) Transgene inheritance in plants genetically engineered by microprojectile bombardment. Mol Biotechnol 6(17):30

12. Pawlowski WP, Torbert KA, Rines HW, Somers DA (1998) Irregular patterns of transgene silencing in allohexaploid oat. Plant Mol Biol 38:597–607

13. Morino K, Olsen OA, Shimamoto K (1999) Silencing of an aleurone-specific gene in transgenic rice is caused by a rearranged transgene. Plant J 17(275):85

14. Tang CM, Sun J, Zhu XF, Guo WZ, Zhang TZ, Shen JL, Gao CF, Zhou WJ, Chen ZX, Guo SD (2000) Inheritance of resistance to *Helicoverpa armigera* of 3 kinds of transgenic *Bt* strains available in upland cotton in China. Chinese Sci Bull 45(363):367

15. Bao-hong Z, Teng-long G, Qing-lian W (2000) Inheritance and segregation of exogenous genes in transgenic cotton. J Genet 79:71–75

16. Sachs ES, Bendict JH, Stelly DM, Taylor JF, Altman DW, Berbrich SA, Devis SK (1998) Expression and segregation of gene encoding Bt insecticidal proteins in cotton. Crop Sci 38:1–11

Part IV

Application

Agrobacterium rhizogenes-Induced Cotton Hairy Root Culture as an Alternative Tool for Cotton Functional Genomics

Hee Jin Kim

Abstract

Although well-accepted as the ultimate method for cotton functional genomics, *Agrobacterium tumefaciens*-mediated cotton transformation is not widely used for functional analyses of cotton genes and their promoters since regeneration of cotton in tissue culture is lengthy and labor intensive. In certain cases, *A. rhizogenes*-induced hairy root culture has been a suitable molecular tool for functional analyses of genes and promoters for plants that are difficult to regenerate by *A. tumefaciens*-mediated transformation. Similarly, *A. rhizogenes*-induced hairy root cultures are an alternative tool for cotton functional genomics. In this chapter, the advantages and disadvantages of using *A. rhizogenes*-induced cotton hairy root culture over *A. tumefaciens*-mediated cotton transformation are discussed. The procedures for transformation, generation, selection, and molecular analyses of transgenic cotton hairy roots are introduced by describing the functional analysis of a cotton promoter in cotton hairy roots generated by *A. rhizogenes*-mediated transformation.

1. Introduction

Two *Agrobacterium* strains have been widely used for plant genetic engineering due to their ability to transfer their own T-DNA into plant genomes. One is *Agrobacterium tumefaciens* that transfers "tumor-inducing (Ti) T-DNA" into plant genomes and develops a crown gall, and the other is *Agrobacterium rhizogenes* that transfers "root-inducing (Ri) T-DNA" into plant genome and produces hairy roots (1). *A. tumefaciens*-mediated transformation has been the method of choice for regenerating transgenic plants, including cotton. Along with the development of convenient binary vector systems, *A. tumefaciens*-mediated transformation has become a standard method for functional analyses of genes and promoters in some plant species that can be transformed by a relatively quick

Baohong Zhang (ed.), *Transgenic Cotton: Methods and Protocols*, Methods in Molecular Biology, vol. 958, DOI 10.1007/978-1-62703-212-4_15, © Springer Science+Business Media New York 2013

transformation method like the floral dip technique. The lack of a quick and simple *A. tumefaciens*-mediated transformation procedure has been a major obstacle for cotton scientists who are trying to characterize cotton genes and their promoter activities. To circumvent lengthy and labor-intensive tissue culture procedures for *A. tumefaciens*-mediated cotton transformation, many cotton scientists have performed functional analyses of cotton genes and their promoters using heterologous plants that can be quickly regenerated (2–4) despite having a limited understanding of the potential similarities/dissimilarities between cotton and heterologous plants.

A. rhizogenes-induced hairy root culture is widely used in functional genomics for other plants, especially plants requiring lengthy and labor-intensive tissue culture procedures through *A. tumefaciens*-mediated transformation. In addition to the morphological similarity of hairy roots with normal roots, the rapid and simple transformation procedure of *A. rhizogenes*-induced hairy root culture is the major advantage of this approach. *A. rhizogenes* can cotransform efficiently both Ri T-DNA and the gene of interest. *A. rhizogenes*-induced hairy root culture was successfully used for RNAi-based functional analyses of genes in both *Arabidopsis* and *Medicago* (5). In addition, *A. rhizogenes*-mediated transformation was used to study tissue-specific expression patterns of promoters (6) and to determine the subcellular localization of proteins in many other plants (7, 8). A comparison of the spatial expression of promoter activity was reported in both hairy roots and normal roots of transgenic plants (9).

A. rhizogenes-mediated transformation has rarely been used for functional analyses of cotton genes and their promoters since a method for regenerating transgenic cotton plants from *A. rhizogenes*-transformed roots has not yet been developed. Cotton hairy roots have been used to produce secondary metabolites having potential pharmaceutical value (10–12) and to serve as hosts for nematodes (13). Recently, our group used *A. rhizogenes*-induced hairy root culture for analyzing the promoter activity of *GhCesA4*, a cellulose synthase subunit that is involved in producing cellulose in the secondary cell walls of cotton xylem tissues as well as cotton fibers (14). The quick and simple *A. rhizogenes*-mediated transformation procedure allowed us to characterize tissue-specific expression patterns of the *GhCesA4* promoter in cotton hairy roots.

There are three advantages to using *A. rhizogenes*-induced hairy root culture over the *A. tumefaciens*-mediated cotton transformation for characterizing cotton genes. The first advantage is the rapid and efficient procedure for generating transgenic cotton tissues. Compared with *A. tumefaciens*-mediated transgenic cotton transformation requiring 6 months to 2 years for regenerating multiple transgenic cotton lines (15), *A. rhizogenes*-mediated transformation resulted in dozens of transgenic cotton hairy root lines within a few weeks (14, 16). The second advantage is the seemingly unlimited host range of *A. rhizogenes*. Unlike *A. tumefaciens*-mediated transformation that limited to a few highly regenerable cotton varieties like *G. hirsutum* cv. Coker 310, 312, and 315 (15), *A. rhizogenes*-mediated transformation can be used for most cotton species and varieties (10). The third advantage is the technical simplicity of the tissue culture since *A. rhizogenes*-induced hairy root culture does not require phytohormones for producing cotton hairy roots (16). The major drawback of using cotton hairy root culture for functional analyses of cotton genes is that the research must be limited to genes that are expressed in cotton root tissues. Therefore, *A. rhizogenes*-induced hairy root culture is a suitable alternative molecular tool for functional analyses of cotton genes that are expressed in cotton roots.

In this chapter, the methods of transforming a binary vector containing the β-*glucuronidase* (*GUS*) reporter gene regulated by the *GhCesA4* promoter (*pGhCesA4::GUS*) into *A. rhizogenes*, producing cotton hairy roots, screening for transgenic cotton hairy roots co-transformed by both *pGhCesA4::GUS* and Ri T-DNA, and analyzing *GhCesA4* promoter activity in hairy cotton roots are described.

2. Materials

1. Cotton seeds (*Gossypium hirsutum*, cv. Texas Marker-1).

2. *Agrobacterium rhizogenes* #15834 (see Note 1).

3. pCAMBIA binary vector 1391z containing a *GUS* reporter fused to a *GhCesA4* promoter (−898/+56) or promoterless pCAMBIA binary vector 1391z (14).

4. Germination medium : a packet of MS salts plus vitamins (Sigma-Aldrich, St. Louis, MO, USA) per L, adjust pH to 6.0 with 1.0 M KOH, add 2.5 g Phytagel. Autoclave for 15–20 min. Cool to 65°C and pour into autoclaved Magenta boxes (Sigma-Aldrich, St. Louis, MO, USA).

5. Hairy Root Induction Media (HRIM): a packet of MS salts plus vitamins (Sigma) per L, adjust pH to 5.8 with 1.0 M KOH, add 30 g/L sucrose and 7 g/L agar. Autoclave 15–20 min. Let cool to 65°C. Pour into 100×25 mm Petri plates.

6. YEB medium: 0.5% sucrose, 0.5% peptone, 0.5% beef extract, 0.1% yeast extract, 0.049 g/L $MgSO_4 \cdot 7H_2O$, pH 7.2.

7. Tris-ethylenediaminetetraacetic acid (TE): 10 mM Tris–HCl, pH 7.5, 1 mM EDTA, pH 8.0.

8. E.Z.N.A. Plant DNA Kit (Omega Bio-Tek, Inc., Norcross, GA, USA).

9. Histochemical staining solution: 0.1 M potassium phosphate buffer pH 7.2, 0.5 mM potassium ferricyanide, 0.5 mM potassium ferrocyanide, 0.1% Triton X-100, and 0.5 g/L X-Gluc (see Note 2).

10. X-Gluc: 5-bromo-4-chloro-3-indolyl-β-D-glucuronide (Gold Biotechnology, St. Louis, MO, USA) (see Note 3).

11. Hygromycin B (Life Technologies).

12. 30% Bleach and 0.01% Triton X-100.

13. 70% Ethanol solution.

14. Whatman #1 filter paper.

15. Sterile toothpicks.

16. Parafilm.

17. Petri dishes (100×15 mm and 100×25 mm).

18. Electrophoresis apparatus.

19. Thermal cycler and PCR accessories.

3. Methods

Multiple transgenic cotton hairy lines that were co-transformed by both Ri T-DNA and *pGhCesA4::GUS* in pCAMBIA 1391z binary vector (17) were generated by *A. rhizogenes*-mediated transformation. The promoter activity was analyzed by a histochemical GUS assay (18).

3.1. Transformation of Binary Vector into A. rhizogenes

Binary vectors containing *pGhCesA4::GUS* were transformed into *A. rhizogenes* by a freeze-thaw transformation method (19). In addition, binary vectors can be transformed into *A. rhizogenes* using an electroporation method (20).

1. Pick up a single colony of *A. rhizogenes* with a sterile loop and inoculate 5 ml of liquid YEB at 28°C for 16–24 h.

2. Inoculate 1 ml of cultured A. *rhizogenes* from step 1 into 100 ml YEB and shake for 5 h at 28°C.

3. Centrifuge *A. rhizogenes* at 3,000×*g* for 20 min at 4°C.

4. Wash the pellet with 30 ml of cold and sterile TE.

5. Centrifuge *A. rhizogenes* at 3,000×*g* for 10 min at 4°C.

6. Resuspend the washed pellet with 10 ml of fresh YEB medium and dispense into 500 µl aliquots.

7. Add 1–2 µg of pCAMBIA 1391z binary vector DNAs into the freshly made *A. rhizogenes* competent cells (see Note 4).

8. Incubate on ice for 5 min.

9. Freeze the mixture in liquid nitrogen for 5 min.

10. Thaw in a water bath at 37°C for 5 min.

11. Add 1 ml of YEB and incubate at 28°C for 8–12 h with gentle shaking.

12. Plate 100–500 µl onto YEB plates with appropriate antibiotic selection and incubate at 28°C for 2–3 days (see Note 5).

13. Select a single colony for PCR to confirm presence of the transformed binary vector (see Note 6).

3.2. Preparation of Sterile Cotton Seedlings

1. Sterilize delinted cottonseeds (*Gossypium hirsutum*, Texas Marker-1) by stirring them with solution containing 30% bleach and 0.01% Triton X-100 for 15 min (see Note 7).

2. Rinse the seeds for 1 min with a 70% ethanol solution (see Note 8).

3. Wash the seeds three times with sterile water.

4. Put three to five sterile seeds on the germination media in each Magenta box.

5. Incubate the seeds at 24–28°C with 16 h light/8 h darkness for 1–2 weeks.

3.3. Inoculation of Cotton Cotyledons with A. rhizogenes

1. Inoculate the selected *A. rhizogenes* containing the binary vectors into 5 ml of YEB and shake at 28°C for 24 h.

2. Centrifuge the cells at $4,000 \times g$ for 5 min and remove most of the media.

3. Resuspend the pellet with 500 µl of YEB and transfer to a sterile microcentrifuge tube.

4. Remove the cotyledonary leaves of 1–2 week old cotton seedlings and place each cotyledon pair in a 100×15 mm Petri dish containing two pieces of autoclaved Whatman filter #1 paper moistened with sterile water.

5. Stab the vascular tissue (leaf veins) of cotyledon eight to ten times using a fresh toothpick dipped in the concentrated *A. rhizogene*. (Fig. 1a).

3.4. Generation of Cotton Hairy Roots

The generated hairy roots are morphologically very similar in structure to wild-type roots; however hairy roots show a higher incidence of lateral branching and grow in an agravitropic manner.

1. Seal the Petri dish with Parafilm and place in a plant growth chamber at 24–27°C with 16 h light for 2–3 days.

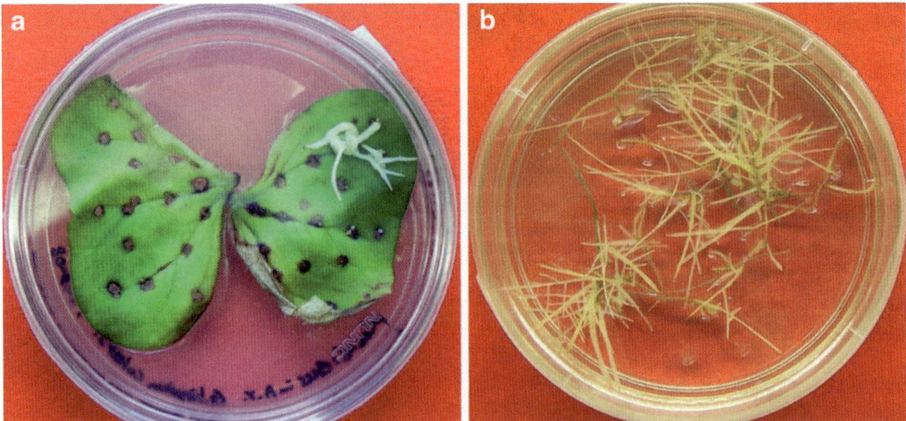

Fig. 1. Production of cotton hairy roots by *A. rhizogenes*-mediated transformation. (**a**) Cotton hairy roots developed at the *A. rhizogenes* infection sites on cotyledonary leaves. (**b**) Selected hairy roots were cultured on HRIM containing no antibiotics and no phytohormones.

2. Transfer the *A. rhizogenes*-inoculated cotyledon pairs to fresh 100×25 mm Petri dishes containing solid HRIM medium (see Note 9).

3. Incubate until hairy roots are visible, usually within 2 weeks (Fig. 1a).

4. Transfer the hairy roots longer than 1 cm onto HRIM medium containing hygromycin B (10–20 mg/L) to select cotton hairy roots containing pCAMBIA1391z (see Note 10).

5. Continue transferring the selected hairy roots every 2–3 weeks to fresh HRIM plates containing no antibiotics (Fig. 1b).

3.5. Verification of Transgenic Cotton Hairy Roots

The selected hairy roots can be confirmed by either PCR analysis or histochemical GUS assay. Since the *rol* (root oncogenic loci) genes located in Ri T-DNA of *A. rhizogenes* are responsible for production of hairy roots (21), PCR amplification of *rolB* or *rolC* is routinely used to confirm transfer of the *A. rhizogenes* Ri T-DNA into hairy roots. In addition, PCR analysis of the gene (*pCesA4::GUS*) in the pCAMBIA 1391z binary vector transformed into hairy roots needs to be conducted. Approximately 30% of the generated hairy roots were co-transformed by Ri T-DNA and *pCesA4::GUS*.

3.5.1. PCR Analysis of Transgenic Cotton Hairy Roots

1. Isolate genomic DNA from the selected hairy roots by the method provided with the E.Z.N.A. Plant DNA Kit.

2. Conduct PCR with 100 ng genomic DNA and appropriate concentrations of primers and dNTPs using the PCR conditions for *rolB* and *rolC* (see Note 11).

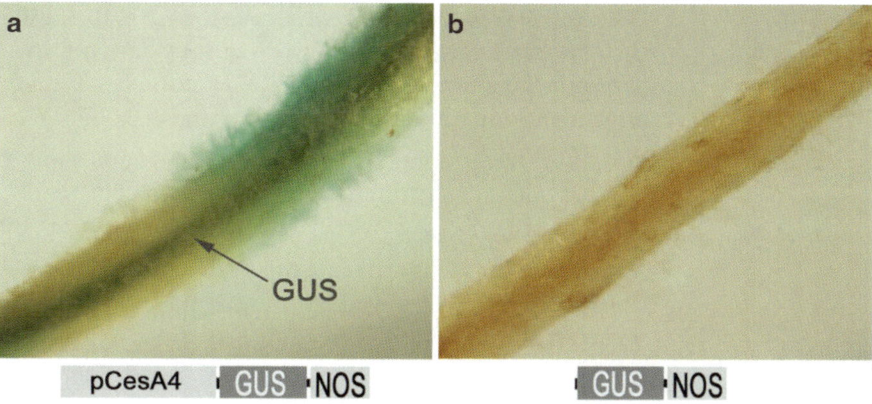

Fig. 2. Histochemical analyses of cotton hairy roots transformed by a *GUS* reporter regulated by a *GhCesA4* promoter and the null promoter construct. (**a**) The blue color represents GUS activity regulated by the *GhCesA4* promoter (−898/+56) in cotton hairy roots. (**b**) As a negative control, the promoterless pCAMBIA 1391z was transformed into cotton hairy roots.

3. Run 10 µl of the PCR reaction on a 0.8% agarose gel and confirm the size of the amplified PCR product (see Note 12).

3.5.2. Histochemical GUS Assays of Transgenic Cotton Hairy Roots

1. Incubate the selected hairy roots with a histochemcial staining solution at 37°C for 16 h.

2. Remove X-Gluc staining solution and add 70% ethanol.

3. Examine the hairy roots under a dissecting microscope (see Note 13) (Fig. 2a, b).

4. Notes

1. *A. rhizogenes* (catalog #15834) was purchased from American Type Culture Collection (ATCC), Manassas, VA or http://www.atcc.org.

2. Potassium ferri/ferrocyanide was added to the histochemical staining buffer to prevent diffusion of the GUS product during staining.

3. X-Gluc is not easily dissolved in aqueous solution. To dissolve X-Gluc, 50 mg of X-Gluc was first dissolved with 1 ml of *N,N*-dimethylformamide and subsequently 99 ml of the histochemical staining buffer was added.

4. *A. rhizogenes* #15834 competent cells that were freshly made always showed good competency. Once frozen, and then thawed *A. rhizogenes* competent cells do not usually show good transformation efficiency.

5. The transformed *A. rhizogenes* with pCAMBIA binary vector 1391z was selected from YEB medium containing kanamycin (50 mg/L) since pCAMBIA 1391z contains a gene for kanamycin resistance for selection in bacteria.

6. For long-term storage of the transformed microbe, freeze-dry 1 ml aliquots of the microbe in YEB liquid culture. *A. rhizogenes* #15834 does not survive well in glycerol stocks at –80°C.

7. Linted cottonseeds are hard to sterilize. Fibers from linted cottonseeds can be removed by incubating with concentrated sulfuric acid followed by rinsing in sterile water.

8. Longer incubation of cottonseeds with 70% ethanol may reduce germination efficiency.

9. Differentiation of cotton hairy roots occurs on HRIM containing no phytohormones.

10. Hygromycin B was added to the HRIM since pCAMBIA 1391z contains a gene for hygromycin resistance for selection in plants. Carbenicillin (500 mg/L) can be added to the HRIM to kill residual *A. rhizogenes* from the medium and plant tissues.

11. The primers used for *rolB* gene were 5′-GCTCTTGCAGTG CTAGATTT-3′/5′-GAAGGTGCAAGCTACCTCTC-3′, and *rolC* 5′-CTCCTGACATCAAACTCGTC-3′/5′-TGCTTCGA GTTATGGGTACA-3′. PCR conditions for *rolB* and *rolC* are 95°C for 30 s, 52°C for 30 s, and 72°C for 60 s.

12. The expected sizes of PCR-amplified *rolB* and *rolC* genes from hairy root are 422 bp and 625 bp, respectively.

13. The blue color represents the promoter activity of *GhCesA4*.

Acknowledgments

This work was supported by United States Department of Agriculture-Agricultural Research Service project 6435-21000-016-00D. The author acknowledges Stephanie Moss, Cheryl Frankfater, Barbara Triplett, and Michael Dowd of USDA-ARS-Southern Regional Research Center who initiated cotton hairy root cultures to produce gossypol compounds. The author also thanks Dr. Barbara Triplett for critically reading manuscript.

References

1. Veena V, Taylor CG (2007) *Agrobacterium rhizogenes*: recent developments and promising applications. In Vitro Cell Dev Biol Plant 43:383–403

2. Wang S, Wang JW, Yu N, Li CH, Luo B, Gou JY, Wang LJ, Chen XY (2004) Control of plant trichome development by a cotton fiber MYB gene. Plant Cell 16:2323–2334

3. Delaney SK, Orford SJ, Martin-Harris M, Timmis JN (2007) The fiber specificity of the cotton FSltp4 gene promoter is regulated by an AT-rich promoter region and the AT-Hook

transcription factor GhAT1. Plant Cell Physiol 48:1426–1437

4. Guan X, Lee JJ, Pang M, Shi X, Stelly DM, Chen ZJ (2011) Activation of *Arabidopsis* seed hair development by cotton fiber-related genes. PLoS One 6:e21301

5. Limpens E, Ramos J, Franken C, Raz V, Compaan B, Franssen H, Bisseling T, Geurts R (2004) RNA interference in *Agrobacterium rhizogenes*-transformed roots of *Arabidopsis* and *Medicago truncatula*. J Exp Bot 55:983–992

6. Nontachaiyapoom S, Scott PT, Men AE, Kinkema M, Schenk PM, Gresshoff PM (2006) Promoters of orthologous *Glycine max* and *Lotus japonicus* nodulation autoregulation genes interchangeably drive phloem-specific expression in transgenic plants. Mol Plant Microbe Interact 20:769–780

7. Moreno-Valenzuela OA, Minero-Garcia Y, Brito-Argaez L, Carbajal-Mora E, Echeverria O, Vazquez-Nin G et al (2003) Immunocytolocalization of tryptophan decarboxylase in *Catharanthus roseus* hairy roots. Mol Biotechnol 23:11–18

8. Marjamaa K, Hilden K, Kukkola E, Lehtonen M, Holkeri H, Haapaniemi P et al (2006) Cloning, characterization and localization of three novel class III peroxidases in lignifying xylem of Norway spruce (*Picea abies*). Plant Mol Biol 61:719–732

9. Grønlund M, Roussis A, Flemetakis E, Quaedvlieg NEM, Schlaman HRM, Umehara Y et al (2005) Analysis of promoter activity of the early nodulin Enod40 in *Lotus japonicas*. Mol Plant Microbe Interact 18:414–427

10. Triplett BA, Moss SC, Bland JM, Dowd MK (2008) Induction of hairy root cultures from *Gossypium hirsutum* and *Gossypium barbadense* to produce gossypol and related compounds. In Vitro Cell Dev Biol Plant 44:508–517

11. Frankfater CR, Dowd MK, Triplett BA (2009) Effect of elicitors on the production of gossypol and methylated gossypol in cotton hairy roots. Plant Cell Tiss Org Cult 98:341–349

12. Verma PC, Trivedi I, Singh II, Shukla AK, Kumar M, Upadhyay SK et al (2009) Efficient production of gossypol from hairy root cultures of cotton (*Gossypium hirsutum* L.). Curr Pharm Biotechnol 10:691–700

13. Wubben M, Callahan FE, Triplett BA, Jenkins JN (2009) Phenotypic and molecular evaluation of cotton hairy roots as a model system for studying nematode interactions. Plant Cell Rep 28:1399–1409

14. Kim HJ, Murai N, Fang DD, Triplett BA (2011) Functional analysis of *Gossypium hirsutum* cellulose synthase catalytic subunit 4 promoter in transgenic *Arabidopsis* and cotton tissues. Plant Sci 180:323–332

15. Trolinder N (2010) Cotton regeneration. In: Stewart JM, Oosterhuis DM, Heitholt JJ, Mauney JR (eds) Physiology of cotton. Springer, New York, pp 379–382

16. Chabaud M, Boisson-Dernier A, Zhang J, Taylor CG, Yu O, Barker DG (2006) *Agrobacterium rhizogenes*-mediated root transformation. In: Mathesius U, Journet EP, Sumner LW (eds) Medicago truncatula handbook. ISBN 0-9754303-1-9. http://www.noble.org/MedicagoHandbook/

17. Roberts CS, Rajagopal S, Smith LA, Nguyen TA, Yang W, Nugroho S, Ravi KS, Cao M, Vijhayachandra K, Patell V, Harcourt RL, Dransfield L, Desamero N, Slamet I, Keese P, Kilian A, Jefferson RA (1998) A comprehensive set of modular vectors for advanced manipulations and efficient transformation of plants by both *Agrobacterium* and direct DNA uptake methods. pCAMBIA vector release manual version 3.05

18. Jefferson RA, Kayanagh TA, Bevan MW (1987) GUS fusion: β-*glucuronidase* as a sensitive and versatile gene fusion marker in higher plants. EMBO J 6:3901–3907

19. Höfgen R, Willmitzer L (1988) Storage of competent cells for *Agrobacterium* transformation. Nucleic Acids Res 16:9877

20. Nagel R, Elliott A, Masel A, Birch RG, Manners JM (1990) Electroporation of binary Ti plasmid vector into *Agrobacterium tumefaciens* and *Agrobacterium rhizogenes*. FEMS Microbiol Lett 67:325–328

21. Nilsson O, Olsson O (1997) Getting to the root: the role of the *Agrobacterium rhizogenes rol* genes in the formation of hairy roots. Physiol Plant 100:463–473

Chapter 16

Overexpression of miR 156 in Cotton via *Agrobacterium*-Mediated Transformation

Baohong Zhang, Min Wang, Xin Zhang, Chengqi Li, and Qinglian Wang

Abstract

microRNAs (miRNAs) are an extensive class of newly identified endogenous small regulatory molecules. Many studies show that miRNAs play a critical role in almost all biological and metabolic progresses through targeting protein-coding genes for mRNA cleavage or translation inhibition. Many miRNAs are also identified from cotton using computational and/or experimental approaches, including the next generation deep sequencing technology. However, the function of the majority of miRNAs are unclear. In this chapter, we describe a detailed method for overexpressing miR 156 in cotton using *Agrobacterium*-mediated genetic transformation. This provides an approach to investigate the function and regulatory mechanism of miRNAs in cotton.

1. Introduction

microRNAs (miRNAs) are an extensive class of newly discovered endogenous small regulatory RNAs (1, 2). miRNAs exist in almost all organisms, including plants and animals (3, 4). Many miRNAs are highly evolutionarily conserved from species to species, such as from moss to high flowering crops in plants (5). miRNAs have become one of the most important gene regulators, which regulate gene expression by targeting mRNAs for degradation or for translation inhibition (1–4, 6). Elevating evidences demonstrate that miRNAs play a critical role in almost all biological and metabolic process, including organ differentiation, developmental timing, phase change from vegetative growth to reproductive growth, signaling transduction, and response to environmental biotic and abiotic stress (1–4, 7–11).

Cotton is the most important textile and cash crops. Recently, many miRNAs are also identified in cotton using a computational

Baohong Zhang (ed.), *Transgenic Cotton: Methods and Protocols*, Methods in Molecular Biology, vol. 958,
DOI 10.1007/978-1-62703-212-4_16, © Springer Science+Business Media New York 2013

approach (12–14), direct cloning (15), microarray and PCR (16, 17), and deep sequencing technology (18–20). Although some miRNAs are differentially expressed in different tissues, the function and regulatory mechanism is unclear. Transgenic technique is a powerful tool for studying miRNA function and its regulatory mechanism. Using this strategy, several miRNA targets have been well studies in model plant species, including *Arabidopsis* and rice (21–24). For example, miR 156 controls leave development and apical dominance through targeting transcription factor Squamosa-promoter binding protein-like protein (SBP). Overexpression of miR156 significantly increase leave initiation and plant biomass in several plant species, including model plant species *Arabidopsis* and a biofuel crop switchgrass (25).

In this chapter, we present a step by step protocol for functionally analyzing miRNAs in cotton using *Agrobacterium*-mediated genetic transformation.

2. Materials

2.1. Vector Construction

1. miR 156. Obtained from *Arabidopsis* (mature sequence: UGACAGAAGAGAGUGAGCAC; precursor sequence: CA AGAGAAACGCAAAGAAACUGACAGAAGAGAGU GAGCACACAAAGGCAAUUUGCAUAUCAUUGCAC UUGCUUCUCUUGCGUGCUCACUGCUCUUUC UGUCAGAUUCCGGUGCUGAUCUCUUU) (see Note 1).

2. Primers for miR156a (Forward primer (FP): ATGC***GGATCC***AAAGAGATCAGCACCGG; Reverse primer (RP): ATGC ***GGATCC***CAAGAGAAACGCAAAGAAAC). Bold letters show the restriction enzyme *Bam*HI cut sites.

3. *Arabidopsis* seedlings or gDNA.

4. PCR machine.

5. PCR mixture with Pfu DNA polymerase.

6. Gel electrophoresis.

7. DNA marker.

8. Nuclease-free water.

9. Restriction enzyme *Bam*HI.

10. Plasmid pBI121 vector carrying the *nptII* gene as the selectable marker gene, which resists to antibiotics.

11. DNA ligase.

12. *E. coli.*

13. 0.1 M $CaCl_2$.

14. Petri dish.

15. Water bath.

16. Ice.

17. LB medium.

18. Amp/X-gal/IPTG.

19. Plasmid extraction kits.

20. *Agrobacterium tumefaciens* strain LBA4404.

2.2. Cotton Transformation

1. Seeds of cotton (*Gossypium hirsutum* L.) cultivar Coker 201.

2. Bleach.

3. Ethanol.

4. Murashige and Skoog (MS) inorganic salts (26).

5. B$_5$ vitamins (27).

6. 2,4-D. Store at 4°C.

7. ZT. Store at –20°C.

8. 1 N KOH.

9. 1 N HCL.

10. Agar.

11. Sucrose.

12. Kanamycin. Store at –20°C.

13. Rifampicin. Store at –20°C.

14. Sterilized deionized water (ddH$_2$O).

15. Glucose.

16. MES.

17. Sodium phosphate buffer pH 5.6.

18. Acetosyringone.

19. ProX 0.22 μm membrane filter.

20. Syringe.

21. Hood.

22. 250×30 mm test tubes.

23. 1.5 mL centrifuge tubes.

24. Filter paper.

25. LB medium.

26. Plant growth incubator with light and temperature control.

2.3. miR156 Expression and Its Regulatory Mechanism on Its Target SPL2

1. *mir*Vana™ miRNA Isolation kit (Ambion, Austin, TX).

2. 100% RNase-free ethanol.

3. Liquid nitrogen.

4. RNase-free water.

5. Primers for miR 156. RT primer: GTCGTATCCAGTGCAG GGTCCGAGGTATTCGCACTGGATACGACGTGC; forward primer: GCGGCGGTGACAGAAGAGAGT; reverse primer: ATCCAGTGCAGGGTCCGAGG.

6. Primers for miR156 target *spl2* gene. Forward primer: GGT CTGGAACGCCGGTTCTGC; reverse primer: GGCGTCTC GCATTGTGGTCG.

7. PCR primers for reference genes GAPDH (forward primer is TGATGCCAAGGCTGGAATTGCTT; reversed primer is GT GTCGGATCAAGTCGATAACACGG).

8. TaqMan® MicroRNA Reverse Transcription Kit (Applied Biosystems, Foster City, CA). Store at −20°C.

9. PCR master mix with SYBR green.

10. PCR and real-time PCR machines.

11. 0.2 and 1.5 mL microcentrifuge tubes.

2.4. Effect of Overexpressed miR156 on Cotton Growth and Development

1. Soil.

2. Pots.

3. Water.

4. Fertilizer.

5. Pesticides.

3. Methods

3.1. Vector Construction

1. Cloning pre-miRNA 156a from *Arabidopsis* using the FP and RP primers listed in Subheading 2.

2. Digest the PCR products using *Bam*HI.

3. Run the digested PCR products using 2% Agarose gel (see Note 2).

4. Extract the digested PCR products (pre-miR 156a) from the gel and resolve in 50 μL nuclease-free water.

5. Digest plasmid pBI121 using *Bam*HI.

6. Insert pre-miR 156a into the pBI121 vector using DNA ligase.

7. Transfer the recombinant pBI121 plasmid into *E. coli* by the heat shock method.

8. Select the recombinant plasmid containing pre-miR156a on LB Amp/X-gal/IPTG plates.

9. Culture the *E. coli* containing transformed pBI121 plasmid overnight in liquid LB medium.

10. Isolate plasmid pBI121 containing pre-miR156a from *E. coli*.

11. Transfer the recombinant pBI121 plasmid into *Agrobacterium tumefaciens* strain LBA4404 by the heat shock method.

12. Select *Agrobacterium tumefaciens* strain LBA4404 containing plasmid pBI121 with pre-miR156a.

3.2. Cotton Transformation

Agrobacterium-mediated genetic transformation is employed to transfer pre-miR 156 into cotton genome. The detailed protocol can be found in Chapter 3.

1. Select cotyledon and hypocotyl explants from 7- to 10-day old seedlings.

2. Cut cotyledons and hypocotyls into small pieces.

3. Coculture cotton explants with *Agrobacterium* pre-induced by 100 µM acetosyringone for 10 min (see Note 3).

4. Place horizontally the hypocotyls segment on the agar-solidified MS medium containing 0.5 mg/L 2,4-D, 0.1 mg/L ZT, and 100 µM acetosyringone (see Note 3).

5. Place the cotyledon disks on the agar-solidified MS medium containing 0.5 mg/L 2,4-D, 0.1 mg/L ZT, and 100 µM acetosyringone (see Note 3). The adaxial sides should be faced down the medium.

6. Coculture them for 48 h in dark at 22°C.

7. Transfer the cotyledon disks or hypocotyls segments to the agar-solidified MS medium containing 0.5 mg/L 2,4-D, 0.1 mg/L ZT, 50 mg/L kanamycin, and 400 mg/L carbenicillin.

8. Culture them at 28 ± 2°C under a 14–16 h photoperiod with a light intensity of approximately 2,000 lux provided by cool white fluorescent lamps.

9. After 4 weeks of culture, the callus is transferred to fresh medium containing 0.5 mg/L 2,4-D, 0.1 mg/L ZT, and 50 mg/L kanamycin.

10. Regular subculture to fresh medium with 0.5 mg/L 2,4-D, 0.1 mg/L ZT, and 50 mg/L kanamycin every 4 weeks.

11. Transfer callus to the agar-solidified MS medium with 0.1 mg/L ZT, 0.1 mg/L IAA, and 150 mg/L for kanamycin (see Note 4)

12. Select and subculture the callus with high embryogenesis to fresh medium every 3 weeks.

13. Transfer embryogenic callus with high embryogenesis into the agar-solidified MS medium containing with 0.1 mg/L ZT, 2 g/L charcoal, and 150 mg/L for kanamycin (see Note 5).

14. Subculture to fresh medium every 3 weeks.

15. Recover plantlets from mature somatic embryos on agar-solidified MS medium with 2 g/L charcoal.

16. Detect whether or not a regenerated plant is transgenic plants using a traditional method (see Note 6).

17. Transplant the transgenic cotton regenerated plants into pots and then to greenhouse (see Note 7).

18. Regularly take care of the transgenic cotton plant.

19. Self-cross the transgenic plants.

20. Harvest the T_1 generation seeds.

3.3. miR156 Expression and Its Regulatory Mechanism on Its Target SPL2

Quantitative real-time PCR (qRT-PCR) technology is employed to monitor the expression level of miR 156 and its target spl2 gene in both transgenic cotton plants and their non-transgenic plants Coker 312. Applied Biosystems TaqMan microRNA Assays is employed to detect and quantify miR 156 using stem loop real-time PCR according to the manufacturer's instructions (28). There are two steps in the TaqMan miRNA Assays: (a) reverse transcription of the mature miRNA to a longer single-stranded cDNA sequence using a stem-looped primer and (b) real-time PCR. The expression level of spl2 gene is detected using a similar method as miR 156, the only difference is the reverse transcription, in which poly(T) primer is used instead of a stem-looped primer.

1. Total RNAs are isolated using mirVana™ miRNA Isolation Kit (Ambion, Austin, TX) according to the manufacture's protocol (see Note 8).

2. A single-stranded miRNA cDNA is generated from 1,000 ng of the total RNA sample by reverse transcription using the Applied Biosystems TaqMan microRNA Reverse Transcription Kit. Each reaction will have three replicates (see Note 9).

3. The 1st strand of cDNA for *spl2* and reference gene GADPH is generated from 1,000 ng of total RNAs. Each reaction will have three replicates.

4. Run qRT-PCR with 40 circles. The detailed protocol for running qRT-PCR can be found in Chapter 3.

5. Analyze the data generated from qRT-PCR.

6. Calculate the relative expression levels of miR 156 and its target *spl2* gene.

7. Compare the expression profile between transgenic plants and non-transgenic plants.

3.4. Effect of Overexpressed miR156 on Cotton Growth and Development

1. Plant the T_1 transgenic seeds in the greenhouse.

2. Standard cultivation practices and insect control measures are used.

3. Compare the growth and development between the transgenic plants and non-transgenic plants.

4. Observe the effect of overexpressed miR 156 on cotton plant growth and development.

5. Self-cross the transgenic plants and harvest the T_2 generation seeds.

4. Notes

1. Here, miR 156 as an example for investigating the function and regulatory mechanism of miRNAs in cotton. WMD3, a web-based miRNA designer also can be used to design any artificial miRNAs for targeting a specific protein-coding gene as well design DNA construct for a specific miRNAs, which can be found in the web site http://wmd3.weigelworld.org/cgi-bin/webapp.cgi (29).

2. Because the PCR products is around 100 nt, high percentage of agarose gel should be used to run the gel electrophoresis. When running the gel, the voltage should be less than 150 V.

3. Pre-inducing *Agrobacterium* using 100 μM acetosyringone and coculturing explants with *Agrobacterium* in medium with 100 μM acetosyringone significantly increase the transformation rate (30–32) because acetosyringone induce the *vir* gene expression and enhance the transformation process.

4. Cotton embryogenic callus is less sensitive to kanamycin exposure than others. Increasing kanamycin concentration will increase the selection rate for transgenic cells and also help to keep the non-embryogenic cells (33).

5. Induction and selection of embryogenic callus is a critical step for cotton tissue culture as well obtaining transgenic cotton plants. It may require multiple subcultures during this process.

6. Many methods can be used to detect transgenic cotton plants. Please refer the related chapter in this book.

7. Many methods can be used to transplant regenerated plants into pots. We suggest using a grafting technology to transplant the regenerated plants because it is high successful rate.

8. mirVana™ miRNA Isolation Kit can be used to extract both small RNAs and protein-coding mRNAs. If you use other methods, please make sure you did not lose small RNAs.

9. This technique can be used to analyze miRNAs with less than 10 ng of total RNAs. However, we suggest use 100–1,000 ng total RNAs for both miRNAs and their target analysis.

References

1. Bartel DP (2004) MicroRNAs: genomics, biogenesis, mechanism, and function. Cell 116: 281–297

2. Carrington JC, Ambros V (2003) Role of microRNAs in plant and animal development. Science 301:336–338

3. Zhang BH, Pan XP, Cobb GP, Anderson TA (2006) Plant microRNA: a small regulatory molecule with big impact. Dev Biol 289:3–16

4. Zhang BH, Wang QL, Pan XP (2007) MicroRNAs and their regulatory roles in animals and plants. J Cell Physiol 210:279–289

5. Zhang BH, Pan XP, Cannon CH, Cobb GP, Anderson TA (2006) Conservation and divergence of plant microRNA genes. Plant J 46:243–259

6. Ambros V (2001) microRNAs: tiny regulators with great potential. Cell 107:823–826

7. Jones-Rhoades MW, Bartel DP, Bartel B (2006) MicroRNAs and their regulatory roles in plants. Annu Rev Plant Biol 57:19–53

8. Chen XM (2005) microRNA biogenesis and function in plants. FEBS Lett 579:5923–5931

9. Chiou TJ (2007) The role of microRNAs in sensing nutrient stress. Plant Cell Environ 30:323–332

10. Lu SF, Sun YH, Shi R, Clark C, Li LG, Chiang VL (2005) Novel and mechanical stress-responsive microRNAs in *Populus trichocarpa* that are absent from *Arabidopsis*. Plant Cell 17:2186–2203

11. Sunkar R, Zhu JK (2004) Novel and stress-regulated microRNAs and other small RNAs from *Arabidopsis*. Plant Cell 16:2001–2019

12. Qiu CX, Xie FL, Zhu YY, Guo K, Huang SQ, Nie L, Yang ZM (2007) Computational identification of microRNAs and their targets in *Gossypium hirsutum* expressed sequence tags. Gene 395:49–61

13. Zhang BH, Wang QL, Wang KB, Pan XP, Liu F, Guo TL, Cobb GP, Anderson TA (2007) Identification of cotton microRNAs and their targets. Gene 397:26–37

14. Khan Barozai MY, Irfan M, Yousaf R, Ali I, Qaisar U, Maqbool A, Zahoor M, Rashid B, Hussnain T, Riazuddin S (2008) Identification of micro-RNAs in cotton. Plant Physiol Biochem 46:739–751

15. Abdurakhmonov IY, Devor EJ, Buriev ZT, Huang LY, Makamov A, Shermatov SE, Bozorov T, Kushanov FN, Mavlonov GT, Abdukarimov A (2008) Small RNA regulation of ovule development in the cotton plant, *G. hirsutum* L. BMC Plant Biol 8:12

16. He XH, Cai YF, Sun Q, Yuan YL, Shi YZ (2011) MicroRNA expression profiling during upland cotton gland forming age by microarray and quantitative reverse-transcription polymerase chain reaction (qRT-PCR). Afr J Biotechnol 10:8695–8702

17. Pang MX, Xing CZ, Adams N, Rodriguez-Uribe L, Hughs SE, Hanson SF, Zhang JF (2011) Comparative expression of miRNA genes and miRNA-based AFLP marker analysis in cultivated tetraploid cottons. J Plant Physiol 168:824–830

18. Kwak PB, Wang QQ, Chen XS, Qiu CX, Yang ZM (2009) Enrichment of a set of microRNAs during the cotton fiber development. BMC Genomics 10:457

19. Pang MX, Woodward AW, Agarwal V, Guan XY, Ha M, Ramachandran V, Chen XM, Triplett BA, Stelly DM, Chen ZJ (2009) Genome-wide analysis reveals rapid and dynamic changes in miRNA and siRNA sequence and expression during ovule and fiber development in allotetraploid cotton (*Gossypium hirsutum* L.). Genome Biol 10(11):R122

20. Ruan MB, Zhao YT, Meng ZH, Wang XJ, Yang WC (2009) Conserved miRNA analysis in *Gossypium hirsutum* through small RNA sequencing. Genomics 94:263–268

21. Chuck G, Meeley R, Hake S (2008) Floral meristem initiation and meristem cell fate are regulated by the maize AP2 genes ids1 and sid1. Development 135:3013–3019

22. Lauter N, Kampani A, Carlson S, Goebel M, Moose SP (2005) microRNA172 down-regulates glossy15 to promote vegetative phase change in maize. Proc Natl Acad Sci USA 102:9412–9417

23. Aukerman MJ, Sakai H (2003) Regulation of flowering time and floral organ identity by a microRNA and its APETALA2-like target genes. Plant Cell 15:2730–2741

24. Chen XM (2004) A microRNA as a translational repressor of APETALA2 in Arabidopsis flower development. Science 303:2022–2025

25. Fu C, Sunkar R, Zhou C, Shen H, Zhang JY, Matts J, Wolf J, Mann DG, Stewart CN, Tang Y, Wang ZY (2012) Overexpression of miR156 in switchgrass (*Panicum virgatum* L.) results in various morphological alterations and leads to improved biomass production. Plant Biotechnol J. doi: 10.1111/j.1467-7652.2011.00677.x

26. Murashige T, Skoog F (1962) A revised medium for rapid growth and bioassays with tobacco tissue cultures. Physiol Plant 15:473–497

27. Gamborg OL, Miller RA, Ojima K (1968) Nutrient requirements of suspension cultures of soybean root cells. Exp Cell Res 50:151–158

28. Chen CF, Ridzon DA, Broomer AJ, Zhou ZH, Lee DH, Nguyen JT, Barbisin M, Xu NL, Mahuvakar VR, Andersen MR, Lao KQ, Livak KJ, Guegler KJ (2005) Real-time quantification of microRNAs by stem-loop RT-PCR. Nucleic Acids Res 33:e179

29. Ossowski S, Schwab R, Weigel D (2008) Gene silencing in plants using artificial microRNAs and other small RNAs. Plant J 53:674–690

30. Jin SX, Zhang XL, Liang SG, Nie YC, Guo XP, Huang C (2005) Factors affecting transformation efficiency of embryogenic callus of Upland cotton (*Gossypium hirsutum*) with Agrobacterium tumefaciens. Plant Cell Tiss Org Cult 81:229–237

31. Sunilkumar G, Rathore KS (2001) Transgenic cotton: factors influencing Agrobacterium-mediated transformation and regeneration. Mol Breed 8:37–52

32. Wu S-J, Wang H-H, Li F-F, Chen T-Z, Zhang J, Jiang Y-J, Ding Y, Guo W-Z, Zhang T-Z (2008) Enhanced *Agrobacterium*-mediated transformation of embryogenic calli of upland cotton via efficient selection and timely subculture of somatic embryos. Plant Mol Biol Rep 26:174–185

33. Zhang B-H, Liu F, Liu Z-H, Wang H-M, Yao C-B (2001) Effects of kanamycin on tissue culture and somatic embryogenesis in cotton. Plant Growth Reg 33:137–149

Development of Transgenic *CryIA(c)* + *GNA* Cotton Plants via Pollen Tube Pathway Method Confers Resistance to *Helicoverpa armigera* and *Aphis gossypii* Glover

Zhi Liu, Zhen Zhu, and Tianzhen Zhang

Abstract

Elite cotton cultivar Sumian16 was transformed with p7RPSBK-mGNA-NPTII containing *Bt* (*CryIA(c)*), *Galanthus nivalis* agglutinin (*GNA*) resistance genes and selectable marker *NptII* gene via the pollen tube pathway method and two fertile transgenic *Bt*+*GNA* plants were obtained in the present study. The integration and expression of the *Bt* and *GNA* genes were confirmed by molecular biology techniques and insect bioassays. Insect bioassays showed that the transformed plants were highly toxic to bollworm larvae as well as obviously retarding development of aphid populations. PCR analyses and identification of resistance to Kanamycin and bollworm showed that the resistance to bollworm for the two transgenic plants was dominantly inherited in a Mendelian manner and the two resistance genes and selectable marker co-segregated from primary transformed parents to the first self-fertilized progeny plants.

1. Introduction

The plant genetic transformation technology to introduced insect-resistance genes into crops genome offers an alternative to conventional breeding for the development of resistant plants. The commercial utilization of transgenic *Bt* cotton varieties in America, Australia, China, etc. has controlled *Helicoverpa* species, the main cotton pests, efficiently and reduced pesticide usage dramatically with concomitant economic and environmental benefits (1–3). *Aphis gossypii* Glover, another major pest in cotton, not only affects the growth and development of cotton, yield and fiber quality seriously, but also acts as a vector for economically important viral diseases. Artificial diet bioassay and transgenic plant showed *Galanthus nivalis* agglutinin (GNA) to be effective against aphids (4, 5). Transgenic tobacco plants expressing snowdrop lectin

Baohong Zhang (ed.), *Transgenic Cotton: Methods and Protocols*, Methods in Molecular Biology, vol. 958, DOI 10.1007/978-1-62703-212-4_17, © Springer Science+Business Media New York 2013

significantly retarded aphid development and reduced its weight and fecundity. The transformed plants inhibited aphid populations by approximately 50% and sometimes by up to 90% (5–7). At present, engineering multiple genes into crops is one of important researches direction in plant gene engineering (8). Development of transgenic crops expressing two or more resistance genes may delay and prevent adaptation in pest species, prolong the useful life span of transgenic plants and broaden their resistance spectrum (9). The introduction of modified *cryIA(b)* and *cryIC* genes of *Bacillus thuringiensis* into tobacco and tomato resulted in protection against *Spodoptera exigua*, *Heliothis virescens*, and *Manduca sexta* (10). Transgenic *CryIA + GNA* tobacco plants had efficacy against *Helicoverpa armigera* and *Myzus persicae* (11) and transgenic tobacco plants expressing both the cowpea trypsin inhibitor (CpTI) and P-lec(Pea Lectin) had enhanced resistance to *Helicoverpa virescens* (12). Laboratory selection indicated that transgenic tobacco harboring double genes (*Bt + CpTI*) may delay significantly the development of pest resistance compared with transgenic *Bt* tobacco plants (13). Transgenic upland cottons plants harboring two insecticidal genes have been developed in America (14) and China (15–17) and the transgenic *Bt + CpTI* cotton cultivar, sGK321, is now being expanded on a large scale in China (18). To date there is no report about transgenic cottons conferring resistance to both *Helicoverpa armigera* and *Aphis gossypii* Glover. In this chapter, we introduced *Bt* and *GNA* insecticidal genes into an elite cotton (*Gossypium hirsutum* L.) cultivar Sumian16 via the pollen tube pathway method and the transgenic plants conferred resistance to *Helicoverpa armigera* and *Aphis gossypii* Glover as identified by insect bioassays.

2. Materials

2.1. Plant Expression Vector

The plant expression vector p7RPSBK-mGNA-NPTII containing *Bt* (*CryIA(c)*), *GNA* resistance genes and selectable marker *NptII* gene was constructed and kindly provided by Professor Z Zhu at the Genetics and Developmental Biology Institute, Chinese Academy of Sciences (Fig. 1). The *Bt* gene is driven by the replication protein promoter of CLCuV(Cotton Leaf Curl Virus) (19), which is strong and constitutive, and modified with the endoplasmic reticulum (ER) retention signal coding sequence *KDEL*.

2.2. Plants and Insects

1. The elite cotton (*Gossypium hirsutum* L.) cultivar Sumian16 with susceptibility to bollworm (*Helicoverpa armigera*) and aphid (*Aphis gossypii* Glover) was used as transformant and control material.

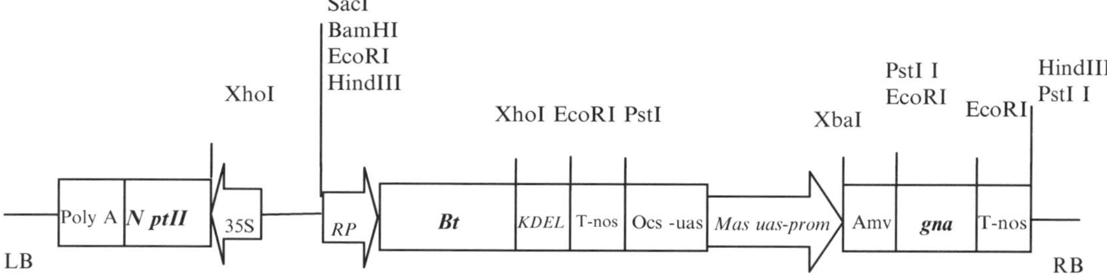

Fig. 1. The map of p7RPSBK-mGNA-NPTII. The vector of p7RPSBK-mGNA-NPTII was constructed by inserting the BglII/PvuII fragment of p7RPSBK into *Bam*HI/*Sac*I site of p7RPG2300. *RP* represents replication protein promoter of CLCuV. *Bt* represents *CryIA(c)* gene. *KDEL* represents endoplasmic reticulum (ER) retention signal coding sequence, T-nos represents nos terminator. Ocs uas represents octopine synthase (ocs) upstream activating sequence (uas) and it repeats for six times. *Mas uas-prom* represents mannopine synthase (mas) upstream activating sequence (uas) and promoter. Amv represents nontranslatable sequence. *GNA* represents snowdrop lectin (*Galanthus nivalis* agglutinin) gene. T-ocs represents ocs terminator.

2. Transgenic Bt cotton line, Shanxi94-24 and transgenic *Bt+ CpTI* cotton line, Shuangkang-1, which are highly toxic to bollworm, were obtained from the Cotton Research Institute, Shanxi Academy of Agricultural Sciences and the Biotechnology Research Institute, Chinese Academy of Agricultural Sciences, respectively (20, 21).

3. The *Helicoverpa armigera* larvae susceptible to Bt (*CryIA*) toxins was kindly provided from the Plant Protection Institute, Jiangsu Academy of Agricultural Sciences.

4. The aphids (*Aphis gossypii* Glover) were maintained on potted plants of cultivar Sumian16 in the greenhouse.

3. Methods

3.1. Genetic Transformation

1. The pollen tube pathway method (22) was adopted.

2. The plasmid (p7RPSBK-mGNA-NPTII) DNA was extracted and purified according to the procedure of Sambrook et al. (23) and then diluted to 20 ng/μl with distilled water.

3. The transformation was conducted during flowering stage. Erased the stigma from the self-crossed bolls developed from 20 to 24 h after anthesis, inserted vertically into the ovary till about 1 cm along the base of stigma with a glass injector in capability of 50 μl volume and then injected 0.2 μg plasmid DNA into the young boll.

4. After injection, the boll was marked and the other squares on the branch were removed and 0.005%(W/V) GA$_3$ was used as well to maintain the injected bolls. About 1,500 young bolls

were injected in total and only 4,400 seeds were harvested from the marked bolls.

3.2. Screening of Putatively Transgenic Plants

1. All resulting seeds were delinted with sulfuric acid and planted in seedling beds at Jiangpu Cotton Breeding Station, Nanjing Agricultural University, in April, 2000.

2. Firstly, absorbent cotton strips moistened with 0.005%(W/V) Kanamycin solution were placed on the true leaves when most seedlings had two true leaves and the seedlings with obviously yellow spots on leaves were removed after about 5 days.

3. Then eight 1st instar bollworm larvae were released for every plant at the tender cotton apexes.

4. The seedlings with no or little damage were selected and transplanted into the field.

5. These putatively transgenic plants were further identified with bollworm bioassay and finally confirmed by molecular biology techniques during the squaring stage.

3.3. Insects Bioassays

1. Bollworm bioassays were performed with the methods described by Sun et al. (24). According to the results of bioassays, the plants with three or four in leaf damage index and 3rd instar larvae survival were classified as "susceptible" and those of one and two in leaf damage index and no 3rd instar larvae survival were classified as "resistant" (20).

2. Aphid bioassays were conducted with whole cotton seedlings potted with four or five true leaves in a greenhouse (T: $26 \pm 2°C$, RH: 60–70%) at Nanjing Agricultural University.

3. The homozygous lines of TL1 and 8-11, TBG and LBG, respectively, and control plants, Shanxi94-24, Shuangkang-1 and Sumian16 were planted in pots in February, 2002, with one plant per pot and five duplications for each material.

4. Five neonate aphid nymphs were transferred to the back of a young leaf on every plant with a soft brush pen. Growth of the aphid populations on each plant was counted every other day for 14 days.

5. The reduction rate was calculated to mirror the efficacy of transgenic plants against aphids and the formula was defined as $((UN - TN)/UN) \times 100\%$ (UN means average number of aphids on untransformed parent Sumian16 and TN means average number of aphids on transgenic plant after infesting 14 days).

3.4. PCR and RT-PCR Analysis

1. Cotton leaf genomic DNA was extracted and purified according to CTAB method (25). PCR analysis was carried out using primers specific for *Bt* and *GNA* genes with 50–100 ng DNA.

2. Total RNA of cotton leaves was extracted using the procedures of Song et al. (26). Synthesis of first cDNA was carried out

according the procedures of the RevertAid™ First Strand cDNA Synthesis Kit (MBI Co) with random hexamer primer and then 2 μl products were used as template for PCR analysis with the specific primers for the *Bt* and *GNA* genes.

3. The primer sequences for PCR were as follows: *Bt* gene, forward primer 5′-GGGCCCGCTGAATCCAACTGGAGAG GC-3′, reverse primer 5′-CCATACAACTGCTTGAGTAACC CAGAAGTTG-3′; *GNA* gene, forward primer 5′-CAAAATGG CTAAGGCAAGTC-3′, reverse primer 5′-CGGTCATTACTT TGCCGTCA-3′. The PCR primers were synthesized in Sangon, Inc. (Shanghai, China).

4. PCR reactions were performed on a GeneAmp PCR system 9600 (Perkin-Elmer). Samples were first denatured for 2 min 95°C then carried through 30 cycles by using the following temperature sequence: 94°C for 0.5 min, 57°C for 1.5 min, and 72°C for 1.5 min and a final extension cycle at 72°C for 10 min.

5. PCR products were electrophoresed through a 1.4% agarose gel, stained with Ethidium Bromide and then observed with UV light.

3.5. Southern Blot Analysis

1. About 20 μg genomic DNA per sample was digested with *Hin*dIII enzyme and separated on a 0.8% agarose gel.

2. Membrane transferring and blotting protocols were performed according to the manual of the Hybond-N + System (Amersham Pharmacia Biotech, Inc.).

3. The labeling of probes from recovery and purification of PCR products of the *Bt* and *GNA* genes, respectively, using expression vector as template followed the instructions of the Prime-a-gene® Labeling System (Promega Inc). α-^{32}P-dCTP was purchased from Beijing Yahui Pharmaceutical Bioengineering, Inc.

4. PCR products from the genomic DNA of the transgenic plants with specific primers of *Bt* and *GNA* genes were separated on a 1.4% agarose gel and also transferred onto membrane and hybridized as the manual of the Hybond-N + System.

---

4. Notes

1. All 4,400 resulting seeds were sown in nursery beds to be analyzed. Half of seedlings with an obvious yellow spot on the treated leaf were removed after the first round of selection using Kanamycin as an indirect marker. When infested by bollworms for 5 days, the true leaves, terminal buds, and even cotyledons of most seedlings were damaged heavily. Seven seedlings with no or little damage were screened and transplanted into field.

Bollworm bioassays and Kanamycin resistance for these plants continued for further characterization during the squaring stage. Finally, two transgenic plants, TL1 and 8-11, conferring high efficacy against bollworm and resistance to Kanamycin were obtained, while others might escape from selection before transplanting, which was confirmed by PCR analysis with specific primers for *nptII* and *Bt* genes.

2. PCR analyses with specific primers of *Bt* and *GNA* genes showed the presence of the amplified 938 and 484 bp fragments corresponding to *Bt* and *GNA* genes, respectively, in genome of TL1 and 8-11 plants, while no target fragments were presented for untransformed parent Sumian16 (Fig. 2a). The results were further confirmed by PCR-Southern blotting using *Bt* and *GNA* gene fragments as probes (Fig. 2b).

   The production of mRNA transcripts from both resistance genes was detected by RT-PCR analyses in total RNA of transgenic plants and no product was observed in that of control sample (Fig. 3a, b).

   Southern blotting using the probe of the *Bt* gene fragment showed the integration of *Bt* gene into plant genome but the positions of the *Bt* gene were different in genomes of TL1 and 8-11 (Fig. 4a, b). When re-probed the stripped blot with the *GNA* gene fragment, we obtained the same band patterns. From Fig. 4b, following *Hin*dIII digestion, we found that the blotted fragments for TL1 and 8-11 were all larger than the expected fragment. The reasons resulting in modification of the enzyme recognition site are unknown and need to be further investigated.

3. All T_1 progeny plants from self-pollination of the primary transgenic plants were tested by bollworm bioassay with the

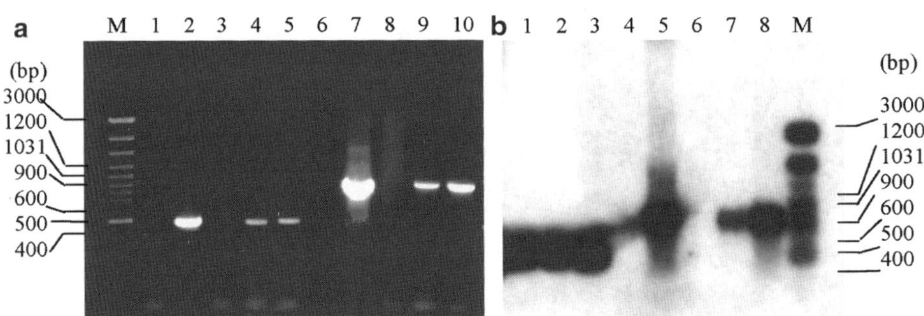

Fig. 2. PCR (**a**) and PCR-Southern (**b**) analyses of transgenic cotton plants. (**a**) Double-distilled water as template (lane 1, 6), positive plasmid (lane 2, 7), non-transformed cotton cultivar "Sumian16" (lane 3, 8), transgenic plant 8-11 (lane 4, 9) and TL1 (lane 5, 10). (**b**) Positive plasmid (lane 1, 5), transgenic plant 8-11 (lane 2, 7) and TL1 (lane 3, 8), non-transformed cotton cultivar "Sumian16" (lane 4, 6).

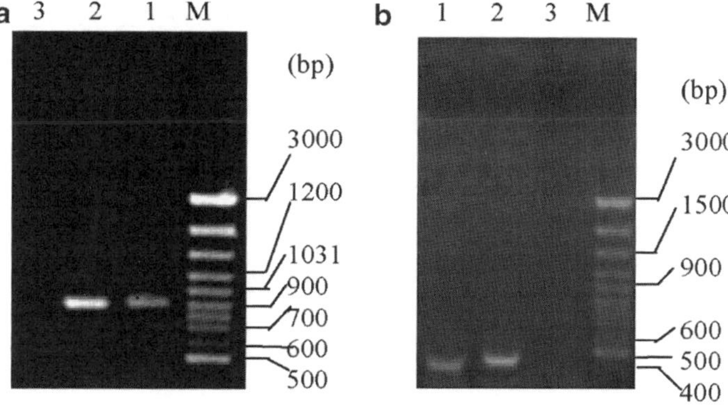

Fig. 3. RT-PCR analyses of transgenic cotton plants with *Bt*-specific primer (**a**) and *GNA*-specific primer (**b**). Lanes 1–3 show transgenic plants 8-11, TL1 and non-transformed cotton cultivar "Sumian16", respectively.

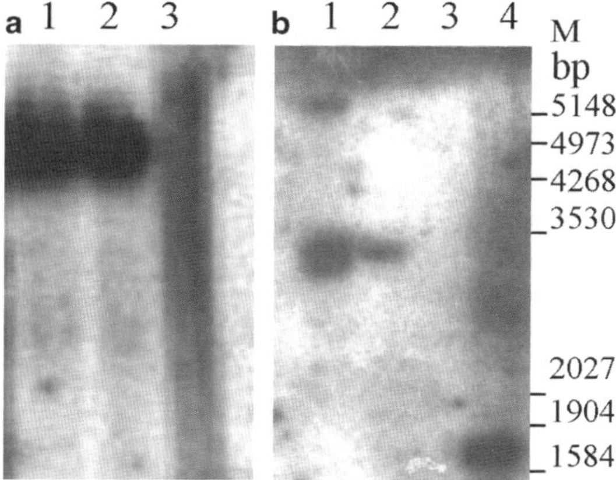

Fig. 4. Southern blot analysis of plant genomic DNA undigested (**a**) and digested by *Hind*III restriction enzyme (**b**) using a probe for the *Bt* gene. Lanes 1–3 represent transgenic plants TL1, 8-11, non-transformed cotton cultivar "Sumian16", respectively, lane 4 is positive plasmid.

just expanded apical leaves during the squaring stage. The segregations of the resistant and susceptible plants in the T_1 populations of both TL1 ($n=113$, $\chi^2=0.0741$, $P=0.750$–0.900) and 8-11 ($n=205$, $\chi^2=0.0758$, $P=0.750$–0.900) fit a 3:1 ratio, which revealed the resistance to bollworm for the two transgenic plants was dominantly inherited as a Mendelian manner. The results were confirmed by PCR analyses of the *Bt* gene and by Kanamycin resistance by placing cotton strips moistened with 0.005% Kanamycin solution on young leaves in the two T_1 progenies during the same developmental stage (data not shown).

Furthermore, the F_1 hybrids crossed the transgenic *Bt*+*GNA* cotton homozygous line TBG pedigree-selected from TL1 with six nontransgenic cotton cultivars showed significant resistance to bollworm and the segregation of the resistant and susceptible plants in F_2 ($n = 3,431$, $\chi^2 = 4.1629$, $P = 0.025–0.050$) and BC_1 ($n = 3,246$, $\chi^2 = 0.0499$, $P = 0.75–0.90$) populations fit 3:1 and 1:1 ratios, respectively, which further confirmed that the bollworm resistance of TBG was controlled by one pair of dominant resistant gene.

We found that both *Bt* and *GNA* DNA fragments by PCR analyses were presented in the T_1 plants with resistance to bollworm and Kanamycin and none was observed for the T_1 plants with susceptible to bollworm and Kanamycin. The two resistance genes and selectable marker of the transgenic plants were co-segregated in transmission.

4. Fifteen plants harboring foreign transgenes from each T_1 progeny plants for bollworm bioassays were screened by PCR analyses with specific primers of the *Bt* and *GNA* genes and identification of Kanamycin resistance. The mortality of the bollworm larvae fed with the just expanded apical leaves of transgenic plants after 5 days was higher than 90% during the peak squaring stage and 80% or so during the peak flowering stage, which indicated the two transgenic plants were highly toxic to bollworm larvae (Table 1). There were no significant differences among the insect-resistance of TL1, 8-11, and Shanxi94-24 (Bt cotton) ($P > 0.05$).

5. The effects of transgenic plants on aphid development were determined by infesting the homozygous lines TBG (TL1) and LBG (8–11) plants with first instar aphid nymphs. The rate of growth of aphid population on TBG and LBG plants was obviously slower than that of aphid population on control plants Sumian16 (Fig. 5). The total nymph population was 50.3% and 35.8% less in TBG and LBG than that of the control, respectively.

Table 1
Resistance of transgenic plants to bollworm larvae with bollworm bioassays

Date	Materials	Mortality (%) ± SE	Survival of 3rd bollworm larvae (%) ± SE	Leaf damage index ± SE
June 20	TL1	95.56 ± 8.82	0	0.70 ± 0.55
July 23	8-11	91.67 ± 14.14	0	0.92 ± 0.70
	Shanxi94-24	93.33 ± 10.0	0	0.84 ± 0.61
	Sumian16	7.40 ± 12.56	72.30 ± 9.60	3.25 ± 0.45
	TL1	83.30 ± 15.70	0	1.11 ± 0.60
	8-11	75.86 ± 16.30	0	1.34 ± 0.83
	Shanxi94-24	80.20 ± 12.50	0	1.20 ± 0.53
	Sumian16	9.20 ± 13.20	81.67 ± 11.50	3.20 ± 0.90

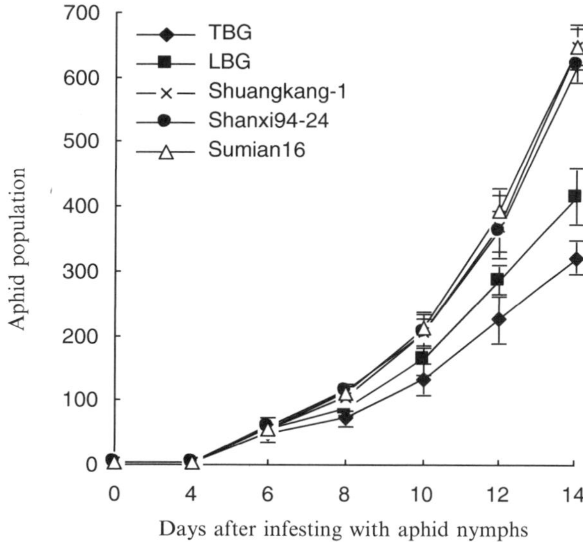

Fig. 5. Aphid populations on transgenic plants during cotton seedling stage. The curves were drawn with the average aphid number on each material during experimental period. Standard errors were indicated as the bar.

However, there was no significant difference among the number of aphids on Shanxi94-24, Shuangkang-1, and Sumian16 during the whole experiment ($P>0.05$), which showed that Bt toxic protein or the combination of Bt and CpTI was not effective against aphid development. The leaf damage and growth of identified materials was also investigated after infesting 14 days.

The young leaves of Shanxi94-24, Shuangkang-1, and Sumian16 curled into a ball shape and their growth was seriously retarded after 14 days infestation, while those of TBG and LBG plants curled slightly and little damage was noted for the transgenic plants.

5. Discussion

Since the pollen tube pathway was reported (22), many transgenic crops have been developed by this genetic transformation method (16, 17, 27–33). Numerous putatively transformed seeds can be obtained and any genotype of flowering crops may be used as recipient because of its simplicity, convenience. At present, many insect resistant transgenic cotton plants have been obtained by the method using elite cultivars as transformants, from which several transgenic cotton cultivars and hybrids were developed and released in China (3). Using an elite cultivar with high yield, super fiber quality and good combining ability as the transformed parent, two fertile transgenic plants with no significantly altered agronomic characters were gained by the pollen tube pathway in the present study. The transformed plants can be released directly as new cultivars after simple selection and the foreign genes can also be easily introduced into other genetic backgrounds by hybridization using them as the father or mother parent in a breeding program. Xie et al. (31) reported the foreign genes in the same expression vector separated and the transformed plants included ones harboring both *NptII* and *Bt* genes along with either the *NptII* gene or *Bt* gene. In the present study, only the transformed plants with high toxicity to bollworm larvae and resistance to Kanamycin were screened by the process of high selection pressure we described and other transformed plants would be removed. Though the transformation efficiency was decreased, the transgenic cotton plants harboring intact *Bt* and *GNA* resistance genes and selectable marker *NptII* gene were linked completely and co-segregated in transmission as confirmed by PCR analyses.

The efficacy of the transgenic plants against bollworm and aphid were identified with insect bioassays. The strength of resistance correlates with the per cent of 3rd instar bollworm larvae, survival and leaf damage index with laboratory bioassays (20). Not like the Bt toxic protein to kill bollworm larvae, however, GNA is known that the toxin has a greater effect on aphid growth, development and fecundity, and no effect on its survival (5–7, 34). It is difficult for aphid bioassay in the laboratory to use detached leaves or branches because the conditions to keep the materials fresh is impossible for the aphids to live a long time. Therefore, we performed the experiment in the greenhouse with whole potted plants suitable for the growth and development of the aphids. Primary field performance showed that the growth of aphid population and leaf damage on transgenic plants TBG and LBG were significantly lower than those of control plants Sumian16 during the squaring stage after infesting 14 days with 50 first instar aphid nymphs for each plant and the same results were observed when the materials were infested with naturally emerging aphids during boll setting stage (data not shown). In order to control cotton aphid efficiently, we think it is necessary to modify the *GNA* gene sequence and expression cassette or isolate novel resistance genes to enhance the toxin in engineered plants or to kill the aphid directly.

Acknowledgments

This project was financially supported in part by the National Research and Development Project of Transgenic Crops of China (2010ZX08009-003, 2011ZX08005-001) and the Priority Academic Program Development of Jiangsu Higher Education Institutions.

References

1. Benedict JH, Sachs ES, Altman WR, Deaon RJ, Kohel RJ, Ring DR, Berberich SA (1996) Field performance of cottons expressing transgenic Cry1A insecticidal protein for resistance to *Heliothis virescens* and *Helicoverpa zea* (Lepidoptera: Nochxidae). J Econ Entomol 89:230–238

2. Fitt GP, Mares CL, Llewellyn DJ (1994) Field evaluation and potential ecological impact of transgenic cottons in Australia. Biocontrol Sci Technol 4:535–548

3. Zhang TZ, Tang CM (2000) Commercial production of transgenic *Bt* insect resistant cotton varieties and the resistance management for bollworm (*Helicoverpa armigera* Hubner). China Sci Bull 45(14):1249–1257

4. Powell KS, Gatehouse AMR, Hilder VA, Gatehouse JA (1993) Antimetabolic effects of plant and fungal enzymes on the nymphal stages of two important rice pest, *Nilaparvata lugens* and *Nephotettix cinciteps*. Entmol Exp Appl 66:119–126

5. Hilder VA, Powell KS, Gatehouse AMR, Gatehouse JA, Gatehouse LN, Shi Y, Hamilton WDO, Merryweather A, Newell CA, Timans JC, Peumans WJ, Van Damme EJS, Boulter D (1995) Expression of snowdrop lectin in transgenic tobacco plant results in added protection against aphids. Transgenic Res 4:18–25

6. Zhou Y, Tian YC, Wu B, Mang KQ (1998) Inhibition effect of transgenic tobacco plants expressing snowdrop lectin on the population of *Myzus persicae*. Chin J Biotechnol 14(1):13–19

7. Yuan ZQ, Zhao CY, Zhou Y, Tian YC (2001) Aphid-resistant transgenic tobacco plants expressing modified *GNA* gene. Acta Bot Sin 43(6):592–597

8. Liu Z, Yuan XL, Zhang TZ (2001) Strategies to develop transgenic crops expressing multiple genes. Hereditas 13(2):182–186

9. McGaughey WH, Gould F, Gelernter W (1998) *Bt* resistance management. Nat Biotechnol 16:144–146

10. Van der Salm T, Bosch D, Honee G, Feng LX, Munsterman E, Bakker P, Stickema WJ, Visser B (1994) Insect resistance of transgenic plants that express modified Bacillus thuringiensis cryIA(b)and cryIC genes: a resistance management strategy. Plant Mol Bio 26:51–59

11. Wang ZB, Guo SD (1999) Transgenic *cryIA+GNA* tobacco plants conferring resistance to bollworm and aphid. China Sci Bull 4(19):2068–2075

12. Boulter D, Edward GA, Gatehouse AMR, Hilder VA (1990) Additive protective effects of different plant-derived insect resistance genes in transgenic tobacco. Crop Prot 9:351–354

13. Zhao JZ, Fan YL, Fan XL, Shi SP, Lu GM (1999) Evaluation of effect of transgenic tobacco plants expressing double resistance genes on delaying resistance of bollworm (*Helicoverpa armigera* Hubner). China Sci Bull 44(15):1635–1639

14. Constable G (1998) Two gene technology in the Australian scene. The Australian Cotton Grower 5:27–29

15. Wang W, Zhu Z, Gao YF, Shi CL, Chen WX, Guo ZC, Li XH (1999) Obtaining a transgenic upland cotton harboring two insecticidal genes. Acta Bot Sin 41(4):384–388

16. Guo SD, Cui HZ, Xia LQ, Wu DL, Ni WC, Zhang ZL, Zhang BL, Xu YJ (1999) Development of bivalent insect-resistant transgenic cotton plants. Sci Agri Sin 32(3):1–7

17. Li FG, Cui JJ, Liu CL, Wu ZX, Li FL, Zhou Y, Li XL, Guo SD, Cui HZ (2000) The study of insect-resistant transgenic cotton harboring double-gene and its insect-resistance. Sci Agri Sin 33(1):46–52

18. Guo SD, Ni WC, Zhao GZ (2001) Development of transgenic cotton plants harboring two insecticidal genes. In: Jia SR, Guo SD, An DQ (eds) Transgenic cotton. Series book of 863 biological high-tech. Science press, Beijing, pp 160–163

19. Xie YQ, Zhu Z, Wu Q, Xu HB, Meng M, Liu YL (2000) Isolation of a strong promoter and its functional study. Acta Bot Sin 42(1):50–54

20. Tang CM, Sun J, Zhu XF, Guo WZ, Zhang TZ, Shen JL, Gao CF, Zhou WJ, Chen ZX, Guo SD (2000) Inheritance of resistance to *Helicoverpa armigera* of 3 kinds of transgenic *Bt* strains available in upland cotton in China. Chin Sci Bull 45(4):363–367

21. Yuan XL, Sun J, Tang CM, Zhang TZ (2001) Temporary fluctuation of resistance of the transgenic cotton harboring *Bt+CpTI* double-gene to *Helicoverpa armigera*. High Tech Lett 10(2):9–12

22. Zhou GY, Weng J, Zheng YS, Huang JQ, Qian SY, Liu GL (1983) Introduction of exogenous DNA into cotton embryos. In: Wu R, Gnssman L, Moldave K (eds) Methods in enzymology, vol 101. Academic, New York, pp 433–481

23. Sambrook J, Friston EF, Maniatis T (eds) (1989) Molecular cloning: a laboratory manual. Cold Spring Harbor Laboratory Press, New York

24. Sun J, Tang CM, Zhu XF, Guo WZ, Zhang TZ, Zhou WJ, Meng FX, Sheng JL (2002) Characterization of resistance to *Helicoverpa*

armigera in three lines of transgenic *Bt* upland cotton. Euphytica 123:343–351

25. Paterson AH, Brubaker CL, Wendel JF (1993) A rapid method for extraction of cotton (*Gossypium* spp.) genomic DNA suitable for RFLP and PCR analysis. Plant Mol Biol Rep 11:122–127

26. Song P, Yamamoto E, Allen R (1995) Differential display of cotton transcripts. Plant Mol Biol Rep 13:174–181

27. Chong K, Bao S, Xu T, Tan K, Liang T, Zeng J, Huang H, Xu J (1998) Functional analysis of the *ver* gene using transgenic wheat. Physiol Plant 102:87–92

28. Luo ZX, Wu R (1988) Simple method for the transformation of rice via the pollen-tube pathway. Plant Mol Biol Rep 6(3):165–174

29. Ni WC, Zhang ZL (1998) Development of transgenic insect-resistant cotton plants. Sci Agri Sin 31(2):8–13

30. Xie DX, Fan YL, Ni PZ (1991) Obtaining transgenic plants by introducing *Bacillus thur-ingiensis* insecticidal crystal protein gene into rice cultivar Zhonghua11 in China. Sci China B 8:830–834

31. Xie DX, Fan YL, Ni WC, Huang JQ (1991) Obtaining transgenic plants by introducing *Bacillus thuringiensis* insecticidal crystal protein gene into cotton. Sci China B 4:367–373

32. Zeng JZ, Wang DJ, Wu YQ, Zhang J, Zhou WJ, Zhu XP, Xu NZ (1993) Obtaining of transgenic wheat plants by pollen tube pathway. China Sci Bull 23(3):256–262

33. Zeng JZ, Wu YQ, Wang DJ, Zhang J, Ma ZR, Zhou ZY (1998) Discussion in transformation mechanism and inheritance of transgenic plants progeny by pollen tube pathway (or vector). China Sci Bull 43(6):561–566

34. Maqbool SB, Riazuddin S, Loc NT, Gatehouse AMR, Gatehouse JA, Christou P (2001) Expression of multiple insecticidal genes confers broad resistance against a range of different rice pests. Mol Breed 7:85–93

Agrobacterium-Mediated Transformation of Cotton (*Gossypium hirsutum*) Shoot Apex with a Fungal Phytase Gene Improves Phosphorus Acquisition

Zhiying Ma, Jianfeng Liu, and Xingfen Wang

Abstract

Cotton is an important world economic crop plant. It is considered that cotton is recalcitrant to in vitro proliferation. Somatic embryogenesis and plant regeneration has been successful by using hypocotyl, whereas it is highly genotype dependent. Here, a genotype-independent cotton regeneration protocol from shoot apices is presented. Shoot apices from 3- to 5-day-old seedlings of cotton are infected with an *Agrobacterium* strain, EHA105, carrying the binary vector pC-KSA contained phytase gene (*phyA*) and the marker gene neomycin phosphotransferase (NPTII), and directly regenerated as shoots in vitro. Rooted shoots can be obtained within 6–8 weeks. Plants that survived by leaf painting kanamycin (kan) were further analyzed by DNA and RNA blottings. The transgenic plants with increased the phosphorus (P) acquisition efficiency were obtained following the transformation method.

1. Introduction

Cotton (*Gossypium hirsutum*) is an excellent source of textile fiber and cultivated in many countries. Genetic transformation plays an important role in modern cotton breeding research. Although many somatic embryogenesis and plant regeneration has been reported in recent years (1–3), the limitations of the used explant types are of low regeneration rate and genotype-dependence, limiting to application in selected group of cultivated varieties (4–6). Hence, an alternative approach using shoot apices as explants was developed for free of genotype-dependence, with time-saving in regeneration of cotton plantlet in vitro. Renfroe and Smith (7) firstly reported the regeneration of cotton plants from shoot meristems, and isolated shoots could be cultured into rooted plants (8). Subsequently, other reports concluded that shoot meristems could be regenerated into cotton plants (9–11). Here, a more

Baohong Zhang (ed.), *Transgenic Cotton: Methods and Protocols*, Methods in Molecular Biology, vol. 958,
DOI 10.1007/978-1-62703-212-4_18, © Springer Science+Business Media New York 2013

efficient and detailed procedure for production of transgenic cotton plants was presented using shoot apices with *Agrobacterium*-mediated transformation. Shoot apex explants were transformed by coculture with *Agrobacterium* strain EHA105 carrying the binary vector pC-KSA contained phytase gene (*phyA*) and the marker gene neomycin phosphotransferase (NPTII). Transgenic plants with increased the phosphorus (P) acquisition efficiency were obtained (12).

2. Materials

2.1. Agrobacterium Strain and Vector

The *Agrobacterium* strain EHA105 contains *phyA* and a plant selection marker, the neomycin phosphotransferase (NPTII) gene. The plasmid construct of pC-KSA2300 (Fig. 1) was kindly provided by Prof. Minggang Li of the College of Life Sciences, Nankai University.

2.2. Cotton Seeds

Cotton (*G. hirsutum*) seeds from the cv. ND94-7 were provided by North China Key Laboratory for Crop Germplasm Resources of Education Ministry, Agricultural University of Hebei (see Note 1).

2.3. Seed Germination Medium

Based on the MS medium (13) solidified with 6.5 g/L agar, the pH of the medium had been adjusted to 5.8 prior to autoclaving at 121°C for 15 min, and dispensed into sterile petri dishes (30 ml/dish).

2.4. Shoot Apex Explants

Shoot apices were isolated from 3- to 5-day-old seedlings with the aid of a dissecting microscope (Fig. 2).

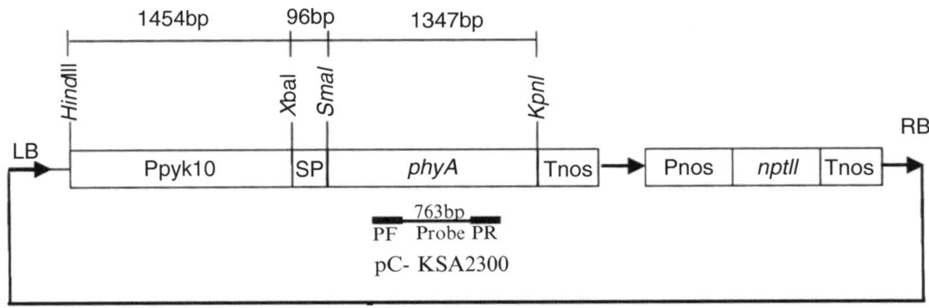

Fig. 1. Map of the recombinant binary vector pC-KSA2300. LB, left border. RB, right border. phyA, phytase gene. nptII, neomycin phosphotransferase. SP, a carrot extracellular targeting peptide sequence. The restriction maps and primers PF and PR used in probe production are shown with their respective locations on the pC-KSA2300 construct.

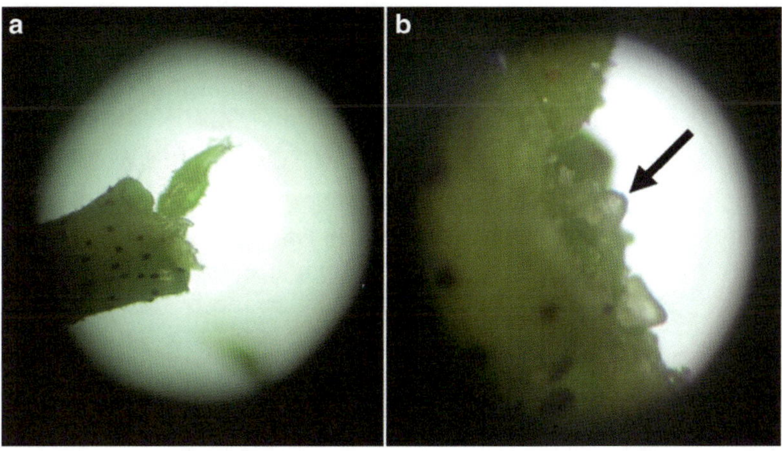

Fig. 2. Isolation of shoot apex of cotton. (**a**) Cotton shoot apex with two true leaf. (**b**) Isolated cotton shoot apex. The *arrow* indicates the specific sites of shoot apex.

2.5. Infection Medium	LB: Bacterial culture media is solidified using 15 g/L agar.

2.5. Infection Medium

LB: Bacterial culture media is solidified using 15 g/L agar.

LB+kan: LB + 50 mg/L kan, used with *Agrobacterium* strain EHA105.

MS + AS: MS, 100 µM acetosyringone (AS).

2.6. Coculture Medium

MS, 200 µM AS.

2.7. Selection Media

MS medium (13), 0.1 mg/L IAA, 0.1 mg/L 6-BA, 50 mg/L kan, solidified with 8 g/L agar, sterilized by autoclaving and dispensed into sterile petri dishes (see Note 2).

2.8. Rooting Media

MS medium, 0.1 mg/L IAA, 0.1 mg/L NAA, carbenicillin (AC) 500 mg/L, 50 mg/L kan, solidified with 8 g/L agar (see Note 3).

2.9. Potting Soil

The mixture soil composed of vermiculite + sand in a ratio of 1:1, mixed well, was prepared to transplant the plants from culture vessels to soil.

2.10. 75 mg/L kan Stock Solution

1. Dissolve 0.75 g kan sulfate in 350 ml of distilled water.
2. Make up volume to 1,000 ml with distilled water.
3. Sterilize by filtration and store at −20°C.

2.11. DNA and RNA Blotting Gel Components

1. Washing buffer: 0.1 M Maleic acid, 0.15 M NaCl, pH 7.5 (20°C), 0.3% (v/v) Tween 20 (see Note 4).
2. Maleic acid buffer: 0.1 M Maleic acid, 0.15 M NaCl, pH 7.5 (20°C) (see Note 5).
3. Blocking solution: Prepare a 1×working solution by diluting the 10×Blocking solution 1:10 in maleic acid buffer (see Note 6).

4. Antibody solution: Centrifuge anti-digoxigenin-AP for 5 min at $560 \times g$ in the original vial prior to each use, and pipet the necessary amount carefully from the surface. Dilute anti-digoxigenin-AP 1:10,000 (75 mU/ml) in blocking solution.

5. Detection buffer: 0.1 M Tris–HCl, 0.1 M NaCl, pH 9.5 (20°C) (see Note 7).

6. 10× MOPS: 0.2 M (41.86 g) mops free acid (MW: 209.3), 0.05 M (6.8 g) sodium acetate (MW136.1), 20 ml 0.5 M EDTA, pH 7.0 with NaOH and bring to a total volume of 1 L with DEPC-treated water (see Note 8).

3. Methods

Carry out all procedures under strict aseptic conditions unless otherwise specified.

3.1. Prepared Agrobacterium Strain

1. Culture the *Agrobacterium* strains in 20 ml of LB medium plus antibiotics (50 mg/L kan), for strain EHA 105 and incubate it in a 100 ml Erlenmeyer flask overnight (about 17 h) on a shaker set for $0.18–0.27 \times g$ at 28°C.

2. Then withdraw 2 ml of the overnight culture and inoculate 50 ml of LB medium without antibiotics.

3. Add AS to the culture with a final concentration of 200 μM. After incubation for 3–4 h at 28°C with shaking, dilute those cultures with additional LB medium (containing 200 μM AS) to a concentration (OD600 0.6) for transformation.

3.2. Seed Sterilization and Germination

1. Treat seeds with H_2O_2 (30%) for 30 min and 0.1% $HgCl_2$ for 10 min, and then wash with sterile distilled water (see Note 9).

2. Germinate 5–10 seeds per Erlenmeyer flask on seed germination medium in the dark at 28°C for 1 day, 16/8 h for 2 days (see Note 10).

3.3. Shoot Apex Isolation and Infection

1. Isolate shoot apices from 3- to 5-day-old seedlings with the aid of a dissecting microscope (Fig. 3a–e). The detailed procedure as follows: firstly, remove the two cotyledons, then removed euphyllia under microscope and expose the shoot apex (see Note 11).

2. Shake the explant gently in the bacterial suspension for about 15 min and blot it on a sterile filter paper for 5 min.

3.4. Coculture

Put a piece filter paper on the coculture medium and transfer the infected shoot apex to the filter paper and coculture under dark

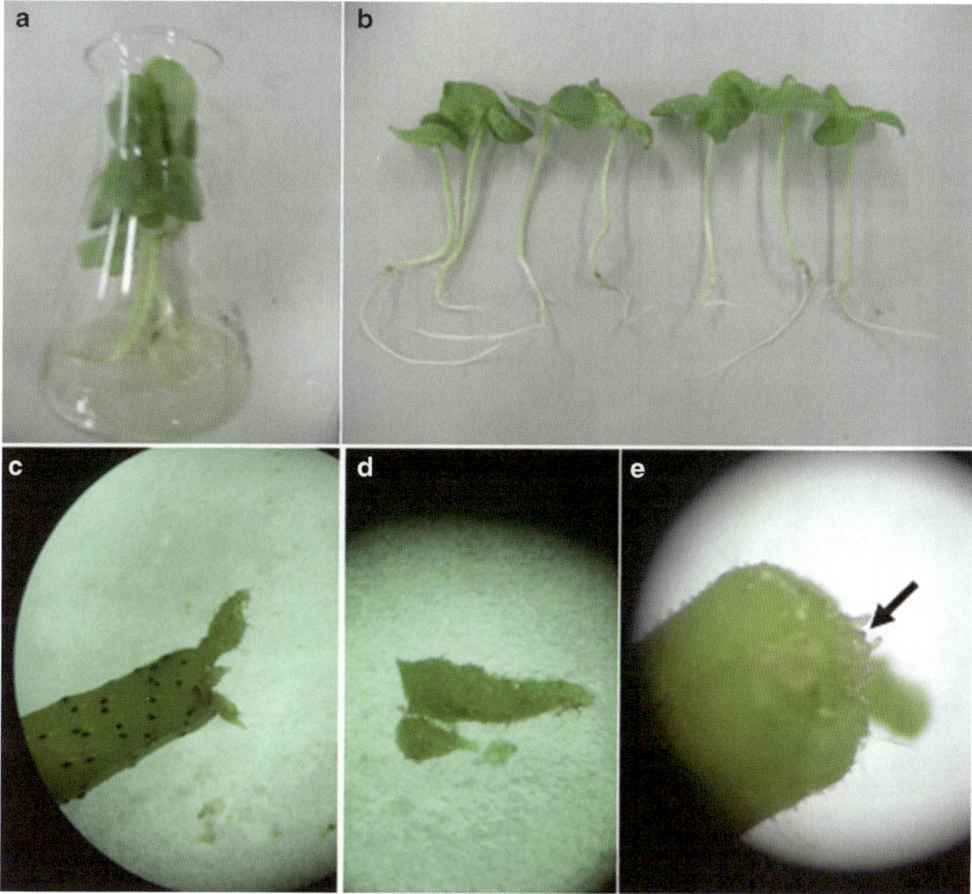

Fig. 3. The procedure of shoot apex isolation. (**a**, **b**) The 4-day-old seedling. (**c**) Removed cotyledon of shoot apex remaining euphyllia under microscope. (**d**) The shoot apex removed euphyllia. (**e**) The exposed shoot apex.

conditions for 2 days at 28°C, 20 shoots in each sterile petri dishes (*see* Note 12).

3.5. Selection

After coculture, wash the explant with sterilized distilled water, blot it and transfer it to selection medium (MS + 50 mg/L kan + 0.1 mg/L 6-BA + 0.1 mg/L IAA). Culture in the selection medium for 2–3 weeks (Fig. 4a, b), and then subculture one time.

3.6. Shoot Rooting Development

1. Root healthy and elongated shoots in rooting media (MS + 500 mg/LAC + 0.1 mg/L IAA + 0.1 mg/L NAA) for 3–4 weeks. Culture 7–10 shoots/plate (Fig. 4c, d) (*see* Note 13).

2. Transfer the individual plantlet after rooting to soilrite mix in plastic cups covered with polythene bags. After 1 week of

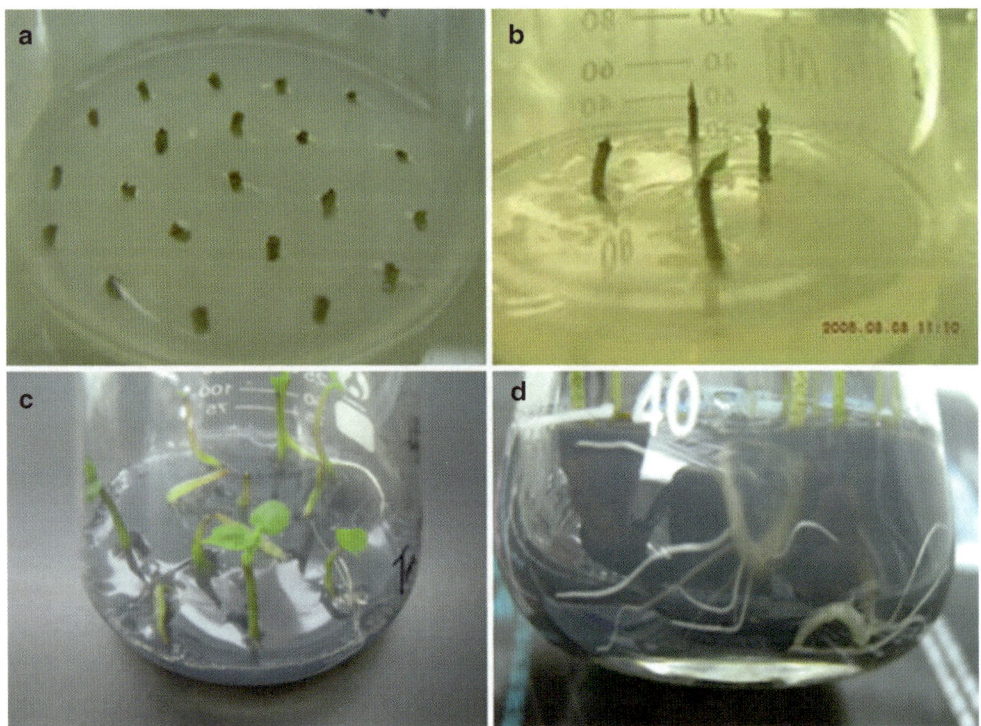

Fig. 4. The selection and rooting of shoot apex. Explants cultured on selection medium for 2 days (**a**) and 2 weeks (**b**) after cocultured. Shoot explants cultured on rooting media for 3 weeks (**c**) and 4 weeks (**d**).

hardening, transfer the plantlets to pots filled with vermiculite and sand (1:1) and later to green house (see Note 14).

3.7. Transfer of Rooted Plantlets to Soil

1. Prepare a soil mixture used for culturing the plants from culture vessels. The mixture is composed of vermiculite and sand medium in a ratio of 1:1, mixed well, and moistened with distilled water (see Note 15).

2. Take the rooted shoots out of the culture vessel, wash with tap water to remove the medium from roots and plant it into the pot filled with soil mixture (see Note 16).

3.8. Flowering and Pollination

1. Flowering of the T_0 plants occurs within 50 days.

2. Cotton plants grown in a greenhouse are self-pollinated (see Note 17).

3.9. Leaf-Painting Assay for Progenies

1. Paint first euphyllia of the healthy normal plantlets using kan at a concentration of 75 mg/L (T_1, T_2 and T_3 generation) (see Note 18).

2. For resistant transgenic plants to kan, its leaf color would remain green, while sensitive untransgenic plants would become yellow (Fig. 5).

3.10. Molecular Characterization of Putative Transgenic Plants

Southern blot is performed according to the procedure described by Sambrook et al. (14).

1. Label a 763 bp fragment (used as a probe) of the *phyA* amplified from pC-KSA2300 using primers PF (5′-TCATTTCAGA GGCCAGCACATC-3′, forward) and PR (5′-AAACCCTTCA CGAAGCTATCCC-3′, reverse) with DIG (DIG High Prime DNA Labeling and Detection Starter Kit II, Roche) according to the manufacturer's instructions (see Note 19) (Fig. 6).

Fig. 5. Kan leaf-painting assay of T$_1$ transgenic plants expressing the new gene (*left*) and a non-transformed plant (*right*), 7 days after kan application. The *arrow* indicates the specific sites of kan application.

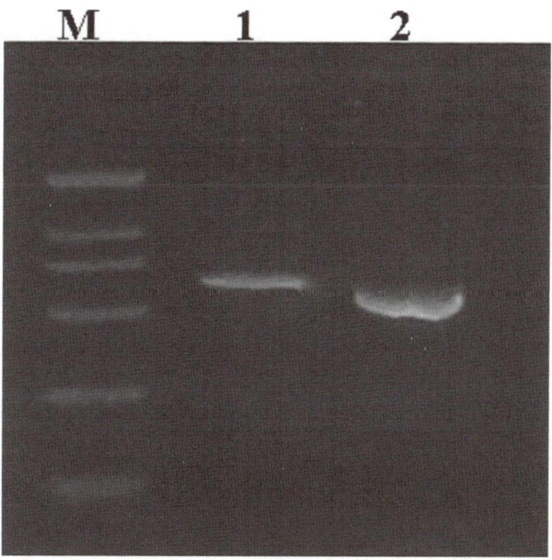

Fig. 6. Detection of DIG-labeled probe. Lane M, DL2000. Lane 1, DIG-labeled probe. Lane 2, control (unlabeled probe).

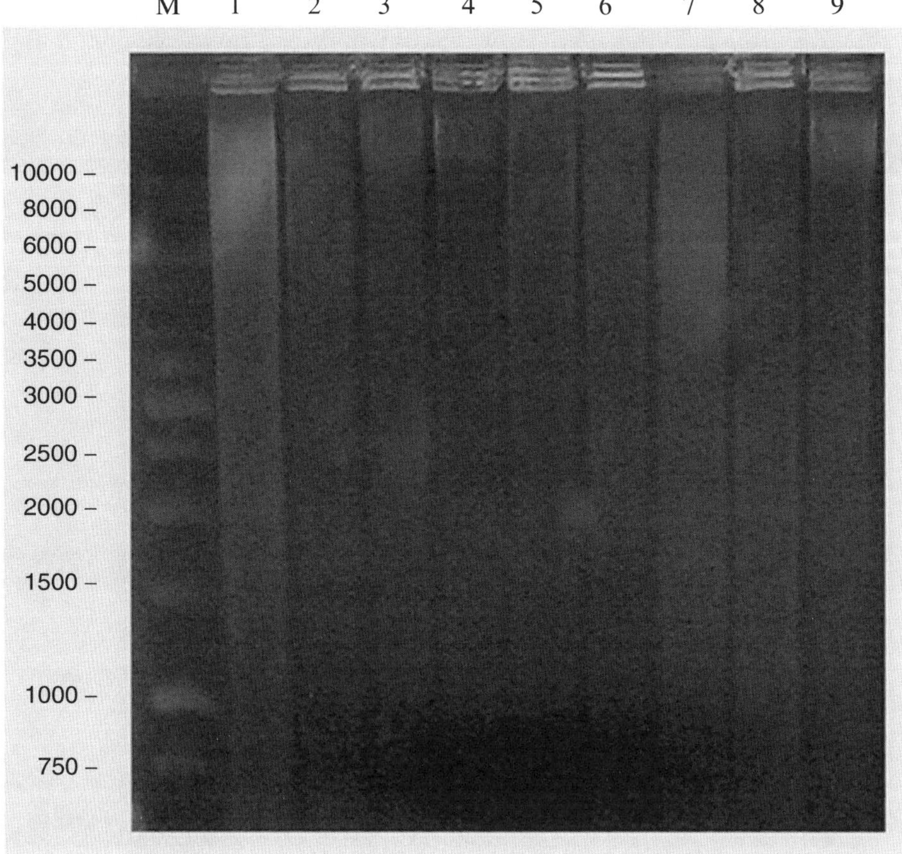

Fig. 7. The result of genomic DNA digested with *Kpn*I. Lane M, 1 kb DNA Ladder. Lane 1, the genomic DNA of wild-type plants. Lane 2–9, the genomic DNA of transgenic plants.

2. Digest genomic DNA samples of cotton leaf (60 μg) with *Kpn*I enzymes (see Note 20) (Fig. 7), then separate digested DNA in 1% (w/v) agarose gels and transfer it to membrane (Hybond-N$^+$ positively CHGD Nylon Transfer Membrane, Amersham Biosciences) (see Note 21).

3. Then perform prehybridization, hybridization, and detection using DIG-High Prime DNA Labeling and Detection Starter Kit II (Roche) according to the supplier's instructions.

4. The number of hybridization signals show that all transformants have a single copy of the transgene in the genome. DNA isolated from wild-type plants does not hybridize with the *phyA* probe (Fig. 8a).

5. Isolate total RNA (25 μg) from roots and then separated by electrophoresis in 1% agarose–formaldehyde denaturing gel using 1 × MOPS as running buffer, and then transfer it to Hybond N$^+$ nylon membrane (Amersham Bioscience) (see Note 22).

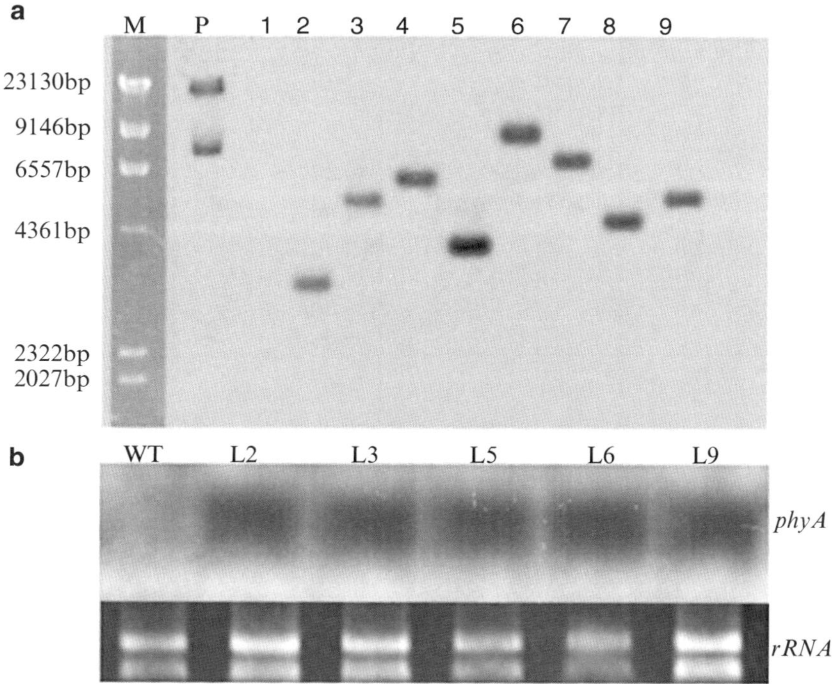

Fig. 8. Detection of *phyA* in T_0 transgenic and wild-type cottons (**a**) Southern blot analysis. Lane M, λDNA/*Hind*III marker. Lane P, plasmid pC-KSA2300. Lane 1, wild-type plant. Lane 2–9, transgenic plants. (**b**) Expression of *phyA* in transgenic plants. Lane WT, wild-type plants; Lane 2, 3, 5, 6 and 9, transgenic plants.

6. Perform probe labeling, hybridization and detection using DIG-High Prime DNA Labeling and Detection Starter Kit II (Roche) according to the supplier's instructions. The *phyA* transcripts are detected in the transgenic plants but not in wild-type plants (Fig. 8b).

3.11. Measurement of Total P

1. Designate the primary transformed plants as T_0 plants, T_1, T_2 and T_3 plants (first, second, and third generations). All the transgenic plants are self-pollinated.

2. Germinate cotton seeds in the greenhouse at a day/night temperature regime of 28/20°C under a 16 h photoperiod.

3. For the measurement of total P, collect dried shoot of T_3 transgenic homozygous lines in 15 mm × 100 mm glass tubes with three replicates (100 mg/replicate), and add 25 ml of 10% $Mg(NO_3)_2$ solution to each tube. Ashes the samples by shaking over strong flames until the brown fumes disappeared.

4. Dissolve the sample ashes in 1.5 ml of 0.5 M HCl for P determination by colorimetric assay at 820 nm wave (see Note 23).

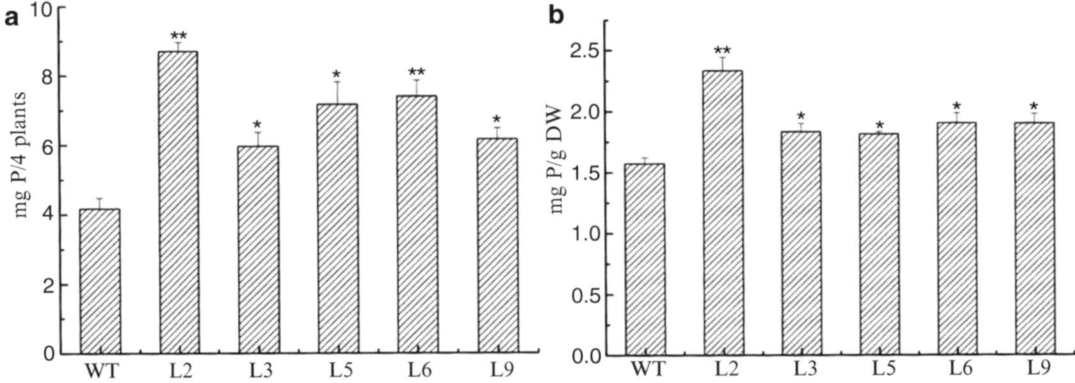

Fig. 9. Phosphorus acquisition analysis of T_3 transgenic and wild-type cottons supplied with phytate (+Po) as the sole source of P. (**a**) Total P content in shoots after 30 days of growth. (**b**) Concentrations of P in shoots after 30 days of growth. WT, wild-type plant. L2, L3, L5, L6, and L9, transgenic lines carrying the *phyA*. Data are presented as means $\pm$ SD ($n=3$) (*$P<0.05$, **$P<0.01$).

5. The result shows that total P contents of all the lines studied are significantly higher than those of wild-type plants. Total P accumulated by the two highest lines, L2 and L6, was 1.8- and 2.1-fold, respectively, of that accumulated by wild type (Fig. 9a) and five transgenic lines show significantly higher concentrations than the wild types (Fig. 9b).

4. Notes

1. The seeds should be fresh and delinted.

2. The pH of the MS medium+0.1 mg/L IAA+0.1 mg/L6-BA+8 g/L agar should be adjusted to 5.8 prior to autoclaving at 121°C for 15 min, and after it cooled to 40–50°C, the 50 mg/L kan was added.

3. The pH of the MS medium+0.1 mg/L IAA+0.1 mg/L NAA+500 mg/L AC+8 g/L agar should be adjusted to 5.8 prior to autoclaving at 121°C for 15 min, and after it cooled to 40–50°C, the 50 mg/L kan was added.

4. It is stable to store at 15–25°C.

5. It is stable to store at 15–25°C.

6. Always prepare fresh.

7. It is stable to store at 15–25°C.

8. Sterile filter and wrap with foil to store in dark at 4°C.

9. Cotton seed is usually difficult to decontaminate. Here, we provide an effective method. Wrap delinted seeds in cheesecloth and soak in water 30 min. Remove water and soak seeds in H_2O_2 (30%) for 30 min. Rinse three times with sterilized water. Soak seeds in 0.1% $HgCl_2$ for 30 min, change the solution every 15 min. Remove $HgCl_2$ and rinse seeds 30 min in sterilized water, change the solution every 10 min.

10. Remove the seed coat prior to germination to reduce contamination.

11. The seedling shoot is embedded in the stem between the cotyledons. Remove one cotyledon by pushing down on it until it snaps off.

12. There would better not exist bubbles between filter paper and the solid medium.

13. Shoots that do not root should be returned to MS for 2–4 weeks, then transferred to rooting media.

14. If shoots haven't rooted, another alternative is to graft the shoot onto a seedling of the same variety. Place under high light, cover with a lid from a plastic Petri plate and keep watered.

15. Filled in an autoclavable bag and autoclaved. After autoclaving the soil composite was cooled and filled in plastic pots having a drain hole at the bottom.

16. 20–50 ml of Hoagland solution was added and pots were covered with a polythene bag. Plants were kept in a growth room at $30° \pm 2°C$ for 16 h, photoperiod (250–300 $\mu M/m^2$ s). After 2 weeks the covers were removed completely. Hoagland solution was added if required. The plants were ready for shifting to glasshouse in earthen pots.

17. Plants need to be hand pollinated to insure self-pollination when determining segregation patterns.

18. To develop a selective growth system for genetically transformed cotton, we first studied the effects of various concentrations of kan on wild-type plants (cv. ND94-7). Leaves of the seedling are pre-painting the solution medium containing 0, 25, 50, 75 and 100 mg/L kan, respectively. After 2 weeks, we examine the leaf color. The selective concentration was used as our selection concentration when 50% of the shoot regeneration survived.

19. The DNA fragment complete denaturation is essential for efficient labeling, longer incubations (up to 20 h) will increase the yield of DIG-labeled DNA and the length of labeled-DNA fragment is longer than unlabeled-DNA.

20. Need to digest enough genomic DNA of cotton (60 μg) and be digested to completion. To avoid partial digests the DNA

should be relatively clean. Performing the digestion in a very large volume (400 μl) with an excess of enzyme also reduces partial digests because it dilutes salts and other potential contaminants.

21. An overnight run at 30–45 V for 14–16 h using TAE gel buffer produces highly resolved bands in the size range of about 500 bp–10 kb.

22. Formaldehyde vapor is toxic and casting this gel should be performed in a fume hood. The gel tank must be covered when in use.

23. The samples should be burned to ashes at 550–600°C of high temperature for 16 h.

References

1. Trolinder NL, Goodin JR (1987) Somatic embryogenesis and plant regeneration in cotton (*Gossypium hirsutum* L.). Plant Cell Rep 14:758–776

2. Rajasekaran K, Grula JW, Hudspeth RL, Pofelis S, Anderson DM (1996) Herbicide-resistant Acala and Coker cottons transformed with a native gene encoding mutant forms of acetohydroxyacid synthase. Mol Breed 2:307–319

3. Zhang BH, Feng R, Liu F, Wang QL (2001) High frequency somatic embryogenesis and plant regeneration of an elite Chinese cotton variety. Bot Bull Acad Sin 42:9–16

4. Feng R, Zhang BH, Zhang WS, Wang QL (1998) Genotype analysis in cotton tissue culture and plant regeneration. In: Larkin PJ (ed) Proceedings of the 4th Asia-Pacific conference on agricultural biotechnology, Darwin, 13–16, July. Canberra, UTC Publishing, pp 161–163

5. Cousins YL, Lyon BR, Llewelly DJ (1991) Transformation of an Australian cotton cultivar: prospects for cotton improvement through genetic engineering. Aust J Plant Physiol 18:481–494

6. Trolinder NL, Goodin JR (1987) Somatic embryogenesis and plant regeneration in cotton (*Gossypium hirsutum* L.). Plant Cell Rep 14:758–776

7. Renfroe MH, Smith RH (1986) Cotton shoot tip culture. In: Proc Beltwide Cotton Prod Res Conf 78–79

8. Gould J, Banister S, Hasegawa O, Fahima M, Smith RH (1991) Regeneration of *Gossypium hirsutum* and *G. barbadense* from shoot apex tissues for transformation. Plant Cell Rep 10:12–16

9. Nasir AS, Zafar Y, Malik KA (1997) A simple procedure of *Gossypium* meristem shoot tip culture. Plant Cell Tiss Org Cult 51:201–207

10. Morre JL, Permingeat HR, Maria VR, Cintia MH, Ruben HV (1998) Multiple shoots induction and plant regeneration from embryonic axes of cotton. Plant Cell Tiss Org Cult 54:131–136

11. Zapata C, Park SH, El-Zik KM, Smith RH (1999) Transformation of a Texas cotton cultivar by using *Agrobacterium* and the shoot apex. Theor Appl Genet 98:252–256

12. Liu JF, Zhao CY, Ma J, Zhang GY, Li MG, Yan GJ, Wang XF, Ma ZY (2011) *Agrobacterium*-mediated transformation of cotton (*Gossypium hirsutum* L.) with a fungal phytase gene improves phosphorus acquisition. Euphytica 181:31–40

13. Murashige T, Skoog F (1962) A revised medium for rapid growth and bioassays with tobacco tissue culture. Plant Physiol 15:473–497

14. Sambrook J, Fritsch EF, Maniatis T (1989) Molecular cloning: a laboratory manual. Cold Spring Harbor Laboratory, New York

Chapter 19

Genetic Transformation of Cotton with a Harpin-Encoding Gene *hpaXoo* Confers an Enhanced Defense Response Against *Verticillium dahliae* Kleb

Weiguo Miao and Jingsheng Wang

Abstract

The soil-borne fungal pathogen *Verticillium dahliae* Kleb causes Verticillium wilt in a wide range of crops including cotton (*Gossypium hirsutum*). To date, most upland cotton varieties are susceptible to *V. dahliae* and the breeding for cotton varieties with the resistance to Verticillium wilt has not been successful. Hpa1Xoo is a harpin protein from *Xanthomonas oryzae* pv. oryzae which induces the hypersensitive cell death in plants. When hpa1Xoo was transformed into the susceptible cotton line Z35 through Agrobacterium-mediated transformation, the transgenic cotton line (T-34) with an improved resistance to *Verticillium dahliae* was obtained. Here, we describe the related research approach, such as Western blot, Southern blot, immuno-gold labeling, evaluation of resistance to *Verticillium dahliae*, and how to detect the micro-hypersensitive response and oxidative burst elicited by harpin$_{Xoo}$ in plant tissue.

1. Introduction

The soil-borne fungal pathogen *Verticillium dahliae* Kleb causes *Verticillium* wilt in a wide range of crops including cotton (*Gossypium hirsutum*). *V. dahliae* can be found in many cotton-growing areas and it has been considered as a major threat to the cotton production worldwide (1). The reduction of cotton biomass caused by *Verticillium* wilt is mainly due to the discoloration of cotton leaves and stems vascular bundles, decreased photosynthesis, and increased respiration (2, 3).

V. dahliae infects cotton roots and then grows into the host vascular system. Symptoms caused by *V. dahliae* in cotton include the necrosis on leaves, wilting, and the discoloration of vascular tissues. Plants infected with *V. dahliae* often develop characteristic mosaic patterns (leaves wilt with interveinal yellowing before

Baohong Zhang (ed.), *Transgenic Cotton: Methods and Protocols*, Methods in Molecular Biology, vol. 958,
DOI 10.1007/978-1-62703-212-4_19, © Springer Science+Business Media New York 2013

becoming necrotic) (4). Light to dark brown vascular discoloration is common in stems and branches of the infected cotton. Pathogenesis of *V. dahliae* is complicated due to the existence of defoliating and non-defoliating strains. The defoliating strains are the most virulent, which can cause typical symptoms of *Verticillium* wilt and lead to the complete defoliation of infected plants (1). Cotton cultivars resistant to *Verticillium* wilt often show decreases in the rate of the disease progress and the symptom severity with a lower percentage of foliar symptoms (4).

Verticillium wilt in cotton is usually controlled by cultural practices, such as the crop rotation (5), biological control with organic amendments (6), and fungicides (7). Although the crop rotation and the application of organic amendments can be successfully in managing *Verticillium* wilt, these methods are not always practical (6). Chemical fungicides are not environment-friendly and tend to raise concerns about the public health and the development of fungicide resistance in pathogens (8). Moreover, none of the available commercial upland cotton varieties is immune to *V. dahliae* (9). Conventional breeding methods for cotton varieties resistant to *Verticillium* wilt have not been successful.

Genetic engineering utilizing plant genes conferring disease resistance offers an alternative to conventional breeding methods for the improved resistance against pathogens, insects, or herbicides (10). Genes encoding antifungal proteins, such as endochitinase (11), β-1,3-glucanases (12), and glucose oxidase (13), or components of signaling pathways involved in the defense response (14–17), have been used to generate transgenic plants with the improved resistance to various plant pathogens.

Several attempts have been made to generate transgenic cottons with a higher tolerance to *Verticillium* wilt. For example, a bean chitinase gene was transformed into cotton and crude leaf extracts from the transgenic cotton lines inhibited the growth of *V. dahliae* in vitro (18). Furthermore, a transgenic cotton line with an over-expressed foreign *Gastrodia* antifungal protein was more resistant to *Verticillum* wilt than the untransformed cotton (19).

Harpins, encoded by *hrp* (hypersensitive response and pathogenicity) genes from Gram-negative plant pathogenic bacteria, are secreted through the Type III protein secretion systems (T3SSs) (20). The T3SSs inject effector proteins directly into the cytosol of eukaryotic cells and thus allows the manipulation of host cellular activities to the benefit of the pathogen. In plant pathogenic bacteria, T3SSs are encoded by *hrp* (for hypersensitive response and pathogenicity) genes, which are capable of inducing host defense responses mediated by different signaling pathways, such as salicylic acid (SA) (21), jasmonic acid (JA) (21), and ethylene-mediated pathways (22). HarpinXoo is a harpin-like protein encoded by *hpa1Xoo* derived from *Xanthomonas oryzae* pv. *oryzae* (Xoo), which belongs to *hpa* (hrp-associated) gene family related to the pathogenicity of *Xanthomonas* and the induction of hypersensitive response (HR) in non-host plants (23–27). *Hpa1Xoo* encodes a 13.69 kDa glycine-rich protein with an amino acid composition similar to harpins from *Pseudomonas syringae* (28) and *Erwinia* species (29). Hpa1Xoo also shares a high sequence similarity to PopA, a harpin-like protein, from *Ralstonia solanacearum* (30). It has been proposed that Hpa could be involved in the secretion of Type III-dependent proteins. HpaA from *X. campestris* pv. *vesicatoria* promotes the secretion of pilus and effector proteins and therefore appears to be an important control protein of the T3SSs (31).

We had shown previously that the transformation of *hpa1Xoo* into tobacco conferred the improved resistance to *Alternaria alternata* and tobacco mosaic virus in the transgenic tobacco (32). Similarly, a high level of resistance to all predominant races of *Magnaporthe grisea* in China was obtained in the rice line transformed with *hpa1Xoo* from *Xanthomonas oryzae* pv. *oryzae* (33). Several other *hrp* genes have also been successfully transformed into different plant species including tobacco (32, 34), potato (35), rice (33), and pear (36). Unfortunately, the defense responses elicited by harpins and their active sites in hosts have not been fully understood.

In this study, a cotton transgenic line resistant to a range of soil-borne pathogens, including *V. dahliae*, was generated through the genetic transformation with *hpa1Xoo* from *Xoo*. The localization of hpa1Xoo in the transgenic cotton line was investigated. Furthermore the defense response and the expressions of defense-related genes in *hpa1Xoo*-expressing cotton line in response to *V. dahliae* were investigated.

2. Materials

2.1. Plant Materials

ZhongMian 35 (Z35) (*Gossypium hirsutum* L.) was used to generate transgenic cotton lines expressing *harpinXoo* using an *Agrobacterium tumefaciens*-mediated method. Hypocotyl segments of Z35 were used as explants for the transformation, and the transformants were selected.

2.2. Fungal Materials

A non-defoliating *V. dahliae* strain Vdps and a defoliating *V. dahliae* strain V151, obtained from Dr. Ling Lin at the Jiangsu Academy of Agricultural Science (China), were used in our the study.

1. *V. dahliae* strains were maintained on potato dextrose agar (PDA) at 25°C.

2. For the preparation of the inoculum, PDA plates were flooded with a conidial suspension of *V. dahliae* and the flooded plates were incubated at 25°C for 7 days.

3. The PDA plates were then flooded with 50 ml sterile distilled water to collect the conidia using the method described by Joost et al. (37).

4. The conidia were washed once with 100 ml sterile distilled water and the suspension was diluted to a concentration of $1–3 \times 10^7$ conidia/ml. The conidial suspension of *V. dahliae* strain Vdps was used to inoculate the roots of cotton plants and for other experiments.

2.3. Western Blot

1. Resolving gel (for 15 ml gel): 7.5 ml of 30% acrylamide, 3.4 ml of ddH_2O, 3.8 ml of 1.5 M Tris–HCl (pH 8.8), add 50 µl of 10% SDS, 150 µl of 10% APS (make fresh each time), and 15 µl of TEMED (see Note 1).

2. Stacking gel (for 5 ml gel): 0.83 µl of 30% acrylamide, 3.4 ml of ddH_2O, 0.63 µl of 1.0 M Tris (pH 8.8), add 50 µl of 10% SDS, 50 µl of 10% APS (make fresh each time), and 10 µl of TEMED (see Note 1).

3. SDS sample loading buffer (40 ml): 16 ml of ddH_2O, 5 ml of 0.5 M Tris (pH 6.8), 8 ml of 50% Glycerol, 8 ml of 10% SDS, 2 ml of 2-βmercaptoethanol (add immediately before use), bromophenol blue, 10% (v/v) acetic acid.

2.4. Immuno-Gold Labeling

1. Phosphate-buffered saline (PBS) buffer: For 1 L of 1× phosphate-buffered saline (1× PBS buffer) use: 8.00 g of NaCl, 0.20 g of KCl, 1.44 g of Na_2HPO_4, 0.24 g of KH_2PO_4. Dissolve in 800 ml of distilled H_2O, adjust the pH to 7.4 with HCl or NaOH, add distilled H_2O to 1 L, autoclave (20 min, 121°C, liquid cycle).

2. Blocking solution (BL): 50 mM PBS (pH 7.0), containing 1% Bovine Serum Albumin (BSA), 0.02% PEG20000, 100 mM NaCl and 1% NaN_3, and then filtered with 0.22 µm filter, packed to 1.5 ml, Store at 2–8°C (see Note 2).

---

3. Methods

3.1. Plant Transformation

ZhongMian 35 (Z35) (*Gossypium hirsutum* L.) was used to generate transgenic cotton lines expressing *harpinXoo* using an *Agrobacterium tumefaciens*-mediated method described by Shao et al. (33).

1. A pGEM-T vector containing *hpa1Xoo* (pGEM-*hpa1Xoo*) was digested with *Bam*HI and *Sac*I. The *Bam*HI- and *Sac*I-digested *hpa1Xoo* fragment was ligated into a pBI121 vector (Clontech, Palo Alto, CA, USA) to generate the recombinant binary vector pBI35S-*hpa1Xoo*-nptII, which contained a neomycin

phosphotransferase II (nptII) with a nopaline synthase (nos) promoter and terminator, a CaMV35S promoter, an *hpa1Xoo* insert, and a nopaline synthase terminator.

2. The binary vector pBI35S-*hpa1Xoo*-nptII was mobilized into *Agrobacterium tumefaciens*-disarmed helper strain LBA4404 by the heat shock method (38). Hypocotyl segments of Z35 were used as explants for the transformation, and the transformants were selected using the method described by Sunilkumar and Rathore (39).

3.2. Kanamycin Resistance Tests

1. Seedlings were screened for kanamycin resistance at the 3- to 4-leaf stage.

2. Kanamycin was applied onto the leaf surface at a concentration of 5,000 mg/L. kanamycin-susceptible seedlings, which changed from green to yellow, were discarded a week after the Kanamycin treatment. The treatment was repeated three times and only kanamycin-resistant plants were retained for further study.

3. From T1 to T6, the transgenic cotton lines were screened for the resistance to kanamycin, the presence of $hpa1_{Xoo}$ insertion, and the expression of $harpin_{Xoo}$. Only plants tested positive for the three attributes and showed an improved resistance to *Verticillium* wilt were selected and used for the further screening.

3.3. DNA Extraction, PCR Analysis, and Southern Blot Analysis

1. Total genomic DNA was extracted from leaves of transgenic cotton line T-34 and the untransformed Z35, using a AxyPrep Multisource Genomic DNA Miniprep Kit.

2. Primers for *hpa1Xoo*, CaMV35S promoter, and NOS terminator (listed in Table 1) were used in the PCR assays. PCR reactions were carried out in a 25 µl reaction volume containing 1× PCR buffer (Applied Biosystem), 1.5 mM $MgCl_2$, 0.2 mM dNTPs, 2.5 mM forward and reverse primers, 0.5U Taq polymerase, and 30 ng sample DNA. Amplifications were performed in a thermal cycler (GeneAmp PCR 9700) using the following temperature profile: initial denaturation at 95°C for 2 min; 35 cycles at 95°C for 30 s, 60°C for 30 s, and 72°C for 1 min; and a final extension at 72°C.

3. Four plants from T-34 line (T6 progeny) were randomly selected and used in the PCR analysis. Bands representing $hpa1_{Xoo}$, 35 S promoter, and NOS terminator (420, 310, and 180 bp, respectively) were detected in all four T-34 plants but they were absent in wild-type Z35 plants (Fig. 1a, b).

4. Results of the PCR analysis were verified by the sequencing of amplification products and BLAST against appropriate sequences in the NCBI database (http://www.ncbi.nlm.nih.gov) (data not shown).

Table 1
Oligonucleotides used in PCR and quantitative RT-PCR

Gene	Primer sequence	Anneal temperature (°C)	Segment length (bp)
hpa1Xoo (EF028092[a])	Forward:5′-TTCGGATCCATGAACTCTTTGAACACACAATT-3′ Reverse:5′-GGTGAGCTCTTACTGCATCGATGCGCT-3′	56	438
35S	Forward:5′-AGAGGCTTACGCAGCAGGTC-3′ Reverse:5′-GCCAGTCTTTACGGCGAGTT-3′	52	310
NOS	Forward:5′-GAACTGACAGAACCGCAACG-3′ Reverse:5′-ACCGAGGGGAATTTATGGAA-3′	50	180
GhAOX 1(DQ250028)	Forward:5′-GCGCCTGGGGATGATGATGAGTCGTG-3′ Reverse:5′-GCGCTTCAGTGATAACCGAGCGGAG-3′	57	1,298
hsr203J(X77136)	Forward:5′-TGTACTACACTGTCTACACGC-3′ Reverse:5′-GATAAAAGCTATGTCCCACTCC-3′	55	618
EF-1α(AJ223969)	Forward:5′-AGACCACCAAGTACTACTGCAC-3′ Reverse:5′-CCACCAATCTTGTACACATCC-3′	58	495
Ghdhg-OMT(GQ303569)	Forward:5′-ATGAATATGGGCAATGCTAAT-3′ Reverse:5′-TCAGGGGTAAACCTCAATGAGA-3′	53	1,087
npr1(U76707)	Forword:5′-GGCCTCGAGATGGCTTATTTGTCTGAGCCATCATCT-3′ Reverse:5′-CGTCTCGAGTCACAATTTCCTATACTTGTAGG-3′	62	1,794
hin1(Y07563)	Forword:5′-GAACGGAGCCTATTATGGCCCTTCC-3′ Reverse:5′-CATGTATATCAATGAACACTAAACGCCGG-3′	55	867

[a]GenBank accession numbers

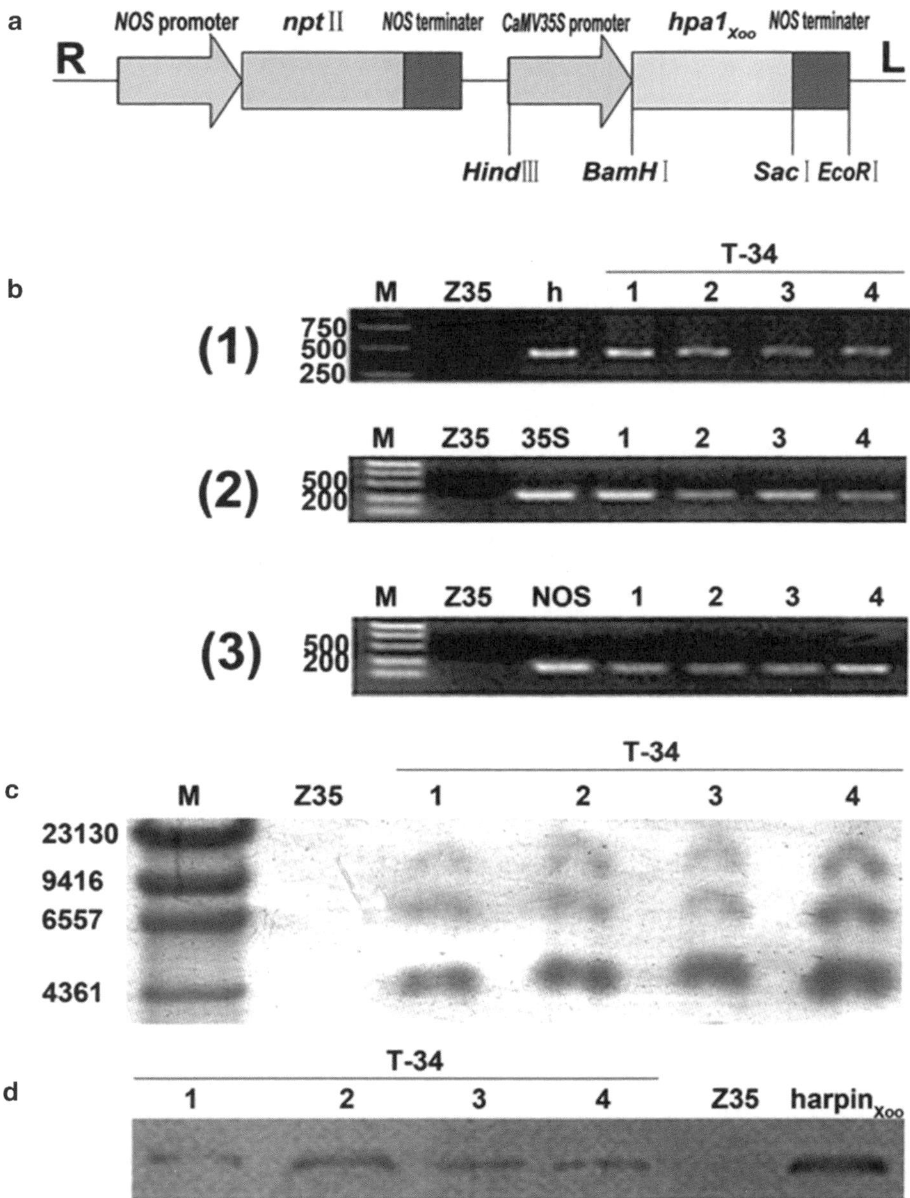

Fig. 1. Molecular analysis of *hpa1Xoo*-transformed T-34 and untransformed cotton (*Gossypium hirsutum*) Z35. (a) The schematic representation of recombinant plasmid pBI35S-*hpa1Xoo*-nptII. R and L represent the right and left borders of T-DNA. (b) Amplifications of *hpa1Xoo* (1), 35S promoter (2), and NOS terminator (3) in PCR; *hpa1Xoo* (h), 35S promoter (35S), and NOS terminator (NOS) represented the DNA fragment amplified from the positive control. M: marker; Four individual plants (1, 2, 3, 4) in T6 progeny of T-34 were tested. (c) Southern blot analysis of *hrp1Xoo* insertions in T-34 and Z35. Ten micrograms of genomic DNA was digested with *Eco*RI and hybridized against a DIG-labeled *hpa1Xoo* probe. M: marker. Four plants (1, 2, 3, 4) in T6 progeny of T-34 were tested. (d) Western blot analysis of harpinXoo in T-34 transgenic lines. Four plants (1, 2, 3, 4) in T6 progeny of T-34 were tested. Purified harpinXoo served as the positive control.

5. For the Southern blot analysis, 3 μg genomic DNA extracted from leaves of the transgenic T-34 and untransformed Z35 was digested with restriction endonuclease EcoR1 in a final volume of 50 μl.

6. The digested genomic DNA was separated on 1.5% (w/v) agarose gel and then transferred onto a hybond- N + nylon membrane after denaturation using the method prescribed by the manufacturer.

7. The probe for hybridizations was amplified from an *hpa1Xoo* fragment and then labeled with digoxigenin using DIG-High Prime DNA Labeling Kit.

8. The hybridization signal was detected using a DIG-High Prime DNA Detection Kit.

The presence of *hpa1*$_{Xoo}$ inserts in transgenic plants was confirmed using Southern blot analysis against a DIG-labeled *hpa1*$_{Xoo}$ probe. Three bands, approximately 4.5, 6.5, and 10.5 kb in length, were detected using the DIG-labeled *hpa1*$_{Xoo}$ probe in the genomic DNA extracted from the four chosen T-34 plants. No positive signal was detected in untransformed Z35 (Fig. 1c).

3.4. Western Blot Analysis

1. Weigh 100 mg fresh cotton leave in a 2 μl tube, freeze the plant tissue in liquid nitrogen, and grind using a rotor starter.

2. Add 300 μl cellytics and 3 μl protease inhibitor cocktail immediately, continue to grind homogenously.

3. Transfer the mixture to 1.5 ml microfuge tube, centrifuge at more than $12,000 \times g$ for 20 min at ambient temperature.

4. Transfer the supernatant into a clean 1.5 ml microfuge tube. The supernatant is the soluble protein of cotton. Concentration determination of the protein extracted by this method should use Braford protein assay instead of BCA protein assay.

5. Total proteins were separated on a 15% sodium dodecyl sulfate-polyacrylamide gel electrophoresis (SDS-PAGE) and then transferred onto a polyvinylidene fluoride (PVDF) membrane.

6. The membranes were blotted with a polyclonal antibody developed against *harpinXoo* and goat anti-rabbit IgG-HRP antibody. The color was developed using DAB. The expression of harpin$_{Xoo}$ in cotton leaves was analyzed using a harpin$_{Xoo}$ polyclonal antibody. The band representing harpin$_{Xoo}$ was observed only in the total proteins extracted from leaves of *hpa1Xoo*-transformed T-34 (Fig. 1d).

All these results indicated that *hpa1*$_{Xoo}$ had been successfully transformed into T-34 and hpa1$_{Xoo}$ was constitutively expressed in the transgenic line T-34.

3.5. Evaluation of Resistance to *Verticillium dahliae*

1. After the surface disinfection for 5 min with a 5% solution of sodium hypochlorite, cotton seeds were sown in a potting mixture (mould and sand, 6:1, v/v).

2. Fifteen 2-week-old cotton seedlings were carefully uprooted and the roots were immersed for 15 min in 100 ml of conidial suspension containing $1-3 \times 10^7$ conidia per ml.

3. Fifteen control plants were immersed in sterile distilled water. All plants were then replanted in a plastic pot (9 cm in diameter) and grown under 12 h of light at 25°C and 70–90% relative humidity.

Pathogenicity was determined based on both external (foliar damage) and internal (vascular discoloration) symptoms 10 and 20 days after inoculation, respectively.

Foliar damage

1. Foliar damage was evaluated by rating the symptom on the cotyledon and leaf of inoculated plant (X) according to the following rating scale: 0 = no foliar symptoms; 1 = yellowing or necrosis of 1–2 cotyledons; 2 = cotyledon fall or yellowing of a leaf; 3 = more than 2 wilted or necrotic leaves; 4 = dead leaf.

2. Foliar alteration index (FAI) was calculated for each inoculated plant : $FAI = 100\Sigma X/(4n)$, where (4) is the maximum score for each plant (maximum score for each plant = 4), (*n*) the total number of inoculated plant.

Vascular discoloration

1. Vascular discoloration was evaluated according to the method described by Yang et al. (40); discoloration was scored (*y*) for every internode using the following scale: 0 = no discoloration; 1 = less than 25% localized brown regions within the vascular tissue of the same internode; 2 = 25–70% localized brown regions within the vascular tissue of the same internode; 3 = more than 70% browning of vessels but not of the adjacent tissues; 4 = browning of both vessels and adjacent tissues.

2. The browning index (BI) was calculated as follows; $BI = 100\Sigma y/4d$, where (*d*) is the total number of seedling internodes including hypocotyls and (4) is the maximum score for an internode. Mean values of FAI and/or BI as Disease severity (DS) were calculated based on four replicates for both inoculated and control plants.

In 2008, the resistance of transgenic cotton was evaluated in a naturally infested *Verticillium* wilt nursery in DaFeng city, Jiangsu province, China. The soil in the nursery is sandy-loam with pH of 8.5. Except for the higher temperature (>30°C) in August, average temperature at the nursery usually ranges between 20 and 25°C during the growing season, which is conducive for the development of *Verticillium* wilt.

Seeds of the transgenic cotton were sown in the field in early May of 2008. Irrigation was provided as needed during the growing season. The experimental plot was divided into four subplots. Each subplot consisted of two rows. Each row is 5 m long and 4 m wide and comprises 15 plants spaced 0.3 m apart. Each treatment was replicated four times, and the replicates were arranged in a randomized complete block design. Untransformed cotton Z35 served as the negative control for *Verticillium* wilt. The testing materials in each replicate were sown randomly in each subplot. The trial plot was separated by at least 50 m from other breeding materials of cotton and sprayed with pesticides to control insect pests. Scoring for disease severity started after the first symptoms appeared on leaves; subsequently, the scoring was conducted on 22 June, 5 August, and 30 August 2008.

The typical symptoms of *Verticillium* wilt first appeared on plants 10 days after the inoculation, and symptoms develop only when the temperature is below 30°C (1). In our study, *Verticillium* wilt resistance of 45 *hpa1Xoo*-transformed T-34 plants inoculated with *V. dahliae* strains Vdps and V151 was assessed 10 days after the inoculation based on the degree of the foliar damage and vascular discoloration as described in the material and method. All plants were individually scored. The susceptible variety, Simian 3, and untransformed Z35 were used as the control. Ten days after the inoculation, only few chlorotic and necrotic spots were visible on leaves of T-34, whereas large chlorotic and necrotic areas were common in leaves of untransformed Z35 and the susceptible line Simian 3 (pictures not shown).

The resistance of T-34 to *Verticillium* wilt was evaluated in the field in 2008. A total of 200 plants were scored. The characteristic mosaic pattern of *Verticillium* wilt was rare in leaves of T-34 and no defoliation occurred during the growing season. In comparison, most Z35 plants showed severe *Verticillium* infections with the characteristic mosaic pattern on leaves and the defoliation occurred 2 or 3 months after the inoculation (Fig. 2a).

The maximum temperature reached 32°C on August 5, 2008 and typical *Verticililum* symptoms were no longer visible. Disease assessment made from 22 June to 5 August showed that *Verticillium* wilt was significantly less severe in *hpa1Xoo*-transformed T-34, compared to untransformed Z35 and the susceptible variety Simian 3 (Fig. 2b). The average *Verticillium* wilt ratings in *hpa1Xoo*-transformed T-34 were 7.32–26.22% lower than those in untransformed Z35.

Although the defoliating strain V151 was more virulent than the non-defoliating strain Vdps (1) on T-34, the disease severity caused by these two stains were both lower on T-34, compared to the untransformed Z35 and the susceptible control Simian 3 (Fig. 2c).

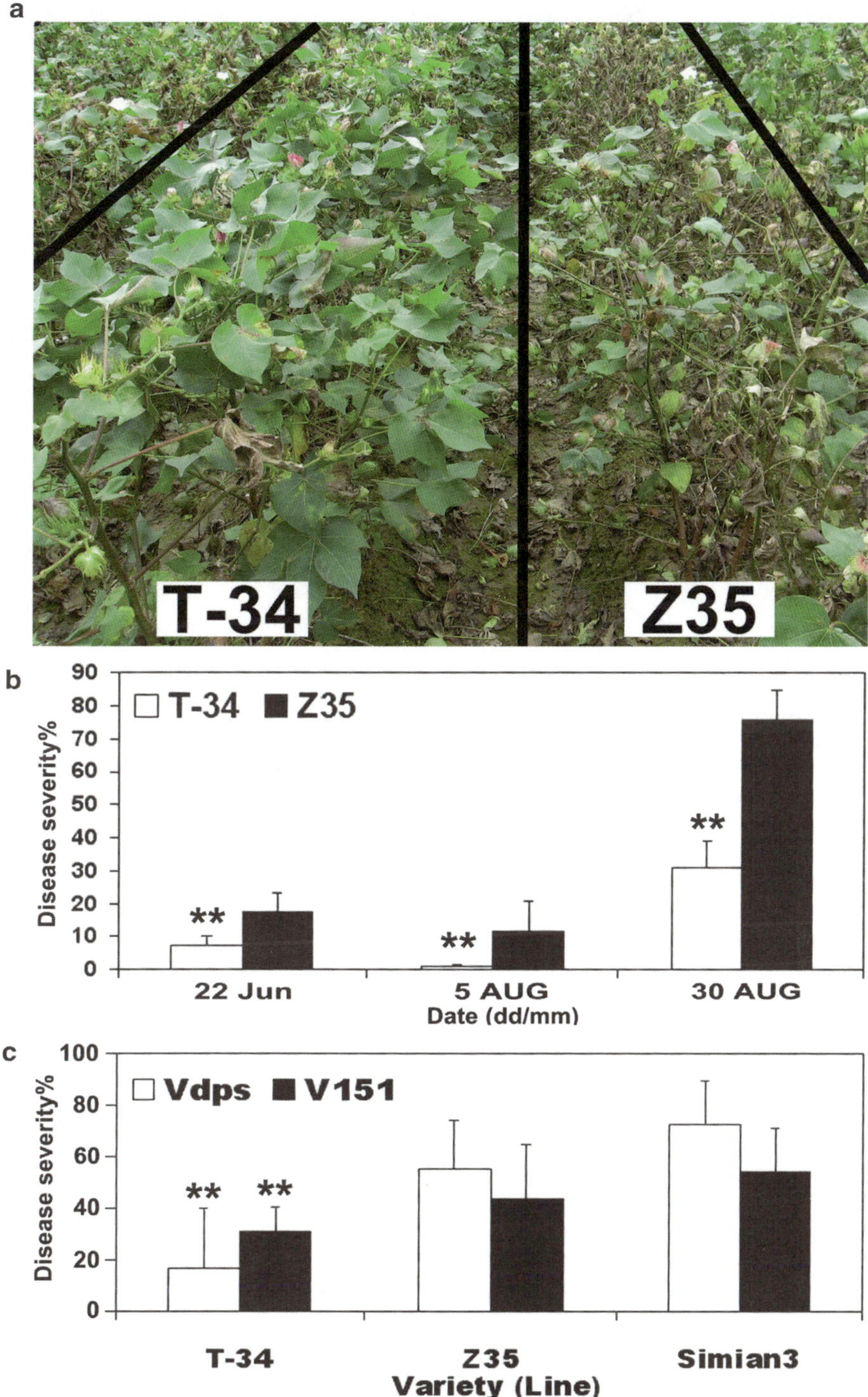

Fig. 2. Resistance of *hpa1Xoo*-transformed T-34 and untransformed Z35 to *Verticillium* wilt. (**a**) Resistance phenotypes of *hpa1Xoo*-transformed T-34 and untransformed Z35 to *Verticillium* wilt in the nursery. (**b**) Disease severity of *Verticillium* wilt on *hpa1Xoo*-transformed T-34 and untransformed Z35 in the nursery. (**c**) Disease severity of *Verticillium* wilt on *hpa1Xoo*-transformed T-34 and untransformed Z35 in plastic pots. Average values and standard errors were calculated from four replicates. Simian 3 was the susceptible control. *Asterisks* represent significant differences at the level of 0.01.

3.6. Preparation of Plant Samples and Immuno-Gold Labeling

1. Samples were collected from the second and the third fresh leaves and stem apex of four T-34 and two Z35 plants at 4–5 leaf stage.

2. The leaf samples were first fixed in a mixture of 3% (v/v) paraformaldehyde and 1% glutaraldehyde in 50 mmol phosphate-buffered saline (PBS), pH 7.2, at 4°C for 3 h.

3. The samples were then washed with the same buffer and dehydrated in 50% ethanol at 4°C for 1 h, followed by washings with 50%, 70%, 90%, and 100% ethanol (3 times each) at –20°C for 2 h.

4. Finally, the samples were embedded in K4M resin and polymerized under UV array at –20°C for 3 days and incubated at the room temperature for 2 days. Ultrathin sections were cut with a diamond knife and collected on Formvar-coated nickel grids.

5. Colloidal gold particles, 15 nm in diameter, were prepared as described by Slot and Geuze (41) and coated with Protein A (PA) at pH 6.0.

6. HarpinXoo antiserum was used for the localization of harpinXoo and the immuno-labeling was performed at 28°C.

7. The ultrathin sections were floated on a drop of double-distilled water for 5 min; the samples were then transferred to the blocking solution (BL) and incubated for 60 min.

8. HarpinXoo antiserum was added to the BL in a dilution of 1:200; the samples were incubated for 60 min; floated first on BL for 60 min and then on PA-gold for 120 min.

9. The samples were thoroughly washed with double-distilled water three times and air-dried. More than 20 ultrathin sections of each sample were examined with a JEM×1200 transmission electron microscope (Nikon, Japan). The experiment was repeated twice.

The localization of harpinXoo in tissues of *hpa1Xoo*-transformed T-34 was investigated using the immuno-gold localization method. The harpinXoo-labeled gold particles were not found in leaf and stem samples collected from the untransformed Z35 (Fig. 3c, f), but they were clearly visible in leaf and stem samples from T-34 (Fig. 3b, e). HarpinXoo-labeled gold particles were mostly seen in clusters along the cell walls of leaves and in apical tissue of stems (Fig. 3b, e). Each cluster contained an average of 10–20 gold particles. Only a few gold particles were found in cell membranes and chloroplasts. None was found in the mitochondria (Fig. 3c).

3.7. Preparations of Cotton Cell Suspension and Assay for Tolerance to Verticillium dahliae

1. Suspensions of Z35 and T-34 cells were prepared using the method described by Wu et al. (42).

2. The cells were cultured in a 500 ml round-bottom flask containing 200 ml modified MS medium (MS; 2,4-D 0.1 mg/L, kinetin 0.1 mg/L, maltose 30 g/L; pH 5.8).

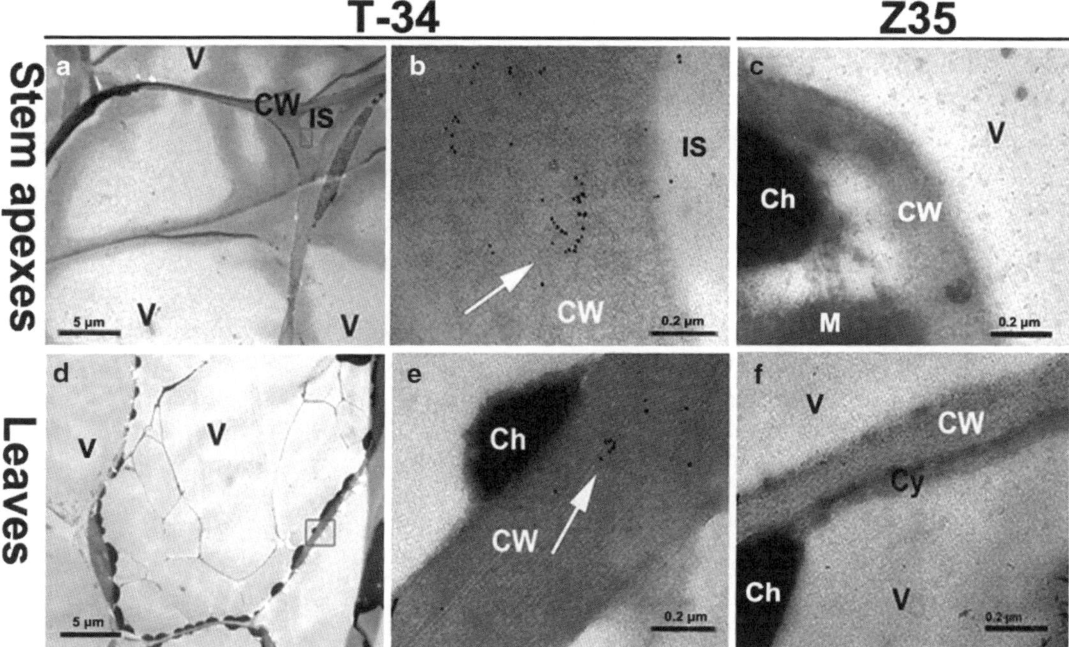

Fig. 3. Immuno-gold localization of harpinXoo in leaves and stem apices of *hpa1Xoo*-transformed T-34 and untransformed Z35. (**a**) and (**b**) Stem apices of *hpa1Xoo*-transformed T-34. (**c**) Stem apices of untransformed Z35. (**d**) and (**e**) Leaves of *hpa1Xoo*-transformed T-34. (**f**) Leaves of untransformed Z35. CW: cell wall, Cy: cytoplasm, V: vacuole, IS: intercellular space, Ch: chloroplasts, M: mitochondria. Bars: a and d = 5 µm; b, c, e, and f = 0.2 µm. *Arrow* points to gold particles labeled with harpinXoo antiserum (15 nm particles). The *squares* indicate the regions of b and e magnified in a and d, respectively. More than 20 ultrathin sections of each sample were examined with a JEM × 1200 transmission electron microscope (Nikon, Japan). The experiment was repeated twice.

3. The flasks were placed under illumination for 12 h with continuous shaking (120 rpm).

4. Cells were subcultured weekly by a fourfold dilution until harvest.

5. The suspensions of cotton cells and of the conidia were then mixed in a ratio of 1:20. The final concentration of both the cells and the conidia in the mix was the same, namely $1–3 \times 10^7$ cells/ml.

6. Death of cotton cells in the mixture was quantified using the FDA stain method described by Amano et al. (43). Immediately before each assay, a stock solution of FDA (0.5% w/v in acetone) was diluted with distilled water to create a fresh FDA working solution 0.01% w/v FDA which was kept in the dark at 4°C.

7. A 20 µl aliquot of this FDA solution was then mixed by gentle stirring with 2 ml of PBS (2.7 mM KCl, 137 mM NaCl, 1.8 mM KH_2PO_4, 4.0 mM Na_2HPO_4) and 0.01% w/v FDA in a quartz cuvette.

8. The measurement of cell activity was initiated by the addition of 100 µl of cotton cells to the cuvette.

The number of viable cells emitting fluorescence was counted under a fluorescence microscope (Olympus, Japan). The percentage of dead cells was calculated as [(total cells – cells emitting fluorescence)/ total cells] × 100. The experiment was repeated three times.

The viability of cotton cells was counted at 3, 6, 9, and 12 h after mixing with *V. dahliae* conidial suspension under a fluorescence microscope by staining with fluorescein diacetate (FDA). Fluorescence emitted from T-34 cells was stronger than that from Z35 cells (Fig. 4a). The percentage of cell death in T-34 cell suspension mixed with *V. dahliae* conidia was significantly lower than that in Z35 cell suspension mixed with *V. dahliae* from 3 to 12 h (Fig. 4b). The viability of Z35 and T34 cells was similar in the absence of *V. dahliae* conidia and almost 100% of untreated Z35 and T34 cells were viable after 12 h (Fig. 4c).

3.8. Microscopy for Micro-hypersensitive Response

1. Roots of transgenic T-34 and untransformed Z35 plants at the 2–3 leaf stage were inoculated with a conidial suspension of *V. dahliae* $(1–3 \times 10^7/\text{ml})$ according to the method described above.

2. Leaves were collected 10 days after the inoculation and stained with Trypan blue using the method described by Lipka et al. (44).

3. Trypan blue staining was performed using 0.1% Trypan blue in lactophenol/ethanol solution.

4. Leaves were cleared by incubation in chloral hydrate (2.5 g/ml).

5. Stained leaf samples were observed under a Leica light microscope (Leica DMRB, Leica Microsystems, Germany) and photographed with a Leica DFC camera (DM2500-3HF-FL, Leica Microsystems, Germany). Leaves (≥4 leaves per plant) without any wound or visible symptom of the disease from ten independent T-34 plants were examined.

6. Leaves were collected from T-34 and Z35 20 days after the root inoculation with *V. dahliae* conidia suspension in the greenhouse and then stained with Trypan blue, which selectively stained dead or dying cells. Leaves inoculated with sterile water

Fig. 4. (continued) (4) Fluorescence emitted from living cells (*red arrow*) of *hpa1Xoo*-transformed T-34 mixed with conidia of *V. dahliae* under a fluorescence microscope. (1), (2), (3), and (4) scale bars = 300 µm. (**b**) Percentage of the cell death in the cotton cell suspension mixed with *V. dahliae* conidia. (**c**) Percentage of cotton cells in the absence of conidia of *Verticillium dahliae*. Cotton cells and *V. dahliae* conidia were mixed in a ratio of 1:20. The percentage of cell death was counted at 3, 6, 9, and 12 h after mixing. Error bars indicate standard error of the mean (*n*=3). Data points marked with *asterisks* are significantly different (Student's *t* test, *p*<0.01). The experiment was repeated three times.

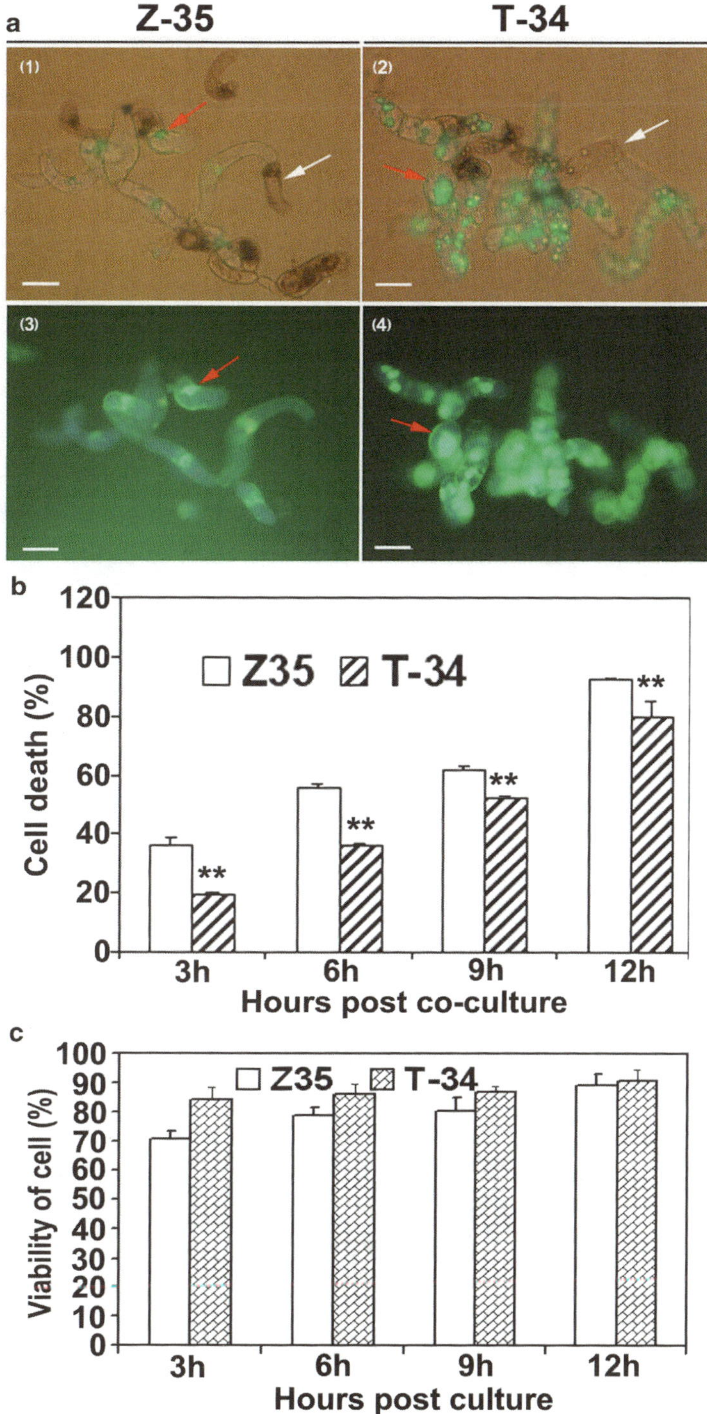

Fig. 4. Viability of cotton cells in the presence of conidia of *Verticillium dahliae*. (**a**) (1) Living (*red arrow*) or dead (*white arrow*) cells in the cell suspension of untransformed Z35 mixed with conidia of *V. dahliae* under a conventional light microscope. (2) Living (*red arrow*) or dead (*white arrow*) cells in the cell suspension of *hpa1Xoo*-transformed T-34 mixed with conidia of *V. dahliae* under a conventional light microscope. (3) Fluorescence emitted from living cells (*red arrow*) of untransformed Z35 mixed with conidia of *V. dahliae* under a fluorescence microscope.

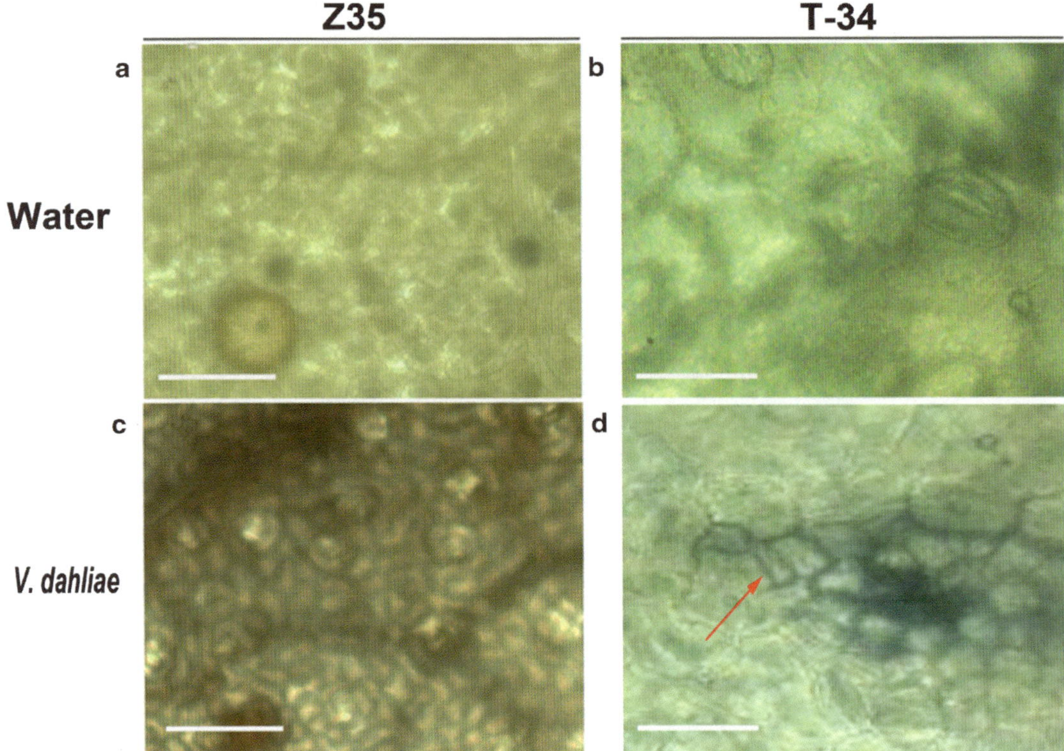

Fig. 5. Microscopic hypersensitive response (HR) in *hpa1Xoo*-transformed T-34 and untransformed Z35 20 days after root inoculations with *Verticillium dahliae*. (**a**) Leaves of uninoculated untransformed Z35. (**b**) Leaves of uninoculated *hpa1Xoo*-transformed T-34. (**c**) Leaves of untransformed Z35 inoculated with *V. dahliae*. (**d**) Leaves of *hpa1Xoo*-transformed T-34 inoculated with *V. dahliae* (*red arrow* indicates microscopic HR). (**a**)–(**d**) scale bars = 1 μm. The experiment was repeated three times.

were used as the control. The results of microscopic examination were shown in Fig. 5.

No Trypan blue stained cells were observed in leaves of T-34 and Z35 treated with water (Fig. 5a, b) and in leaves of Z35 inoculated with *V. dahliae* (Fig. 5c). In comparison, large regions of Trypan blue stained cells were observed in leaves of T-34 inoculated with *V. dahliae* indicating the occurrence of microscopic hypersensitive response (HR) (Fig. 5d). Micro HR was observed in all leaves (≥4 leaves per plant) collected from ten T-34 plants infected with *V. dahliae* (100%), whereas it was not observed in the controls (T-34 and Z35 un-inoculated) (0%).

3.9. Observation of Oxidative Burst and Quantification of H_2O_2 in Cotton Leaves

1. The second and third cotton leaves with freshly cut petioles and no visible wounds were collected when the plants were at the 5 leaf stage.

2. Two-third of the petiole was immersed into 10 ml of conidial suspension ($1–3 \times 10^7$ ml) for 0, 1, or 3 h.

3. To make the accumulation of H_2O_2 in the cotton leaves visible, fresh inoculated leaf samples were incubated in 1 mg/ml of 3,3′-diaminobenzidine (DAB)–HCl (pH 3.8) (a low pH is necessary in order to solubilize DAB) and incubated in the growth chamber for 8 h prior to sampling, and then decolorized in 96% ethanol.

4. The samples were mounted on slides with 60% glycerol and examined under an Olympus light microscope (BH-2).

The accumulation of H_2O_2 was visible as a reddish or brown discoloration. Furthermore, the production of H_2O_2 in leaves was quantified. Leaves dipped in sterile water were used as the negative control. The production of H_2O_2 in leaves was measured 0, 1, and 3 h after inoculation with the method described by Brennan and Frenkel (45) and expressed as a percentage of fresh weight. Hydroperoxides form a specific complex with titanium (Ti^{4+}) which can be measured by colorimetry.

1. Peroxides were extracted by homogenizing 100 g leaf tissue in 200 ml cold acetone.

2. The homogenate was filtered and the filtrate brought to 300 ml with water.

3. Two ml of a titanium reagent (20% titanic tetrachloride in concentrated HCl, v/v) were added to 20-ml samples of the peroxide extract, followed by the addition of 4 ml concentrated NH_4OH to precipitate the peroxide–titanium complex.

4. After centrifugation (5 min at $10,000 \times g$), the supernatant was discarded and the precipitate solubilized in 15 ml of 2 N H_2SO_4, washed repeatedly with acetone, and brought to a final volume of 20 ml with water.

5. The absorbancy of the obtained solutions was read at 415 nm against a water blank.

The concentration of the peroxide in the extracts was determined by comparing the absorbancy against a standard curve representing titanium–H_2O_2 complex from 0.1 to 1 mm. The experiment was repeated three times.

3,3′-Diaminobenzidine tetrahydrochloride (DAB) was used to detect the production of reactive oxygen intermediates (ROI) (46). No reddish or brown spots representing the accumulation of H_2O_2 were observed in T-34 and Z35 leaves dipped in water. After the inoculation, visible reddish or brown spots were only observed in T-34 leaves collected 3 h after dipping in the conidial suspension of *V. dahliae* (Fig. 6a, b). The basal level of H_2O_2 was higher in leaves of transgenic T-34 than in leaves of Z35 prior to dipping. The level of H_2O_2 increased dramatically in leaves of transgenic T-34 3 h after dipping and such increase in H_2O_2 content was not observed in the treated Z35 leaves (Fig. 6c).

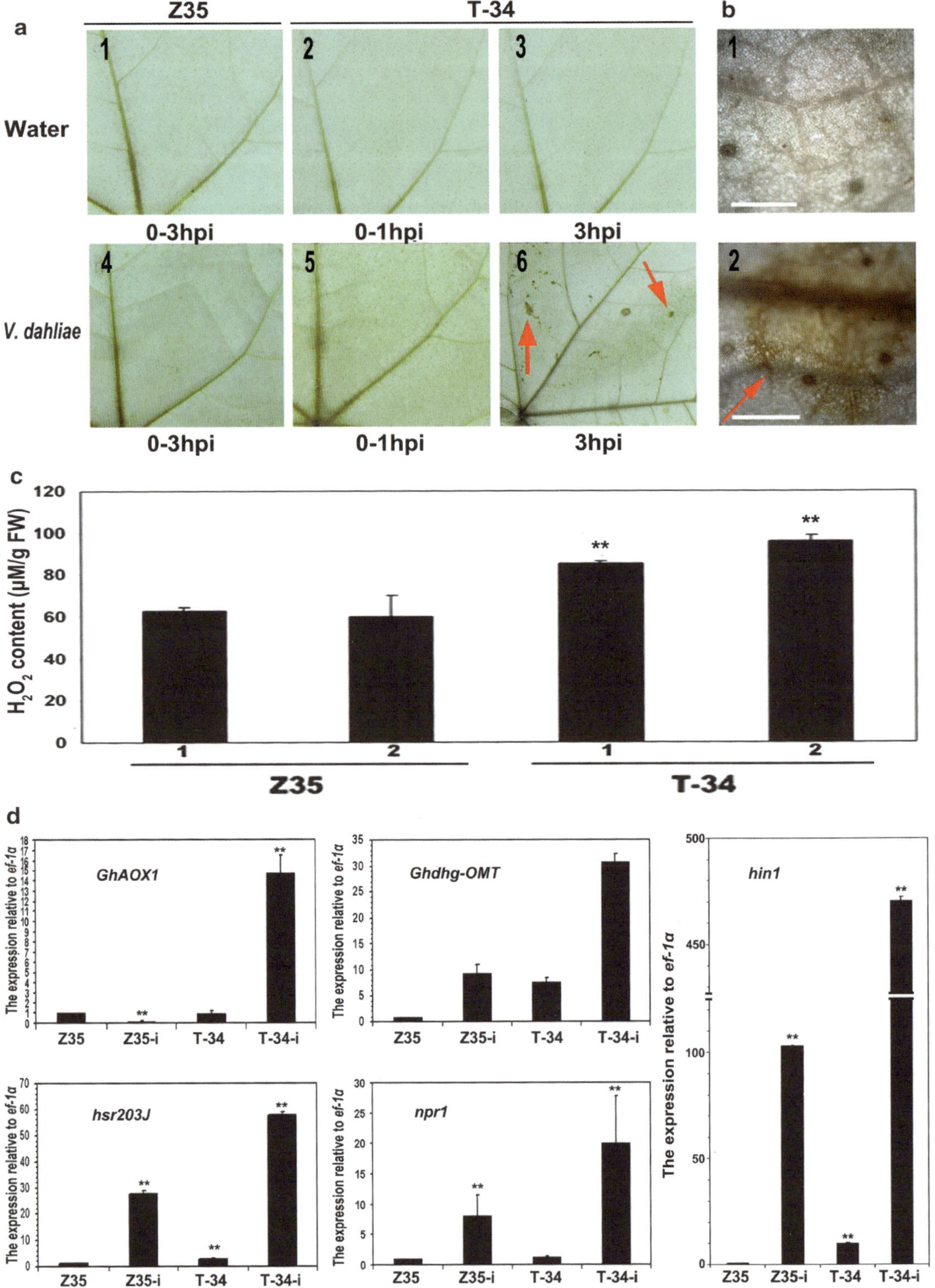

Fig. 6. Generation of active oxygen species (AOS) in leaves of *hpa1Xoo*-transformed T-34 and untransformed Z35. (**a**) Oxygen burst in cotton leaves dipped in the conidial suspension of *Verticillium dahliae* collected from 0 to 3 h after inoculation (*red arrow* points to the location of oxygen burst).

3.10. Quantitative RT-PCR

1. The second and third fully grown leaves of cotton were harvested when the plants were at the 5 leaf stage and inoculated as described above. Leaves treated with sterile distilled water served as control.

2. The leaf blades were frozen in liquid nitrogen immediately after the sampling.

3. Total RNA was extracted from the leaves using a commercial reagent, namely RNAiso. RNA concentrations were quantified using a biophotometer (Eppendorf AG, Hamburg, Germany).

4. Denaturation of template RNA and reverse transcription

   Prepare the following reaction mixture

Reagents	Volume
Template RNA	1 μg (see Note 3)
Oligo d(T) 18 Primer (50 μM)	1 μl
dNTP Mixture (10 mM each)	1 μl
RNase free dH$_2$O	up to 10 μl

5. Place tubes in the Thermal Cycler and set the parameters according to the following conditions for denaturation and annealing (see Note 4).

65°C	5 min	4°C

6. Prepare the reagent mixture for use in step 5 by combining the reagents in the proportion shown as below.

Reagent	Volume
Reaction mixture used for denaturation and annealing from step 5	10 μl
5× PrimeScriptTM Buffer	4 μl
RNase inhibitor(40 U/μl)	0.5 μl
PrimeScriptTM Reverse Transcriptase (200 U/μl)	1 μl
RNase-free dH$_2$O	up to 20 μl

Fig. 6. (continued) (**b**) Light microscopy of the oxygen burst in leaves of untransformed Z35 (1) and *hpa1Xoo*-transformed T-34 (2) 3 h after the inoculation. (**c**) H$_2$O$_2$ content (μg/g fresh weight) in leaves of *hpa1Xoo*-transformed T-34 and untransformed Z35 dipped in the conidial suspension of *V. dahliae* (mean values and standard errors calculated from three replicates). 1, non-inoculated; 2, inoculated. (**d**) Quantitative RT-PCR analysis of *ghAOX1*, *hin1*, *npr1*, *ghdhg-OMT*, and *hsr203J* expression in leaves of *hpa1Xoo*-transformed T-34 (T-34-i) and untransformed Z35 (Z35-i) dipped in the conidial suspension of *V. dahliae* compared with that of *hpa1Xoo*-transformed T-34 (T-34) and untransformed Z35 (Z35) dipped in water (*error bars* indicate standard error). (**b**) (1, 2) scale bars = 0.01 mm. The experiment was repeated three times. *Asterisks* represent significant differences at the level of 0.01.

7. Place the tubes in a Thermal Cycler and perform reverse transcription using the following program.

42°C (~50°C)	15–30 min
95°C	5 min
4°C	

8. Proceed to qPCR analysis of the cDNA. Alternatively, the reactions may be stored frozen for future use.

9. Prepare PCR reaction mixture

Reagents	Volume	Final content
SYBR® Premix Ex TaqTM (2×)	25.0 µl	1×
PCR Forward Primer (10 µM)	1.0 µl	0.2 µM *1 (see Note 5)
PCR Reverse Primer (10 µM)	1.0 µl	0.2 µM *1 (see Note 5)
ROX Reference Dye or Dye II *3 (see Note 7) (50×)	1.0 µl	1×
Template DNA	4.0 µl	*2 (see Note 6)
dH$_2$O	18.0 µl	
Total	50.0 µl	*4 (see Note 8)

10. Start the reaction.
    Shuttle PCR standard protocol is recommended. Try this protocol firstly, and optimize the reaction condition if needed. Try 3 step PCR protocol, when shuttle protocol is difficult (e.g., Tm value of primers is low.)
    Shuttle PCR Standard Protocol:
    Stage 1: Initial denaturation
    Reps: 1, 95°C 30 s.
    Stage 2: PCR
    Reps: 40, 95°C 5 s, 60°C 30–34 s. (see Note 9)
    Stage 3: Melt curve

All PCR reactions were repeated three times and the data were normalized to constitutively expressed ef-1α using the 2-ΔΔCT method described by Livak and Schmittgen (47). The primer sequences used in the quantitative RT-PCR are listed in Table 1.

The expressions of *ghAOX1* [GenBank accession number DQ250028], *hsr203J* [GenBank accession number X77136], *hin1* [GenBank accession number Y07563], and *npr1* (48) were quantified using the real-time RT-PCR. *GhAOX1* is a key gene involved in the production of active oxygen species (AOS) in plants (49, 50) and *hsr203J* and *hin1* are marker genes for HR which express specifically in plant tissues undergoing HRs (51, 52). The data was normalized to a constitutive expressed *ef-1α*.

No up-regulations of *npr1*, *hsr203J*, *hin1*, and *ghdhg-OMT* were observed in the un-inoculated T-34 and Z-35 plants. The basal expression level of *GhAOX1* was higher in the un-inoculated T-34, compared to that in wild-type Z35. *Npr1*, *hsr203J*, *hin1*, and *GhAOX1* were all upregulated in T-34 and Z35 after plants were dipped in the conidial suspension of *V. dahliae*. Nevertheless, the upregulations of these genes were stronger in leaves of transgenic T-34 in response to the dipping treatment (Fig. 6d). In addition, the upregulation of *dhg-OMT* (53) encoding hemigosspol was only observed in T-34 after the dipping treatment (Fig. 6d).

3.11. Statistical Analysis

For quantitative determination, the data were analyzed by the *t* test at $P \leq 0.05$ or 0.01 using the Microsoft Analysis Tool. For differences in disease severity, each transgenic plant was compared with an untransformed plant.

4. Notes

1. Tetramethylethylenediamine (TEMED) is used with ammonium persulfate to catalyze the polymerization of acrylamide when making polyacrylamide gels, used in gel electrophoresis, for the separation of proteins or nucleic acids. Although the amounts used in this technique may vary from method to method, 0.1–0.2%v/v TMEDA is a "traditional" range. In SDS page, the resolving gel and the stacking gel should be immediately made after TEMED was added according to the protocol.

2. Block for at least 10 min to ensure adequate blocking.

3. Template RNA: use up to 8 μl. When total RNA is used, it can be added up to 5 μg. (recommended amount: 100 pg to 1 μg).

4. This denaturation/annealing process facilitates reverse transcription by efficient denaturation of template RNA and efficient annealing of reverse transcription primer to template RNA.

5. The final concentration of primers can be 0.2 μM in most reactions. When it does not work, determine the optimal concentrations within the range of 0.1–1.0 μM.

6. Final template concentration varies depending on the copy number of target present in the template solution. Optimal amount should be determined by preparing the dilution series. It is recommended to apply DNA template in less than 100 ng per 20 μl of reaction mixture. When the RT reactant (cDNA) is used as a template, it should be added in less than 10% volume of PCR reaction mixture.

7. The ROX Reference Dye/Dye II is supplied for performing normalization of fluorescent signal intensities among wells when used with real-time PCR instruments that have option. For ABI PRISM® 7000/7700 and Applied Biosystems 7300 Real-Time PCR System, the use of ROX Reference Dye (50×) is recommended. For Applied Biosystems 7500 Real-Time PCR System and 7500 Fast Real-Time PCR System, the use of ROX Reference Dye II is recommended.

8. Prepare in accordance with the recommended volume for each instrument.

9. This step should be 30 s. with ABI PRISM® 7700, 31 s. with ABI PRISM® 7000 and 7300 Real-Time PCR System, and 34 s. with 7500 Real-Time PCR System.

References

1. Bell AA (1992) Verticillium wilt. In: Hillocks RJ (ed) Cotton diseases. C.A.B. International, Wallingford, UK, pp 87–126

2. Hampton RE, Wullschleger SD, Oosterhuis DM (1990) Impact of Verticillium wilt on net photosynthesis, respiration and photorespiration in field-grown cotton (*Gossypium hirsutum* L.). Physiol Mol Plant Pathol 37:271–280

3. Paplomatas EJ et al (1992) Incidence of Verticillium wilt and yield losses of cotton cultivars (*Gossypium hirsutum*) based on soil inoculum density of *Verticillium dahliae*. Phytopathology 82:1417–1420

4. Pegg GF (1989) Pathogenesis in vascular disease of plants. In: Tjamos EC, Beckman C (eds) Vascular wilt diseases of plants. Springer, Berlin, Germany, pp 51–94

5. Xiao CL et al (1998) Effects of crop rotation and irrigation on *Verticillium dahliae* microsclerotia in soil and wilt in cauliflower. Phytopathology 88:1046–1055

6. Huang J, Li H, Yuan H (2006) Effect of organic amendments on Verticillium wilt of cotton. Crop Prot 25:1167–1173

7. Gray FA, Koch DW (2004) Influence of late season harvesting, fall grazing, and fungicide treatment on Verticillium wilt incidence, plant density, and forage yield of alfalfa. Plant Dis 88:811–816

8. Kurt S, Dervis S, Sahinler S (2003) Sensitivity of *Verticillium dahliae* to prochloraz and prochloraz-manganese complex and control of Verticillium wilt of cotton in the field. Crop Prot 22:51–55

9. Colson-Hanks ES, Deverall BJ (2000) Effect of 2,6-dichloroisonicotinic acid, its formulation materials and benzothiadiazole on systemic resistance to alternaria leaf spot in cotton. Plant Pathol 49:171–178

10. Punja ZK (2001) Genetic engineering of plants to enhance resistance to fungal pathogens: a review of progress and future prospects. Can J Plant Pathol 23:216–235

11. Gentile A et al (2007) Enhanced resistance to *Phoma tracheiphila* and *Botrytis cinerea* in transgenic lemon plants expressing a *Trichoderma harzianum* chitinase gene. Plant Breed 126:146–151

12. Wrobel-Kwiatkowska M et al (2004) Expression of beta-1,3-glucanase in flax causes increased resistance to fungi. Physiol Mol Plant Pathol 65:245–256

13. Lee YH et al (2002) Enhanced disease resistance in transgenic cabbage and tobacco expressing a glucose oxidase gene from *Aspergillus niger*. Plant Cell Rep 20:857–863

14. Komatsu K et al (2007) QTL mapping of pubescence density and flowering time of insect-resistant soybean. Genet Mol Biol 30:635–639

15. Yang KY et al (2003) Overexpression of a mutant basic helix-loop-helix protein HFR1, HFR1-deltaN105, activates a branch pathway of light signaling in Arabidopsis. Plant Physiol 133:1630–1642

16. Cao H, Li X, Dong X (1998) Generation of broad-spectrum disease resistance by overexpression of an essential regulatory gene in systemic acquired resistance. Proc Natl Acad Sci USA 95:6531–6536

17. Chern M et al (2005) Overexpression of a rice NPR1 homolog leads to constitutive activation of defense response and hypersensitivity to light. Mol Plant Microbe Interact 18:511–520

18. Tohidfar M, Mohammadi M, Ghareyazie B (2005) Agrobacterium-mediated transformation of cotton (*Gossypium hirsutum*) using a heterologous bean chitinase gene. Plant Cell Tiss Org Cult 83:83–96

19. Wang YQ et al (2004) Over-expression of Gastrodia anti-fungal protein enhances Verticillium wilt resistance in coloured cotton. Plant Breed 123:454–459

20. Alfano JR, Collmer A (1997) The type III (Hrp) secretion pathway of plant pathogenic bacteria: trafficking harpins, Avr proteins, and death. J Bacteriol 179:5655–5662

21. Dong H et al (1999) Harpin induces disease resistance in Arabidopsis through the systemic acquired resistance pathway mediated by salicylic acid and the NIM1 gene. Plant J 20:207–215

22. Dong HP et al (2004) Downstream divergence of the ethylene signaling pathway for harpin-stimulated Arabidopsis growth and insect defense. Plant Physiol 136:3628–3638

23. Li P et al (2004) A novel member of avrBs3 gene family from *Xanthomonas oryzae* pv. *oryzae* has a dual function. Process Nat Sci 14:767–773

24. Zou LF et al (2006) Elucidation of the *hrp* Clusters of *Xanthomonas oryzae* pv. *oryzicola* that control the hypersensitive response in non-host tobacco and pathogenicity in susceptible host Rice. Appl Envir Microbiol 72:6212–6224

25. Gough CL et al (1993) Homology between the HrpO protein of *Pseudomonas solanacearum* and bacterial proteins implicated in a signal peptide-independent secretion mechanism. Mol Gen Genet 239:378–392

26. Huguet E et al (1998) hpaA mutants of *Xanthomonas campestris* pv. *vesicatoria* are affected in pathogenicity but retain the ability to induce host-specific hypersensitive reaction. Mol Microbiol 29:1379–1390

27. Zhu W, Magbanua MM, White FF (2000) Identification of two novel hrp-associated genes in the *hrp* gene cluster of *Xanthomonas oryzae* pv. *oryzae*. J Bacteriol 182:1844–1853

28. He SY, Huang HC, Collmer A (1993) *Pseudomonas syringae* pv. *syringae* harpinPss: a protein that is secreted via the hrp pathway and elicits the hypersensitive response in plants. Cell 73:1255–1266

29. Gaudriault S, Brisset MN, Barny MA (1998) HrpW of *Erwinia amylovora*, a new Hrp-secreted protein. FEBS Lett 428:224–228

30. Arlat M et al (1994) PopA1, a protein which induces a hypersensitivity-like response on specific *Petunia* genotypes, is secreted via the Hrp pathway of *Pseudomonas solanacearum*. EMBO J 13:543–553

31. Lorenz C et al (2008) HpaA from Xanthomonas is a regulator of type III secretion. Mol Microbiol 69:344–360

32. Peng JL et al (2004) Expression of HarpinXoo in transgenic tobacco induces pathogen defense in the absence of hypersensitive cell death. Phytopathology 94:1048–1055

33. Shao M et al (2008) Expression of a harpin-encoding gene in rice confers durable nonspecific resistance to *Magnaporthe grisea*. Plant Biotechnol J 6:73–81

34. Sohn S et al (2007) Transgenic tobacco expressing the hrpNEP Gene from *Erwinia pyrifoliae* triggers defense responses against *Botrytis cinerea*. Mol Cells 24:232–239

35. Li R, Fan Y (1999) Reduction of lesion growth rate of late blight plant disease in transgenic potato expressing harpin protein. Sci China C Life Sci 42:96–101

36. Malnoy M, Venisse JS, Chevreau E (2005) Expression of a bacterial effector, harpin N, causes increased resistance to fire blight in *Pyrus communis*. Tree Genet Genomes 1:41–49

37. Joost O et al (1995) Differential induction of 3-hydroxy-3-methylglutaryl CoA reductase in two cotton species following inoculation with *Verticillium*. Mol Plant Microbe Interact 8:880–885

38. An G et al (1988) Binary vectors. In: Gelvin SB, Schilperoort RA (eds) Plant molecular biology manual. Kluwer Academic Publishers, Dordrecht, Netherlands, pp 1–19

39. Sunilkumar G, Rathore KS (2001) Transgenic cotton: factors influencing Agrobacterium-mediated transformation and regeneration. Mol Breed 8:37–52

40. Yang C et al (2008) QTLs mapping for Verticillium wilt resistance at seedling and maturity stages in *Gossypium barbadense* L. Plant Sci 174:290–298

41. Slot JW, Geuze HJ (1985) A new method of preparing gold probes for multiple-labeling cytochemistry. Eur J Cell Biol 38:87–93

42. Wu J et al (2005) High-efficiency transformation of *Gossypium hirsutum* embryogenic calli mediated by *Agrobacterium tumefaciens* and regeneration of insect-resistant plants. Plant Breed 124:142–146

43. Amano T et al (2003) A versatile assay for the accurate, time-resolved determination of cellular viability. Anal Biochem 314:1–7

44. Lipka V et al (2005) Pre- and postinvasion defenses both contribute to nonhost resistance in Arabidopsis. Science 310:1180–1183

45. Brennan T, Frenkel C (1977) Involvement of hydrogen peroxide in the regulation of senescence in pear. Plant Physiol 59:411–416

46. Thordal-Christensen H et al (1997) Subcellular localization of H_2O_2 in plants. H_2O_2 accumulation in papillae and hypersensitive response during the barley powdery mildew interaction. Plant J 11:1187–1194

47. Livak KJ, Schmittgen TD (2001) Analysis of relative gene expression data using real-time quantitative PCR and the 2-CT method. Methods 25:402–408

48. Wang X et al (2006) Cloning full-length cDNA of GbNPR1 gene from *Gossypium barbadense* and its expression in transgenic tobacco. Sci Agric Sin 39:886–894

49. Li F et al (2008) Molecular cloning and expression characteristics of alternative oxidase gene of cotton (*Gossypium hirsutum*). Mol Biol Rep 35:97–105

50. Maxwell DP, Wang Y, McIntosh L (1999) The alternative oxidase lowers mitochondrial reactive oxygen production in plant cells. Proc Natl Acad Sci USA 96:8271–8276

51. Pontier D et al (1994) Hsr203J, a tobacco gene whose activation is rapid, highly localized and specific for incompatible plant. Plant J 5:507–521

52. Pontier D et al (2001) Identification of a novel pathogen-responsive element in the promoter of the tobacco gene HSR203J, a molecular marker of the hypersensitive response. Plant J 26:495–507

53. Liu J et al (2002) Cloning and expression of desoxyhemigossypol-6-O-methyltransferase from cotton (Gossypium barbadense). J Agric Food Chem 50:3165–3172

Chapter 20

Development of Insect-Resistant Transgenic Cotton with Chimeric TVip3A* Accumulating in Chloroplasts

Jiahe Wu and Yingchuan Tian

Abstract

An optimized *vip3A* gene, designated as *vip3A** was chemically synthesized and a *thi1* gene chloroplast transit peptide coding sequence was attached to its 5′ end to produce the *tvip3A**. *vip3A** and *tvip3A** genes were transformed into *Gossypium hirsutum* cv. Zhongmiansuo35 mediated by *Agrobacterium tumefaciens*. Four independent transgenic T1 lines with single-copy insertions and unchanged phenotypes (CTV1 and CTV2 for *tvip3A**, and CV1 and CV2 for *vip3A**) were selected by Polymerase chain reaction (PCR), Reverse transcription (RT)-PCR, Southern blotting, enzyme-linked immunosorbent assay (ELISA), and insect bioassay. As expected, the Vip3A* protein of CTV1 and CTV2 were transported to the chloroplasts, where they accumulated. Our results suggest that the two *tvip3A** transgenic lines (CTV1 and CTV2) can be used to develop insect-resistant cultivars and could be used as a resource for raising multi-toxins-expressing transgenic cotton.

1. Introduction

The first commercially available transgenic cotton expressing an insecticidal protein (Cry1Ac from *Bacillus thuringiensis* (*Bt*)) was produced in the USA in 1995 (1). The planting area of *Bt* cotton cultivars has steadily increased since then, especially in China and India (2). Although, *Bt* cotton exerts substantial pressure against many lepidopteran pests of cotton, insects have evolved resistance to Bt δ-endotoxins, and some lepidopteran insects are not as sensitive to Cry1A as the cotton bollworm. Many cases of insect tolerance against Bt δ-endotoxins have been reported (3–7). Therefore, the isolation and characterization of new insecticidal toxins is necessary to widen the scope of pest control programs, including delaying the development of resistant insects on transgenic cotton containing Bt δ-endotoxins proteins. To achieve this goal, we have used a new insecticidal protein gene, *vip3A*, to create a new type of transgenic plant.

Baohong Zhang (ed.), *Transgenic Cotton: Methods and Protocols*, Methods in Molecular Biology, vol. 958,
DOI 10.1007/978-1-62703-212-4_20, © Springer Science+Business Media New York 2013

Vip3A (Vegetative insecticidal protein 3A), a novel insecticidal protein, is secreted during the vegetative phase of *B. thuringiensis* development; in contrast to the crystal δ-endotoxins proteins, which are produced during the reproductive phase of bacterial development (8). The Vip3A protein shares no sequence homology with the known crystal δ-endotoxin proteins (9) and exhibits a broader insecticidal spectrum against a wide variety of lepidopteran and coleopteran insects (9–12). The mode of Vip3A's action in insects' midgut is also different from that of δ-endotoxin proteins, which minimizes any potential insect cross-resistance to δ-endotoxins (13–15). Initial reports of Vip3A's insecticidal activity by Estruch et al. (9) indicated 100% mortality of black cutworm (BCW, *Agrotis ipsilon*), beet armyworm (BAW, *Spodoptera exigua*) and fall armyworm (FAW, *Spodoptera frugiperda*) when added at 140 ng/ml in the diet. By contrast, the δ-endotoxins Cry1Ac and Cry1Ab exerted relatively lower insecticidal impact on BCW, BAW, and FAW than Vip3A (9, 16). However, Vip3A shows similar activities to the Bt δ-endotoxins against various insects. Although Vip3A showed insecticidal activity against the main cotton insects, tobacco budworm (TBW, *Heliothis virescens*), and CBW, 70-fold more Vip3A than Cry1Ac was required to achieve an LD_{50} against TBW, and 25-fold higher to achieve an LD_{50} against CBW. The expression of such large amounts of Vip3A using the CaMV35S promoter in transgenic cotton is theoretically a difficult task to FAW and BAW achieve, and alternative strategies for increasing transgenic expression of Vip3A are required. Thus, we developed a new approach to achieve higher expression of the Vip3A protein in plants. Based on the results reported by Chabregas et al. (13) and Chabregas et al. (17), we chose the coding sequence of the chloroplast transit peptide of THI1 protein to construct a fusion protein gene with *vip3A** gene.

Arabidopsis thaliana THI1 protein, a thiazole enzyme (encoded by the *thi1* gene), is targeted simultaneously to chloroplasts and mitochondria by a posttranslational mechanism (13, 17). Molecular characterization confirmed that this protein contains a typical chloroplast transit peptide and a mitochondrial presequence-like structure at the N-terminus, enabling dual organelle targeting. In fact, most of the THI1 protein is targeted to the chloroplasts by the transit peptide; only a small amount of protein is targeted to the mitochondria (13, 17). The use of the THI1 chloroplast transit peptide to direct Vip3A to the chloroplast could increase the intracellular concentration of Vip3A in transgenic cells.

To increase Vip3A expression in transgenic cotton, two strategies were adopted. First, a novel *vip3A* gene (*vip3A**) was designed and synthesized to increase the GC nucleotide content to enhance mRNA stabilization and cotton-preferred codens were used to favor its translation, based on the Vip3A sequence reported by several laboratories (9, 18, 19). Second, DNA encoding the chloroplast transit peptide of THI1 was fused to *vip3A** to target the expressed Vip3A* protein to the chloroplasts, allowing accumulation of Vip3A* in the chloroplasts. This gene was referred to as *tvip3A**.

The *vip3A** gene and *tvip3A** were separately introduced into *Gossypium hirsutum* via *Agrobacterium tumefaciens*-mediated transformation. The Vip3A* expression level and the insecticidal activity against FBW, BAW, and CBW were investigated in transgenic *tvip3A** or *vip3A** plants. The results indicated that the Vip3A* expression level in transgenic *tvip3A** lines was at least three-fold higher than in transgenic *vip3A** lines. Transgenic *tvip3A** plants also showed higher mortality against CBW than transgenic *vip3A** plants. Thus, the optimization of the nucleotide sequence of *vip3A* gene and application of the chloroplast signal were highly effective for increasing Vip3A expression levels and mortality against insects in transgenic cotton plants.

2. Materials

The materials of molecular analysis, PCR, RT-PCR, Southern blot, Western blot, ELISA, are not listed as they are routine regents and referred in other chapters. However, some important regents and kits are shown in Subheading 3.

2.1. Cotton Cultivar and Insect Pests

1. The cotton variety, *Gossypium hirsutum* cv. Zhongmiansuo35, is used as genetic transformation. Zhongmiansuo35 is widely planted in China, which was provided by Prof. Zhengde Liu.
2. The *cry1Ac* transgenic line BR-98-2 previously produced in our laboratory.
3. The egg masses of FAW and CBW were donated by the Plant Protection Institute of the Chinese Academy of Agricultural Sciences, egg masses of BAW were collected from the cotton plants grown in the field, and all were allowed to hatch in the laboratory.

2.2. Vectors and Bacterium

The plant expression vector pBin438 is constructed by Li et al. (20). *E. coli* DH5α and *Agrobacterium tumefaciens* LBA4404 are purchased from TransGen Biotech, Inc.

2.3. Media, Antibiotics, and Supplementary

1. LB medium:
   Weight 10 g Bacto-Tryptone, 5 g Bacto-yeast extract, and 10 g NaCl, and dissolved in the ddH_2O, then add ddH_2O to 1 L and correct the pH to 7.0.

2. MS medium: see ref. (21).

(a) Salt ingredients: Ammonium nitrate (NH_4NO_3), 1,650 mg/L; Boric acid (H_3BO_3), 6.2 mg/L; Calcium chloride ($CaCl_2 \cdot 2H_2O$), 440 mg/L; Cobalt chloride ($CoCl_2 \cdot 6H_2O$), 0.025 mg/L; Magnesium sulfate ($MgSO_4 \cdot 7H_2O$), 370 mg/L; Cupric sulfate ($CuSO_4 \cdot 5H_2O$), 0.025 mg/L; Potassium phosphate (KH_2PO_4), 170 mg/L; Ferrous sulfate ($FeSO_4 \cdot 7H_2O$), 27.8 mg/L; Potassium nitrate (KNO_3), 1,900 mg/L; Manganese sulfate ($MnSO_4 \cdot 4H_2O$), 22.3 mg/L; Potassium iodide (KI), 0.83 mg/L; Sodium molybdate ($Na_2MoO_4 \cdot 2H_2O$), 0.25 mg/L; Zinc sulfate ($ZnSO_4 \cdot 7H_2O$), 8.6 mg/L; $Na_2EDTA \cdot 2H_2O$, 37.2 mg/L.

(b) Organic additives: i-Inositol, 100 mg/L; Niacin, 0.5 mg/L; Pyridoxine·HCl, 0.5 mg/L; Thiamine·HCl, 0.1 mg/L; Glycine, 2.0 mg/L.

3. B5 medium: see ref. (22)

(a) Salt ingredients: Ammonium sulfate ((NH_4)$_2SO_4$), 134 mg/L; Boric acid (H_3BO_3), 3 mg/L; Calcium chloride ($CaCl_2 \cdot 2H_2O$), 150 mg/L; Cobalt chloride ($CoCl_2 \cdot 6H_2O$), 0.025 mg/L; Cupric sulfate ($CuSO_4 \cdot 5H_2O$), 0.025 mg/L; Ferrous sulfate ($FeSO_4 \cdot 7H_2O$), 27.8 mg/L; Magnesium sulfate ($MgSO_4$), 122 mg/L; Manganese sulfate ($MnSO_4 \cdot H_2O$), 10 mg/L; Potassium nitrate (KNO_3), 2,500 mg/L; Potassium iodide (KI), 0.75 mg/L; Sodium molybdate ($Na_2MoO_4 \cdot 2H_2O$), 0.25 mg/L; Sodium phosphate monobasic (NaH_2PO_4), 130 mg/L; Zinc sulfate ($ZnSO_4 \cdot 7H_2O$), 2 mg/L; $Na_2EDTA \cdot 2H_2O$, 37.2 mg/L.

(b) Organic additives: i-Inositol, 100 mg/L; Niacin, 1 mg/L; Pyridoxine·HCl, 1 mg/L; Thiamine·HCl, 10 mg/L.

4. MSB medium: MS salts and B5 organic components, plus 30 g/L glucose, pH = 5.8.

5. MSB1 medium: MSB supplemented 0.1 mg/L 2.4-D, 0.1 mg/L KT, 500 mg/L cefotaxime, and 100 mg/L kanamycin, pH = 5.8.

6. MSB2 medium: add 100 mg/L kanamycin in MSB medium, pH = 5.8.

7. MSB3 medium: MSB medium supplemented 1.0 g/L glutamine and 0.5 g/L Asparagine, pH = 5.8.

8. Antibiotics: 50 µg/mL of Kanamycin, 25 µg/mL of Rifampicin, 25 µg/mL of Rifampicin, 500 mg/L cefotaxime.

9. Phytohormones: 0.1 mg/L 2.4-D and 0.1 mg/L KT.

10. Transformation buffer: 10 mM $MgCl_2$, 10 mM MES and 200 µM acetosyringone, pH = 5.7.

11. The nylon membrane (Hybond-N+) is purchased from Amersham, Buckinghamshire, UK.

12. α-[^{32}P] using a random primer labeling kit is purchased from Promega, Madison, WI, USA.

13. Trizol reagent is purchased from Invitrogen, Life Technologies, Carlsbad, CA.

14. RNase-free DNase I is purchased from Promega, Madison, WI.

15. The SuperScript™ First-Strand Synthesis system for RT-PCR is purchased from Invitrogen, Life Technologies, Carlsbad, CA.

16. A alkaline phosphatase-conjugated goat anti-rabbit immuno-globulin (Ig)G is purchased from Promega, Madison, WI, USA.

17. The Chloroplast Isolation Kit is purchased from Sigma-Aldrich (P. 4937).

3. Methods

3.1. The Modification and Chemical Synthesis of vip3A*

1. The sequences of wild-type *vip3A* genes were compared using DNAman Software. The genes were highly similar and five variable sites were found, causing variations in the amino acid sequence: Q284K, T291P, E406G, K742G, and P770S. Moreover, wild-type *vip3A* gene was very A+T-rich, which could cause lower expression of *vip3A* genes in the plant expression system.

2. Thus, to optimize the expression of *vip3A* in cotton plants, the sequence of the wild-type *vip3A* gene was modified in three ways: (1) reduction of the amount of A+T nucleotides, (2) deletion of unstable factors in the mRNA transcript, and (3) optimization of the codon usage bias. The modified gene was chemically synthesized and designated as *vip3A** (see ref. 23). The unstable structures of wild-type *vip3A* gene are shown in Table 1.

3.2. The Synthesis of the Nucleotides of Chloroplast Signal

Arabidopsis thaliana THI1 protein, a thiazole enzyme, is targeted to chloroplasts by a posttranslational mechanism. The use of the THI1 chloroplast transit peptide to direct Vip3A to the chloroplast could increase the intracellular concentration of Vip3A in transgenic cells. The chloroplast transit signal sequence of the *thi1* gene is chemical synthesized or obtained by PCR methods (see Note 1).

3.3. The Construction of Plant Expression Vectors

1. DNA encoding the chloroplast transit signal sequence of the *thi1* gene was added to the 5′ end of the modified *vip3A** gene using PCR, to generate a chimeric gene designated as *tvip3A**.

Table 1
The unstable structures of the sequence of *vip3A**

The unstable structures in transcription	Nucleotides	Beginning nucleotide of the unstable structures
mRNA unstable structures	attta	67, 725, 929, 951, 1173, 1528, 1643, 1749, 1823, 1922, 1937, 2321
Poly A signals	aataaa	1564, 1897
Splice site similar sequence of intron	aggtaa	187, 1309

2. The *vip3A** and *tvip3A** were introduced into the plant expression vector pBin438 using *Bam*HI and *Sal*I sites to construct pBVip3A* and pBTVip3A*, respectively. The two vectors are confirmed by enzyme digestion and sequencing, and the sequence of *thi1* is also separated by enzyme digestion (see Note 2).

3. The *vip3A** and *tvip3A** genes in the T-DNA region were placed under the control of a strong constitutive CaMV35S promoter (see ref. (23)). The pBVip3A* and the pBTVip3A* are transferred into *Agrobacterium tumefaciens* LBA4404 by electroporation.

3.4. The Transformation of Cotton Mediated by Agrobacterium tumefaciens

Cotton transgenic via *Agrobacterium tumefaciens*-mediated transformation is difficult mainly due to the genotype limitation, low frequency of somatic embryogenesis, long time and complexity of procedure. We develop a simple, rapid, and high effect protocol of cotton genetic transformation under consecutive optimization of the factors of procedure for 16 years. This protocol is shown below.

3.4.1. The Development of Germplasm of High Frequency Somatic Regeneration

The 100% somatic embryogenesis germplasm has developed through continuously selecting seeds by cultured for somatic embryogenesis. The detail protocol has been shown in our study (24). Nowadays, six commercial available cultivars with 100% somatic embryogenesis have been produced.

3.4.2. The Development of Asepsis Seedling of Cotton

1. The fiber of cotton seeds are denuded with 98% sulfate acid, the naked seeds are rinsed with tap water to get rid of the acid.

2. The hulls of seeds are taken off with scissors and forceps.

3. The seeds without hulls are placed in the half strength of MS medium with 15 g/L glucose.

4. Cultured jars are transferred in the environment control chamber with $26 \pm 1°C$ in dark. The seeds geminate and growth for 7 days or so, the hypocotyls reach about 12 cm long for genetic transformation.

<table>
<tr><td>3.4.3. Perform Genetic Transformation</td><td>

1. Three days before transformation, streak frozen glycerol stocks of *Agrobacterium tumefaciens* LBA4404 containing pBVip3A* or pBTVip3A* on LB agar plates containing 50 μg/mL of Kanamycin and 25 μg/mL of Rifampicin. Incubate the plates at 28°C for 24 h.

2. Two days prior to transformation, pick a single colony for each construct from the above plates and inoculate it into 5 mL of LB medium supplemented with 50 μg/mL of Kanamycin and 25 μg/mL of Rifampicin; Grow the bacterial culture at 28°C for overnight in a shaker at a speed of 50 rpm.

3. Transfer the above culture to a flask with 50 mL of LB medium supplemented with 50 μg/mL of Kanamycin and 25 μg/mL of Rifampicin. Grow the culture at 28°C for overnight in a shaker with a speed of 50 rpm.

4. On the following day, spin down the agrobacterial cells at $3,300 \times g$ for 5 min; resuspend the culture in the transformation buffer. Adjust the OD 600 of the culture to 0.5.

5. Prior to *Agrobacterial* transformation, the hypocotyls are cut into segments, about 1 cm long, with the asepsis scissors in the clean bench.

6. Add *Agrobacterial* culture suspension buffer into the plates with hypocotyls segments, fully mix, and leave it on the clean bench for 15 min.

7. Place the hypocotyls segments with bacterium on the sterilized paper on the surface of the MSB medium. The hypocotyls and *Agrobacterium tumefaciens* are cocultured at 23°C in dark for 48 h. The temperature of coculture of explants and bacterium should be below 25°C (see Note 3).

8. Rinse the segments with sterilized water two to three times, absorb the water on the surface of the segments with asepsis paper, and place them on the plates with medium MSB1 for calli induction. Induce the calli at 28°C, photoperiod of 14 h/10 h, light intensity of 1,200 lux.

9. Subculture the segments with calli every 30 days until the calli grow 1 cm in diameter on the MSB1 (see Note 4), and the calli are cultured on the MSB2 medium for the somatic embryogenesis. The explants without calli cultured for 50 days should be discarded (see Note 5).

10. Pick the embryogenesis calli from the surface of calli (see Note 6), and culture on the MSB3 for somatic embryos maturation and germination. The cultural condition is the same as the above.

11. Pick the 10–15 mm embryos (see Note 7), and plant them in the jar with the MSB3 medium for the plantlet regeneration under 28°C, photoperiod of 14 h/10 h, light intensity of 2,000 lux.

</td></tr>
</table>

12. The complete regenerating plantlets with the 2–3 leaves and healthy roots are acclimatized from asepsis condition by taking off the cover of the jars for 2–3 days, and then transplant in the plastic plot with a mixture of 4 parts peat, 3 parts vermiculite, and 3 parts sand. It takes long time to develop the transgenic plantlets (see Note 8). Transfer the pots in the greenhouse under the 28°C, photoperiod of 14 h/10 h, light intensity of 2,000 lux.

3.5. PCR Detection, Southern Blot Analysis, and Selection of Transgenic Plants

1. Cotton genomic DNA was extracted from young leaves using methods described by Paterson et al. (25).

2. PCR analysis was performed using the primer pair VipF (5′-ctcacgtaagggatgacgc-3′) (forward) and VipR (5′-ttgaatt-gaatacgcatcttc-3′) (reverse), generating a 495 bp amplicon from the *vip3A** gene.

3. The PCR reaction comprised 94°C for 5 min, 35 cycles of 94°C for 1 min, 58°C for 1 min, and 72°C for 1 min, followed by 72°C for 5 min.

4. Southern blot analysis was carried out according to the methods suggested by Sambrook and Russell (26). The vector DNA was used as a positive control.

5. Approximately 20 μg of total genomic DNA was digested with *Hind*III, separated by electrophoresis on a 0.8% agarose gel, and transferred to a nylon membrane (Hybond-N+).

6. The amplicon of the *vip3A** gene was used as a probe, which was labeled with α-[^{32}P] using a random primer labeling kit.

7. T_1 transgenic cotton plant families derived from self-fertile T_0 (T_0 stands for the original transgenic plants) was grown on the experimental farm.

8. No insecticide was sprayed during the entire cotton growth period, to allow investigation of the insect resistance of the transgenic plants.

9. Highly insect-resistant lines with normal phenotypes were selected by visual observation and by infestation on detached leaves in the laboratory.

10. Mature seeds were harvested from individual high resistance plants with a single-copy insertion. Homozygous transgenic cotton lines were selected by molecular and genetic analyses of T_2 lines.

3.6. RNA Preparation and Transcriptional Analysis

1. Total RNA was extracted from young leaves of the transgenic and non-transformed (NT) control plants using Trizol reagent according to the manufacturer's instructions, and the sample was then treated with RNase-free DNase I.

2. Complementary DNA (cDNA) was synthesized from 2 µg of total RNA using the SuperScript™ First-Strand Synthesis system for RT-PCR.

3. One microliter of cDNA from the reverse transcription reaction was amplified using the same primer pair used to amplify the *vip3A** probe.

4. A fragment of the cotton *GhUBI1* gene (GenBank, accession number EU604080) was amplified from the same cDNA sample as an internal control using primers UBI-F: 5′-ctgaatcttcgctttcacgttatc-3′ and UBI-R: 5′-gggatgcaaatcttcgttaagac-3′.

3.7. Detection of Vip3A* Protein in Transgenic Plants

1. Total protein was extracted from fully expanded young leaves of transgenic homozygous lines according to Sambrook and Russell (26).

2. The protein concentration in the samples was determined by the Bradford method (27).

3. To identify Vip3A* expression, western blot analyses were performed. The Vip3A* protein content in transgenic plants was further quantified by ELISA according to Sambrook and Russell (26).

4. Approximately 20 µg of soluble protein was loaded in each well for western blot analysis, and about 1 µg of soluble protein was applied in each well for the ELISA assay.

5. Rabbit antiserum against Vip3A* (1:3,000 v/v) prepared in our laboratory and alkaline phosphatase-conjugated goat anti-rabbit immunoglobulin (Ig) G (1:5,000 v/v) were used as primary and secondary antibodies, respectively, in both protein assays.

6. Chloroplast isolation was carried out according to procedure of the Chloroplast Isolation Kit with minor modification, mainly, the fully expanded young leaves of cotton plants were kept in the dark overnight before isolating the chloroplasts, and the leaf homogenate was centrifuged for 3 min at $250 \times g$.

7. Protein extraction from chloroplasts and leaf remnants (leaf homogenate subtracted chloroplasts) was the same with the method used to extract proteins from leaves and were assayed by ELISA analysis as described above.

8. The Vip3A* content in chloroplast and leaf remnant were expressed as ng/µg chloroplast total soluble protein and ng/µg leaf remnant total soluble protein, respectively.

3.8. Evaluation of Insect-Resistance of Transgenic Plants in Laboratory

1. The insect bioassay on transgenic plants was performed according to Guo et al. (28) with slight modifications.

2. Fully expanded young leaves were detached from transgenic plants and cut into discs of uniform diameter. Each leaf disc

was placed in a plastic box containing a piece of wet filter paper of 30 mm diameter. They were inoculated with four first-instars larvae of FAW, BAW, and CBW, respectively.

3. Four days later, surviving larvae were counted and the insect mortality was recorded. The insect mortality was used to verify the level of insect-resistance of a transgenic plant.

3.9. Evaluation of Insect Resistance of Transgenic Cotton in the Field

1. Evaluation of the insect resistance of transgenic plants in the field was carried out by artificial infestation and natural infestation of CBW at the experimental farm.

2. The materials tested included four homozygous *vip3A** and *tvip3A** transgenic lines, *cry1Ac* transgenic line BR-98-2 as a positive control, and the NT cultivar Zhongmiansuo35 as a susceptible control. No insecticide was applied for lepidopteran pest control during the whole growing season.

3. Each cotton plant was infested with five first-instar larvae of CBW at the flowering stage. The number of squares and bolls damaged by CBW were counted in the field.

4. Notes

1. The transit signal sequence of *thi1 gene* is also cloned by PCR methods from the *Arabidopsis thaliana* exclude the chemical synthesis.

2. The transit signal nucleotide sequence of *thi1* gene can be separated from the vector pBTVip3A* by *Bam*HI and *Eco*RV.

3. The low temperature, 23 °C, is important in coculture of explants and *Agrobacterium tumefaciens*.

4. The kanamycin is kept in the MSB1 until the embryogenesis calli is induced.

5. The explants are discarded if they have not yet formed the calli on the MSB1 for 50 days because the no plant cells are transformed by *Agrobacterium tumefaciens*.

6. It is important how to select the calli to promote the somatic embryogenesis. To pick the pale-yellow, granular, and small group calli to subculture, it facilitates embryogenesis calli formation.

7. The maturation somatic embryos should be immediately picked up from the embryogenesis calli avoid to form abnormal embryos or calli.

8. It takes 120–150 days to produce a complete transgenic plantlets from explants inoculated with bacterium.

References

1. EPA US (1995) Use of the benchmark dose approach in health risk assessment. Office of Research and Development, Washington, DC, EPA/630/R-94/007

2. James C (2009) Global status of commercialized biotech/GM crops. http://www.isaaa.org

3. Cao J, Zhao JZ, Tang D, Shelton M, Earle D (2002) Broccoli plants with pyramided cry1Ac and cry1C Bt genes control diamondback moths resistant to Cry1A and Cry1C proteins. Theor Appl Genet 105(2-3):258–264

4. Ferre J, Van Rie J (2002) Biochemistry and genetics of insect resistance to Bacillus thuringiensis. Annu Rev Entomol 47:501–533

5. Shelton AM, Zhao JZ, Roush RT (2002) Economic, ecological, food safety, and social consequences of the deployment of Bt transgenic plants. Annu Rev Entomol 47:845–881

6. Tabashnik BE, Carriere Y, Dennehy TJ, Morin S, Sisterson MS, Roush RT, Shelton AM, Zhao JZ (2003) Insect resistance to transgenic Bt crops: lessons from the laboratory and field. J Econ Entomol 96(4):1031–1038

7. Zhao JZ, Li YX, Collins HL, Shelton AM (2002) Examination of the F2 screen for rare resistance alleles to Bacillus thuringiensis toxins in the diamondback moth (Lepidoptera: Plutellidae). J Econ Entomol 95(1):14–21

8. Micinski S, Waltman B (2005) Efficacy of VipCOT for control of the bollworm/tobacco budworm complex in Northwest Louisiana. In: Proceedings of 2005 Beltwide cotton conferences New Orleans, LA National Cotton Council Memphis, TN, January 2005. pp 1239–1242

9. Estruch JJ, Warren GW, Mullins MA, Nye GJ, Craig JA, Koziel MG (1996) Vip3A, a novel Bacillus thuringiensis vegetative insecticidal protein with a wide spectrum of activities against lepidopteran insects. Proc Natl Acad Sci USA 93(11):5389–5394

10. Doss VA, Kumar KA, Jayakumar R, Sekar V (2002) Cloning and expression of the vegetative insecticidal protein (vip3V) gene of Bacillus thuringiensis in Escherichia coli. Protein Expr Purif 26(1):82–88

11. Mesrati LA, Tounsi S, Jaoua S (2005) Characterization of a novel vip3-type gene from Bacillus thuringiensis and evidence of its presence on a large plasmid. FEMS Microbiol Lett 244(2):353–358

12. Selvapandiyan A, Arora N, Rajagopal R, Jalali SK, Venkatesan T, Singh SP, Bhatnagar RK (2001) Toxicity analysis of N- and C-terminus-deleted vegetative insecticidal protein from Bacillus thuringiensis. Appl Environ Microbiol 67(12):5855–5858

13. Chabregas SM, Luche DD, Farias LP, Ribeiro AF, van Sluys MA, Menck CF, Silva-Filho MC (2001) Dual targeting properties of the N-terminal signal sequence of Arabidopsis thaliana THI1 protein to mitochondria and chloroplasts. Plant Mol Biol 46(6):639–650

14. Lee MK, Walters FS, Hart H, Palekar N, Chen JS (2003) The mode of action of the Bacillus thuringiensis vegetative insecticidal protein Vip3A differs from that of Cry1Ab delta-endotoxin. Appl Environ Microbiol 69(8): 4648–4657

15. McCaffery A, Capiro M, Jackson R, Marcus M, Martin T, Dickerson D, Negrotto D, O'Reilly D, Chen E, Lee M (2006) Proceedings of effective IRM with a novel insecticidal protein, Vip3A. Beltwide cotton conferences, San Antonio, TX. National Cotton Council Memphis, TN, 3–6 Jan 2006. pp 1229–1235

16. MacIntosh SC, McPherson SL, Perlak FJ, Marrone PG, Fuchs RL (1990) Purification and characterization of Bacillus thuringiensis var. tenebrionis insecticidal proteins produced in E. coli. Biochem Biophys Res Commun 170(2):665–672

17. Chabregas SM, Luche DD, Van Sluys MA, Menck CF, Silva-Filho MC (2003) Differential usage of two in-frame translational start codons regulates subcellular localization of Arabidopsis thaliana THI1. J Cell Sci 116(Pt 2):285–291

18. Chen JW, Tang LX, Tang MJ, Shi YX, Pang Y (2002) Cloning and expression product of vip3A gene from Bacillus thuringiensis and analysis of inseceicidal activity. Sheng Wu Gong Cheng Xue Bao 18(6):687–692

19. Yu CG, Mullins MA, Warren GW, Koziel MG, Estruch JJ (1997) The Bacillus thuringiensis vegetative insecticidal protein Vip3A lyses midgut epithelium cells of susceptible insects. Appl Environ Microbiol 63(2):532–536

20. Li TY, Tian YC, Qing XF, Mang KQ, Li WG, He YG, Shen L (1994) Transgenic tobacco plants with efficent insect resistance. Science in China, Ser B 37:1479–1488

21. Murashige T, Skoog F (1962) A revised medium for rapid growth and bioassays with tobacco tissue cultures. Physiol Plant 15: 473–497

22. Gamborg OL (1970) The effects of amino acids and ammonium on growth of plant cells in suspension culture. Plant Physiol 45:372–375

23. Wu J, Luo X, Zhang X, Shi Y, Tian Y (2011) Development of insect-resistant transgenic cotton with chimeric TVip3A* accumulating in chloroplasts. Transgenic Res 20:963–973

24. Wu J, Zhang X, Luo X, Xiao J (2003) Selection of somatic embryogenesis pure lines of two upland cotton (*Gossypium hirsutum* L) cultivars. Cotton Sci 15:254–256

25. Paterson AH, Brubaker CL, Wendel JF (1993) A rapid method for extraction of cotton (*Gossypium* spp.) genomic DNA suitable for RFLP or PCR analysis. Plant Mol Biol Rep 11(2):122–127

26. Sambrook J, Russell DW (2001) Molecular cloning: a laboratory manual, 3rd edn. Cold Spring Harbor Laboratory Press, New York

27. Bradford MM (1976) A rapid and sensitive method for the quantitation of microgram quantities of protein utilizing the principle of protein-dye binding. Anal Biochem 72: 248–254

28. Guo HN, Wu JH, Chen XY, Luo XL, Lu R, Shi YJ, Qin HM, Xiao JL, Tian YC (2003) Cotton plants transformed with the activated chimeric cry1Ac and AP1-B genes. Acta Botanica Sin 45:108–113

Part V

Risk Assessment

Chapter 21

Determining Gene Flow in Transgenic Cotton

Xiaoping Pan

Abstract

Gene flow is one of the major concerns associated with the release of transgenic plants into the environment. Unrestricted gene flow can results in super weeds, reduction in species fitness and genetic diversity, and contamination of traditional plants and foods. Thus, it is important and also necessary to evaluate the extent of gene flow in the field for transgenic plants already released or being considered for a release. Transgenic cotton is among the first transgenic crops for commercialization, which are widely cultivated around the world. In this chapter, we use transgenic insect resistant cotton and herbicide-tolerant cotton as two examples to present a field practice method for determining transgene flow in cotton. The procedure includes three major sections: (1) field design, (2) seed collection, and (3) field and lab bioassay.

1. Introduction

Gene flow is the movement of genes from one plant to another, a natural genetic drift process for the selection of advantageous traits during species evolution. For transgenic plants, it means that the movement of transgene from transgenic plants to non-transgenic plants or from one type of transgenic plants to another type of transgenic plants. Gene flow is a common phenomenon during plant evaluation (1). Although it plays an important role in plant biodiversity and species evolution, gene flow also cause some significant issues (2), particularly for transgenic plants (3, 4). There are three major consequences caused by transgene flow, which include reduction in species fitness and genetic diversity, development of resistance and becoming super weeds, and contamination of traditional plants and foods (3, 4). Transgene flow-caused contamination of non-transgenic crops has caused significant concerns (5), resulting in product recalls, lawsuit (6), and even collapse of the product market (7). Thus, it is important and also necessary to investigate the gene flow before releasing the transgenic plants into

Baohong Zhang (ed.), *Transgenic Cotton: Methods and Protocols*, Methods in Molecular Biology, vol. 958,
DOI 10.1007/978-1-62703-212-4_21, © Springer Science+Business Media New York 2013

the environment. Pollen dispersal is the major cause of gene flow. Many biotic and abiotic factors facilitate pollen dispersal-mediated gene flow; wind and insects are two dominant factors.

Transgenic cotton is among the first genetically modified commercial crops (8). In the past decades, transgenic cotton has been widely adopted by cotton farmers around the world (5, 9). Insect-resistant and herbicide-tolerant cotton are two major types of transgenic cotton in the field (9). Although less attention was paid on transgene flow in cotton because it is a primarily self-pollinating crop, several studies shows transgene flow occurred in cotton at different geographical conditions (10–13). However, all tested transgene flow occurred within a short distance nearby the transgenic cotton pool; the transgene flow rate was significantly declined as distance increases. For all tested fields, no transgene flow was observed beyond 50 m of distance from the border of the transgenic cotton field. One recent study showed that the transgene flows from a cultivated upland cotton into wild populations at its center of origin, which may influence cotton biodiversity (14). Insects, particularly bees, are the major players causing gene flow in cotton (15).

2. Materials

2.1. Field Trials and Seed Collection

1. Two transgenic cotton events (NewCott 33B and TFD) are used in this method for investigating the transgene flow from transgenic cotton to non-transgenic cotton. Newcott 33B is a commercial transgenic insect-resistant cotton cultivar, which are widely adopted in China as well in the USA since later 1990s. Newcott 33B carries a Bt gene and expresses cry IA insecticidal protein for resistance to *Heliothis virescenes* and *Helicoverpa zea* (Lipidoptera: Noctuidae). TFD is a transgenic herbicide-resistant cotton line, which carries the tfd A gene and expresses 2,4-dichlorophenoxyacetic acid (2,4-D) monooxygenase for resistance to the herbicide 2,4-D (see Note 1).

2. Non-transgenic cultivar CCRI 19 was obtained from the Cotton Research Institute, Chinese Academy of Agricultural Sciences.

3. Field suitable for growing cotton.

4. Pesticides and herbicides potentially used for pest, disease and weed control.

2.2. Field Test of Seeds Obtaining from Field Trials

1. Seeds collected from Subheading 2.1.

2. 2,4-D.

3. *Heliothis virescenes.*

4. Field suitable for growing cotton.

5. Pesticides and herbicides potentially used for pest, disease and weed control.

2.3. Laboratory Bioassay Test of Insect-Resistant Plants from Field Test

1. 100×20 mm Petri Dish.

2. Filter papers (9 cm diameter).

3. *Heliothis virescenes.*

4. Sterilized deionized H_2O (ddH_2O).

2.4. Verification of Resistant Seedlings by Molecular Analyses

1. DNA isolation.

   (a) Pestle and Mortar.

   (b) 1.5 mL Microcentrifuge tubes.

   (c) Tris–HCl.

   (d) EDTA.

   (e) NaCl.

   (f) CTAB.

   (g) Glucose.

   (h) Chloroform:isoamyl alcohol (24:1).

   (i) Isopropanol.

   (j) 80% ethanol.

   (k) 100% ethanol.

   (l) TE buffer: 100 mM Tris–HCl pH 8, 1 mM EDTA.

   (m) Nuclease-Free water.

   (n) Liquid nitrogen.

   (o) Cotton leaves.

   (p) NaoDrop or regular spectrophotometer for check DNA concentration and quality.

   (q) Centrifuge.

2. PCR.

   (a) Thermal cycler (PCR machine).

   (b) Pipettes and tips.

   (c) Primers for cry IA and tfd A. Store at –20°C.

   (d) dNTP. Store at –20°C.

   (e) Taq DNA Polymerase. Store at –20°C.

   (f) 0.2 mL Eppendorf microcentrifuge tubes.

   (g) Nuclease-free water.

   (h) Vortex mixer.

3. Gel electrophoresis.

    (a) 20× TAE buffer (see Note 2).

    (b) Agarose.

    (c) DNA molecular size marker (e.g., 100 bp DNA ladder).

    (d) Gel loading buffer.

    (e) Ethidium bromide (EB) (see Note 3) or GelRed DNA Stains.

    (f) Horizontal gel electrophoresis equipment.

    (g) UV transilluminator.

3. Methods

3.1. Field Trials and Seed Collection

1. All transgenic materials need to be purified by consecutive two-generation self-crossing to ensure that each line is homozygous for the transgenes.

2. Transgenic and non-transgenic cotton are sown by conventional methods at a rate of approximately 15 seed/m of row on a 1.0 m row spacing. Transgenic cotton plants and non-transgenic cotton plants are sown in the same field. Transgenic cotton plants are sown in a 6 m × 6 m square in the middle of a 210 m × 210 m square field according to the scheme presented in Fig. 1. The area surrounding this central plot is planted with the non-transgenic cotton variety CCRI 19 (see Note 4).

3. Final plant density is about 5–7 plants/m^2.

4. Standard cultivation practices and insect control measures are used in an attempt to optimize yields.

5. During the flowering season, the wind condition and insect population should be recorded.

6. At the harvest season, 50 cotton bolls are chosen randomly from each direction (east, south, west, and north) of the transgenic cotton at distances of 1, 2, 5, 10, 20, 50, 60, and 100 m away from the border of the transgenic cotton field, respectively.

7. The seeds are isolated and ready for field test.

3.2. Field Test of Seeds Obtaining from Field Trials (see Note 5)

3.2.1. Gene Flow of Transgenic Herbicide-Resistant Cotton

1. Seeds obtained from the trial with transgenic herbicide-resistant tfd A-cotton are sown in a new field with a 30-cm row spacing and 10-cm plant spacing.

2. Standard cultivation practices and insect control measures are used.

3. At the 4–5-leaf stage of cotton, 2–5 mL of 200 mg/L 2,4-D is sprayed to each plants.

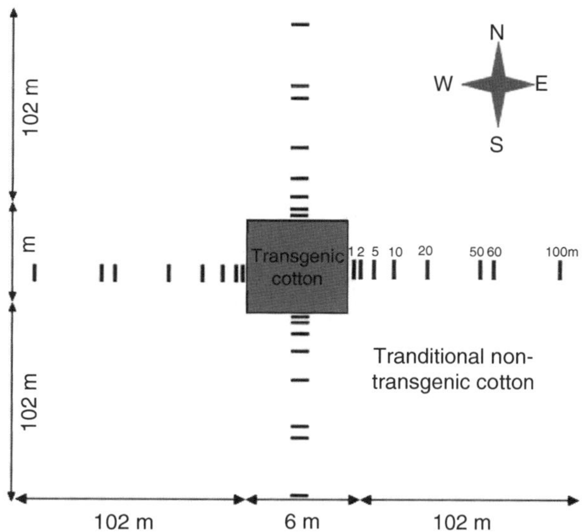

Fig. 1. Field trial design to measure gene flow from transgenic cotton to non-transgenic cotton. The transgenic cotton test plots are planted in a 6-by 6-m square in the middle of a 210-by 210-m square field of non-transgenic cotton plants. Bars indicate sampling points at 1, 2, 5, 10, 20, 50, 60, and 100 m in distance away from the border of the transgenic cotton field.

4. Two weeks after 2,4-D spraying, the growth of cotton plants are observed, and the damage to the cotton leaves are recorded.

5. The total number of resistant seedlings and non-resistant seedlings per sampling point is recorded, respectively, according to the 2,4-D damage on cotton leaves. Plants with twisted, damaged, or absent leaves are identified as non-resistant plants. Plants with normal leaves are identified as resistant plants.

6. Herbicide-resistant plants are likely carry the transgene tfd A as a result of gene flow.

7. Ratios of gene flow are calculated according to the following formula: gene flow ratio = (the number of herbicide-resistant plants)/[(the number of herbicide-resistant plants) + (the number of non-herbicide-resistant plants)].

3.2.2. Gene Flow of Transgenic Insect-Resistant Cotton

1. Seeds, obtained from the transgenic insect-resistant Bt-cotton, are sown in a new field with a 100-cm row spacing and 15–20-cm plant spacing.

2. Standard cultivation practices and insect control measures are used.

3. At the ten-leaf stage of cotton, three to five bollworms are placed on the shoot of all tested plants.

4. Three days later, bollworm mortality is recorded and damage to cotton plants is measured and recorded. Plants without bollworm damage are identified as pest-resistant plants. Plants with bollworm damage are identified as susceptible plants.

5. The total number of resistant seedlings and non-resistant seedlings per sampling point is recorded, respectively.

6. Insect-resistant plants are likely carry the transgene Bt as a result of gene flow. Ratios of gene flow are calculated accordingly the following formula: gene flow ratio = (the number of insect-resistant plants)/[(the number of insect-resistant plants) + (the number of non-insect-resistant plants)].

3.3. Laboratory Bioassays of Insect-Resistant Plants from Field Test

The lab bioassay is employed to further test undamaged plants and confirm that they are real pest-resistant plants (see Note 6).

1. One fresh top opened leave is obtained from each undamaged cotton plant collected from the field test.

2. The leaves are marked and bring to lab using a cooler.

3. Each leave is placed into a Petri dish with two layers of filter papers wet by diH_2O (see Note 7).

4. Two young bollworm larvae are placed on the leaves.

5. After 2 days of exposure, the leaves are observed and damages are recorded. If a damage present, the plant is identified as susceptible. Leaves without damage are identified as resistant plants, which caused by gene flow.

3.4. Verification of Resistant Seedlings by Molecular Analyses

Bioassays of resistant phenotypes in the field and lab as described above are not sufficient to determine the event of gene flow, although previous molecular analyses showed that there was a high correlation between the presence of the resistance gene and the phenotypic response of cotton plants to 2,4-D or bollworms. Molecular evidence is also required to further confirm the field and lab resistance screening. There are many strategies for detecting specific genes in a sample. Here, we only present a conventional PCR method for verification of gene flow in resistant seedlings.

1. Genome DNA isolation from cotton leaves

    Many methods, including commercial kits, have been developed for isolating genome DNA from cotton tissues. Any of those methods can be used in this section. Here, we describes a method from a previous report (16) with some modifications.

    (a) Make extraction buffer (EB), which contains 100 mM Tris–HCl pH 8, 20 mM EDTA, 1.4 M NaCl, 2% CTAB, 0.5 M glucose.

    (b) The resistant cotton plants are marked in the field.

(c) The first unopened leave is collected (see Note 8), stored in liquid nitrogen and then brought to the lab. Alternatively, the cotton leaves can be stored in –80°C freezer if not extracting DNA immediately.

(d) Transfer the cotton leaves (~0.1–0.2 g) from liquid nitrogen and place them in a pre-cooled mortar (see Note 9).

(e) Add liquid nitrogen slowly, grind the cotton leave tissue sample into a fine powder using a pre-cooled pestle.

(f) Transfer the fine powder, making sure it does not thaw, to a 1.5 mL microcentrifuge tube containing 500 μL Extraction Buffer.

(g) Vortex mix the tissue sample with extraction buffer and then sonicate for 15 s.

(h) Homogenize the tissue sample and incubate at 60°C for 30 min with gentle mixing.

(i) Add 3.33 mL choloroform:isoamyl alcohol (24:1) mixture and gently inverting the tube.

(j) Centrifuge the tube at $5,000 \times g$ for 10 min at 4°C.

(k) Carefully remove the supernatant and transfer to a new 1.5 mL tube.

(l) Add 0.8 volume of ice-cold isopropanol to the collected supernatant.

(m) Gently mix by vortex.

(n) Place the tube with mixture at –20°C for 60 min.

(o) Centrifuge the tube at $10,000 \times g$ at 4°C for 10 min.

(p) Discard the supernatant.

(q) Resuspend the pellet in 100 μL TE buffer or RNAase-free water.

(r) The quality and quantity of DNAs are measured using NanoDrop ND-1000 (NanoDrop Technologies, Wilmington, DE).

(s) DNA samples are store in a –80°C freezer until further use.

2. PCR.

(a) In a PCR tube (0.2 mL tube), add the following amount of reagents in order for one reaction: 10.0 μL 2× PCR Buffer, 1.0 μL forward primer, 1.0 μL reverse primer, 0.2 μL dNTP mix (100 mM), and 1.00 μL Taq DNA Polymerase enzyme and 500 ng DNA. Using nuclease-free water to make up a total of 20 μL reaction (see Note 10).

(b) Mix the reagents gently by tapping the tube or briefly centrifuge.

(c) Load all sample tubes into a thermal cycler.

    (d) Program the thermal cycler as follows: 95°C for 10 min, followed by 30 cycles of 95°C for 15 s, 60°C for 30 s, and 72°C for 60 s.

    (e) If not proceeding to gel electrophoresis, store the PCR samples at 4°C or in a −20°C freezer.

3. Gel electrophoresis.

    (a) Prepare 1× TAE buffer from the 20× stock. To make 300 mL 1× TAE buffer, you need to add 15 mL of the 20× TAE buffer to 285 mL deionized water.

    (b) Add 0.6 g agarose to 50 mL 1× TAE buffer for making a 1.2% gel (see Note 11).

    (c) Using a microwave to heat the solution to dissolve the agarose (see Note 12).

    (d) Add 0.5–1.0 µL GelRed to the dissolved agarose and mix (see Note 13).

    (e) Setup the gel tray and pour the gel into the tray (see Note 14).

    (f) Solidify the gel at room temperature.

    (g) Load the PCR products (5 µL per sample) with loading dye and also DNA markers into separate gel wells.

    (h) Run the gel at 100 V for 60 min.

    (i) Remove the gel from the tray and observe the bands under a UV transilluminator (see Note 15).

    (j) Take a picture if needed (Fig. 2).

3.5. Data Analysis

Rate of gene flow are calculated accordingly the following formula: gene flow rate = (the number of resistant plants carrying the transgene)/(the total number of plants tested). The gene flow is mainly caused by pollen outcrossing. Usually, transgenic cottons (both

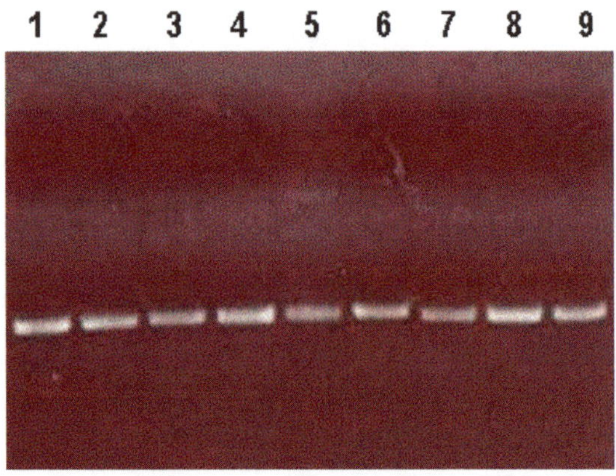

Fig. 2. PCR analysis to detect the Bt gene in cotton plants resistant to bollworms. All the nine cotton plants contain the Bt gene caused by gene flow.

insect-resistant and herbicide-tolerant) cause transgene flow with low level of outcrossing. The outcrossing rate declines sharply as distance into adjacent plots of non-transgenic cotton increased (Figs. 3, 4, 5, and 6).

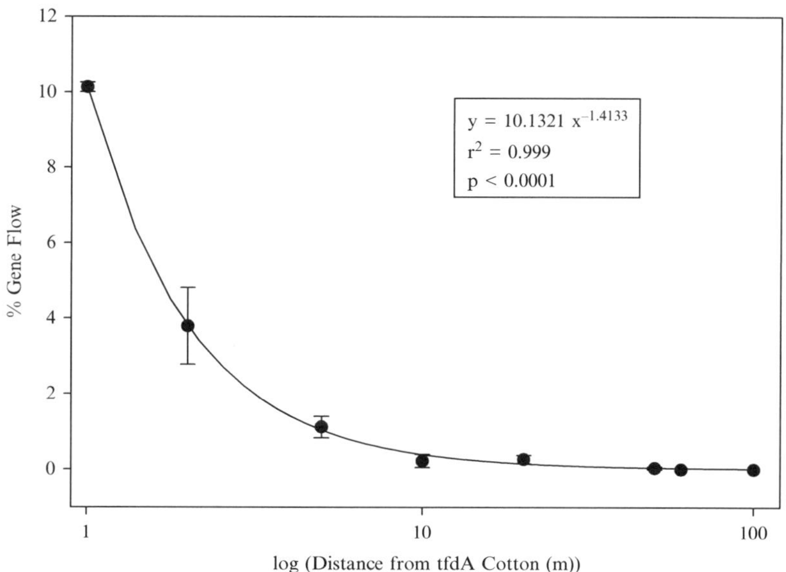

Fig. 3. Gene flow from transgenic *tfd*A cotton into surrounding non-transgenic cotton. *Filled circle* represents observed value and the *continuous line* represents those predicted by the nonliner model of power curve.

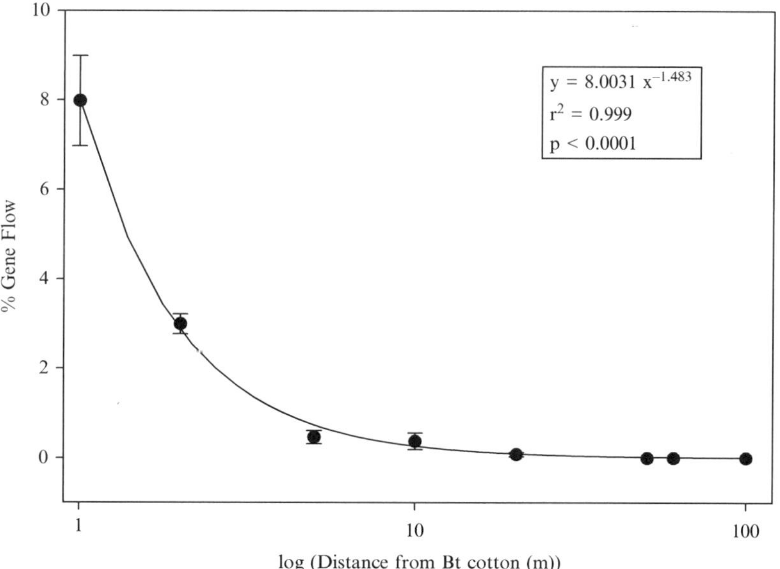

Fig. 4. Gene flow from transgenic Bt cotton into surrounding non-transgenic cotton. *Filled circle* represents observed value and the *continuous line* represents those predicted by the nonliner model of power curve.

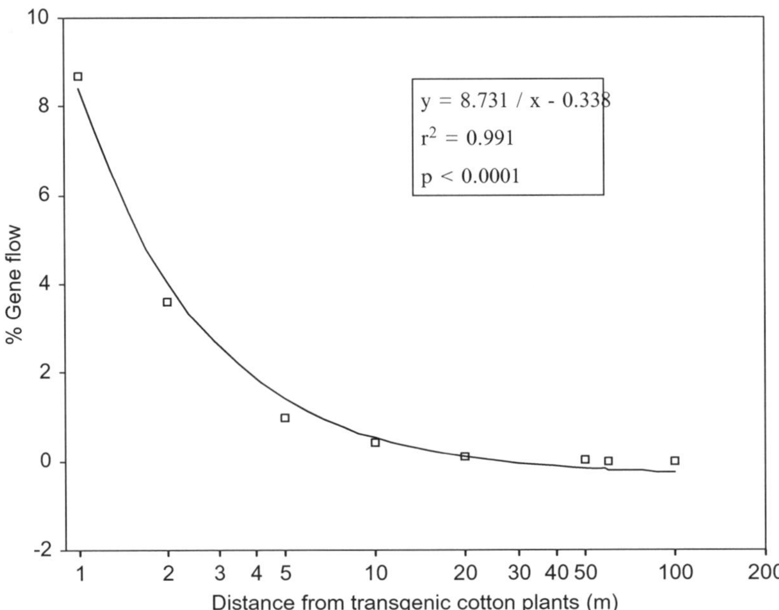

Fig. 5. Gene flow from transgenic cotton (tfdA and Bt) into surrounding non-transgenic cotton. *Squares* represent observed values and the *continuous line* represents those predicted by the nonliner model.

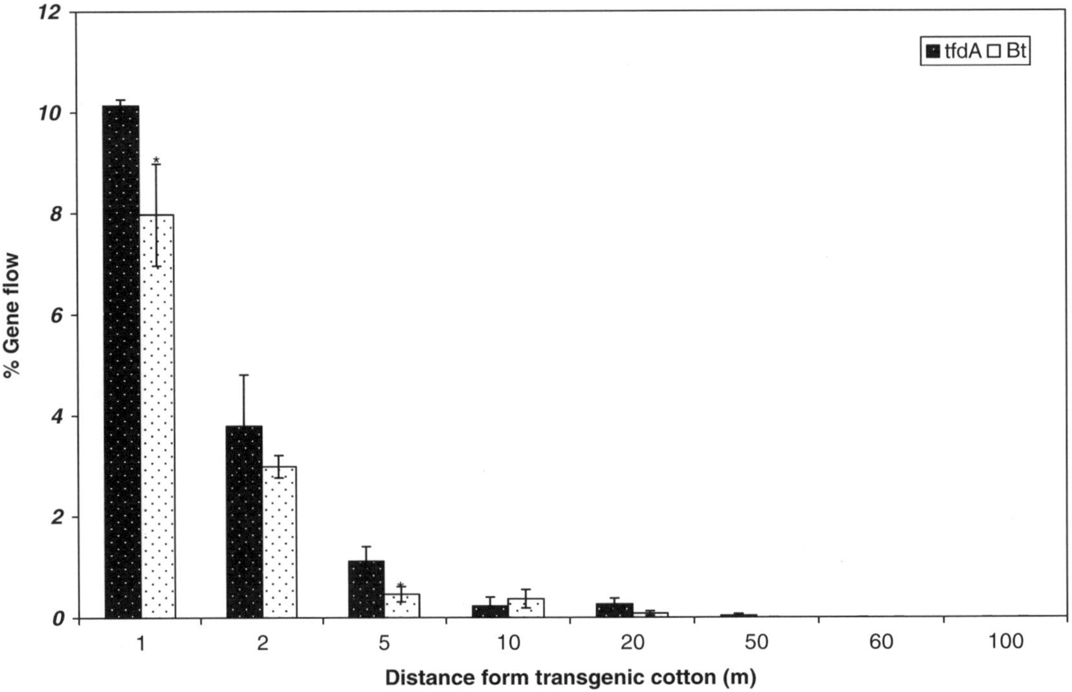

Fig. 6. Mean gene flow in cultivation of transgenic cotton via pollen dispersal. *Error bars* represent standard errors of the means. Chi-square test indicated that there is significant difference in pollen-mediated gene flow between tfdA cotton and Bt cotton ($p < 0.0001$).

4. Notes

1. In this method, NewCott 33B and TFD are chosen as two examples. Other transgenic cultivars can also be used.

2. To make 1 L of 20× TAE buffer, add 96.8 g Tris-base, 22.8 mL of glacial acetic acid, 40 mL of 0.5 M EDTA (pH 8.0) into 800 mL distilled deionized water, dissolve solids and adjust pH with HCl to 7.6–7.8, then bring up to 1 L with distillated deionized water.

3. Ethidium bromide (EB) is a carcinogen, handle with care, and wear appropriate protection.

4. The size of transgenic and non-transgenic cotton areas can be adjusted accordingly to specific experimental location and also specific purpose. Different transgenic events can be tested at the same time. But the locations of each event must be separated by at least 1 km to avoid the cross-contamination. Similarly, no transgenic cotton should be planted around the test area.

5. Due to different transgenic traits, different practices may be adapted to grow different transgenic plants. For herbicide-resistant cotton, they can be sown in a high density because they will be terminated after spraying herbicide. For insect-resistant cotton, they should be grown under normal conditions because they still can be continuously grown and harvested after test.

6. If bollworms crawl away from a certain test plant in the field, false positive may result. Therefore, all undamaged plants need to be further tested in lab.

7. Adding water to filter papers keep humidity in the Petri Dishes, which is required for leaves and bollworms survivals.

8. It is relatively difficult to extract DNA from cotton tissues because cotton is rich in polyphenolic compounds. Compared with mature leaves, it is easier to obtain high yield and quality of DNA from young leaves.

9. Pestle and Mortar should be pre-cooled in −80°C freezer for at least half an hour, which will prevent breakage of the mortar and pestle upon sudden cooling by liquid nitrogen.

10. Always add the DNA template last and the enzyme second last. Also need to set up positive and negative controls.

11. The agarose concentration should be based on the actual size of PCR products; usually the smaller the PCR product, the higher concentrations of the agarose is needed.

12. When using microwave, care should be taken to avoid boiling of the gel solution. Also need to make sure the solution become

clear and the agarose is completely dissolved. Otherwise, distorted bands will be observed.

13. GelRed is a safe DNA stain dye. If no GelRed is available, ethidium bromide (EB) also can be used. However, EB is carcinogen, so always wear gloves to prevent contact with it.

14. Make sure no air bubbles in the gel. If bubble occurs, remove it using a pipette tip.

15. Avoid exposure to UV light directly by using a UV light shield and wear appropriate protection.

References

1. Slatkin M (1985) Gene flow in nature populations. Annu Rev Ecol Syst 16:393–430

2. Lenormand T (2002) Gene flow and the limits to natural selection. Trends Ecol Evol 17:183–189

3. Chevre AM, Eber F, Baranger A, Renard M (1997) Gene flow from transgenic crops. Nature 389:924

4. Snow AA (2002) Transgenic crops – why gene flow matters. Nat Biotechnol 20:542

5. Showalter AM, Heuberger S, Tabashnik BE, Carriere Y (2009) A primer for using transgenic insecticidal cotton in developing countries. J Insect Sci 9:22

6. Vermij P (2006) Liberty Link rice raises specter of tightened regulations. Nat Biotechnol 24:1301–1302

7. Smyth S, Khachatourians GG, Phillips PWB (2002) Liabilities and economics of transgenic crops. Nat Biotechnol 20:537–541

8. Zhang BH, Feng R (2000) Cotton resistance to insect and transgenic cotton. China Agricultual Science and Technology Press, Beijing

9. Zhang BH, Liu F, Yao CB, Wang KB (2000) Recent progress in cotton biotechnology and genetic engineering in China. Curr Sci 79:37–44

10. Heuberger S, Ellers-Kirk C, Tabashnik BE, Carriere Y (2011) Pollen- and seed-mediated transgene flow in commercial cotton seed production fields. PLoS One 5(11):e14128

11. Zhang BH, Pan XP, Guo TL, Wang QL, Anderson TA (2005) Measuring gene flow in the cultivation of transgenic cotton (*Gossypium hirsutum* L.). Mol Biotechnol 31:11–20

12. Llewellyn D, Fitt G (1996) Pollen dispersal from two field trials of transgenic cotton in the Namoi Valley, Australia. Mol Breed 2:157–166

13. Umbeck PF, Barton KA, Nordheim EV, McCarty JC, Parrott WL, Jenkins JN (1991) Degree of pollen dispersal by insect from a field test of genetically engineered cotton. J Econ Entomol 84:1943–1950

14. Wegier A, Pineyro-Nelson A, Alarcon J, Galvez-Mariscal A, Alvarez-Buylla ER, Pinero D (2011) Recent long-distance transgene flow into wild populations conforms to historical patterns of gene flow in cotton (*Gossypium hirsutum*) at its centre of origin. Mol Ecol 20:4182–4194

15. Free JB (1970) Insect pollination of crops. Academic, London, p 544

16. Permingeat HR, Romagnoli MV, Sesma JI, Vallejos RH (1998) A simple method for isolating DNA of high yield and quality from cotton (*Gossypium hirsutum* L.) leaves. Plant Mol Biol Rep 16:89

Baohong Zhang (ed.), *Transgenic Cotton: Methods and Protocols*, Methods in Molecular Biology, vol. 958,
DOI 10.1007/978-1-62703-212-4, © Springer Science+Business Media New York 2013

Printed by Printforce, the Netherlands